TRANSMITTERS AND IDENTIFIED NEURONS IN THE MAMMALIAN CENTRAL NERVOUS SYSTEM

TRANSMITTERS AND IDENTIFIED NEURONS IN THE MAMMALIAN CENTRAL NERVOUS SYSTEM

ANDRIS K. TEBĒCIS

Department of Physiology,
The John Curtin School of Medical Research,
Australian National University,
Canberra, Australia.

BRISTOL: SCIENTECHNICA (Publishers) LTD.
1974

ISBN 0 85608 008 X

PRINTED IN GREAT BRITAIN BY HENRY LING LTD., A SUBSIDIARY OF JOHN WRIGHT AND SONS LTD., AT THE DORSET PRESS, DORCHESTER

TO MAIJA, FOR BEARING WITH ME

PREFACE

A NUMBER of reviews and monographs have been written about chemical substances suspected of being transmitters in the nervous system. Some of these have dealt with various aspects (the physiology, pharmacology, biochemistry, histochemistry) of neurons in both peripheral and central nervous systems of different animal species, whereas others have analysed one particular aspect of neurotransmission in detail. All approaches have been of great value. However, knowledge in each particular discipline is accumulating so rapidly that it is no longer possible to treat all these related fields adequately in a single review. Conversely, a detailed analysis of a single discipline will be of value mainly to the specialist working in that field, since those in related fields will have difficulty in extricating and appreciating the major points discussed. A satisfactory compromise is difficult to achieve.

This book is an attempt critically to review and summarize the main facts pertaining to the identification of synaptic transmitters in the mammalian central nervous system, viewed predominantly from a pharmacological aspect. One approach to find some order in the vast, ever-increasing store of neuropharmacological data is to examine the pharmacology of neurons *identified* as belonging to a particular functional or anatomical type, as these have a relatively consistent pharmacology and therefore enable one to make conclusions about transmitter probability with more certainty. Reports on the pharmacology of various types of unidentified neurons have often left all but the specialist working in the same field confused as to the significance of the findings. However, pharmacological studies of unidentified neurons are also discussed in some detail in this book if they have significantly contributed towards the identification of synaptic transmitters.

It is assumed that the reader has at least a general understanding of the fundamental principles of synaptic transmission. These will not be expounded here since several excellent reviews are available (Hodgkin, 1964; Eccles, 1964, 1969; Katz, 1966; Weight, 1971; Krnjević, 1971; Werman, 1972a). Similarly, the generally-accepted criteria for establishing a compound as a transmitter have been frequently reviewed (Paton, 1958; Florey, 1960; Curtis, 1961; Werman, 1966, 1972a; Phillis, 1970; McLennan,

1970a; Curtis and Johnston, 1970; Steiner, 1971) and will not be amplified here. In addition, some knowledge of neuroanatomy, histochemistry, and neurochemistry is useful.

The types of neurons discussed in this book are grouped into chapters according to the anatomical structures in which they are located. Each chapter begins with some general remarks on the particular structure and the neurons within it, and the pharmacological findings pertaining to transmitter action are then described in detail. Although the detailed text often requires the reader to have had experience in the field for a complete understanding, the main findings are stressed in the text and summarized at the end of each chapter. It is hoped that this approach will provide both well-informed neuropharmacologists and research workers in related fields, as well as graduate students new to such studies, with a reasonably comprehensive idea of what has been done and what remains to be done in the identification of central transmitters. For other expositions of chemical transmission, recent reviews should be consulted (Curtis and Crawford, 1969; Vogt, 1969; Hebb, 1970; Phillis, 1970; McLennan, 1970a; Curtis and Johnston, 1970; Cooper, Bloom and Roth, 1970; Steiner, 1971; Krnjević, 1971; Werman, 1972a, b).

October, 1973 A.K.T.

ACKNOWLEDGEMENTS

I GREATLY appreciate Professor G. A. Kerkut's invitation to write this book, which was begun shortly after the very enjoyable and stimulating few weeks I spent in his Department in mid-1970. It is a pleasure to express gratitude to Professor P. O. Bishop for encouragement and providing facilities for the production of this book. I am particularly indebted to Professor J. W. Phillis and Dr. G. A. R. Johnston for the valuable comments they made on an early version of the manuscript. Sincere thanks are extended to the following people, who sent me either their unpublished results or material in press: F. E. Bloom (Washington, D.C.), G. ten Bruggencate (München), G. Burnstock (Melbourne), D. R. Curtis (Canberra), T. Hökfelt (Stockholm), J. S. Kelly (Cambridge), K. Krnjević (Montreal), I. McCance (Melbourne) H. McLennan (Vancouver), K. Obata (Tokyo), J. W. Phillis (Winnipeg), R. W. Ryall (Cambridge), T. W. Stone (Aberdeen), J. Storm-Mathisen (Kjeller), M. Straschill (München), and R. Werman (Jerusalem). The preparation of this book was greatly facilitated by the excellent library and photography department of the John Curtin School of Medical Research, the skilled technical assistance of Ms. Adrienne Carpenter, and the efficient typing of Mrs. Patricia Attwater and Mrs. Mary Rigg. I thank the following Publishing Companies for their permission to reproduce illustrations: Academic Press, Inc., *Acta Physiologica Scandinavica*, The American Association for the Advancement of Science (*Science, N.Y.*, copyright 1969 and 1971), Birkhäuser Verlag, Cambridge University Press, Elsevier Publishing Company, *Federation Proceedings*, Paul B. Hoeber, Inc., The Japan Academy, The Johns Hopkins University Press, *Journal of Neurophysiology*, MacMillan Journals Ltd., Microforms International Marketing Corporation, C. V. Mosby Company, National Research Council of Canada, Plenum Publishing Corporation, Springer-Verlag, and The Williams and Wilkins Company.

CONTENTS

INTRODUCTION

THE term 'identified neuron' is used rather loosely in this book. In most cases it refers to neurons which have been activated antidromically by stimulating their axons, although neurons with characteristic orthodromic discharges (Renshaw cells, for example) are also included. Cells of a particular nucleus, all of which from anatomical studies are known to have a similar projection, are also considered to be 'identified'.

The chemical compounds discussed have either been generally accepted as transmitter substances in the nervous system, or are currently considered as likely candidates for a transmitter role. To this category belong acetylcholine, certain amino-acids, and monoamines. It is perhaps useful to mention at the outset that not one substance postulated to be a central transmitter meets all the rigid criteria formulated (Werman, 1966; Phillis, 1970; Steiner, 1971), which, none-the-less are all-important in the design of experiments. Direct pharmacological evidence that the depressant amino-acids, glycine and gamma-aminobutyric acid, are inhibitory transmitters in various areas of the central nervous system has been obtained in recent years by the use of certain convulsants which are relatively specific antagonists of these amino acids. Although there is still some controversy about the specificity of these convulsants and all their actions, for the purposes of this monograph strychnine is considered to be a relatively specific antagonist of glycine, whereas picrotoxin and bicuculline are considered to be relatively specific antagonists of gamma-aminobutyric acid.

Little is said about a wide range of transmitter possibilities other than amino-acids, acetylcholine, and monoamines (such as prostaglandins, substance P, and others), as the evidence that they may be central transmitters is meagre or doubtful. However, it is entirely reasonable to expect that substances which have hitherto received little attention as likely transmitters, and substances which have not yet been identified in brain extracts, may prove to be central transmitters. Recently there has been renewed interest in depressant amino-acids other than glycine and gamma-aminobutyric acid (including taurine, serine, L-α-alanine, imidazole acetic acid, cystathionine, gamma-aminobutyrylcholine, gamma-amino-β-hydroxybutyrate) as possibly being inhibitory transmitters (*see* Chapter IX). A tryptamine derivative other than 5-hydroxytryptamine may be released from a system of fibres which originate in the brain-stem and project to many areas of the central nervous system (Björklund, Falck, and Stenevi, 1971). The evidence for adenosine triphosphate as a transmitter in the peripheral nervous system (Burnstock, 1972) raises the possibility that this may also occur in the central nervous system. No doubt other transmitter candidates will be 'discovered' in the future.

Almost all the studies reviewed deal with the effects of substances ejected electrophoretically from multibarrel glass micropipettes into the immediate environment of neurons during recording of their action potentials. In most cases the firing frequency of the neuron has been recorded extracellularly, although a few studies involving intracellular recording have also been made. The technique of microelectrophoresis has been described and critically evaluated in detail and requires no further elaboration here (Curtis, 1964; Krnjević, 1964, 1971; Salmoiraghi and Stefanis, 1967; Curtis and Crawford, 1969; Phillis, 1970; Steiner, 1971). Most of these studies have been made of cats, although some have also been of rats, rabbits, guinea-pigs, and monkeys. Studies involving systemic or topical application of drugs are mentioned only when directly relevant to the results obtained with the electrophoretic method, or when pharmacological studies involving microelectrophoresis have not been made of the particular region.

Abbreviations: acetylcholine (ACh), acetylcholinesterase (AChE), adrenocorticotrophic hormone (ACTH), adenosine triphosphate (ATP), adenosine 3′, 5′-phosphate (cyclic AMP), amino-oxyacetic acid (AOAA), central nervous system (C.N.S.), dihydro-β-erythroidine (DHβE), 2, 4-dinitrophenol (DNP), dopamine (DA), L-3, 4-dihydroxy-phenylalanine (L-DOPA), electroencephalogram (E.E.G.), excitatory postsynaptic potential (EPSP), gamma-aminobutyric acid (GABA), DL-homocysteate (DLH), 5-hydroxytryptamine (5-HT), 5-hydroxytryptophan (5-HTP), inhibitory postsynaptic potential (IPSP), lysergic acid diethylamide (LSD), noradrenaline (NA).

Chapter

1

SPINAL CORD

ALTHOUGH the spinal cord has been the centre of physiological and pharmacological investigations far longer than most other structures of the central nervous system, only two types of spinal neurons have been *routinely* identified by electrophysiological techniques for neuropharmacological studies. These are the *motoneurons* and *Renshaw cells*. As the name implies, the function of motoneurons is to initiate muscle contractions. Antidromic action potentials may be recorded from motoneurons in the ventral horn following stimulation of the appropriate peripheral muscle nerves, or more simply, ventral roots. Although motoneurons can be identified relatively easily in this way, their large size has increased the difficulty in altering their firing frequency by electrophoretically administered substances. On the other hand, the large size of motoneurons has facilitated studies involving intracellular recording during extracellular drug administration. The spinal motoneuron has thus been invaluable for investigating the ionic mechanisms which underly the actions of likely transmitter substances in the spinal cord.

Renshaw cells are usually located ventromedially in the ventral horn, and are characterized by a short-latency (generally less than 1 msec.), high-frequency, repetitive synaptic discharge evoked by a single antidromic volley in a motor nerve (Renshaw, 1946; Eccles, Fatt, and Koketsu, 1954). Although the existence of Renshaw cells had been questioned on the grounds of lack of clear anatomical evidence (Scheibel and Scheibel, 1966; Weight, 1968; cf. Erulkar, Nichols, Popp, and Koelle, 1968), unequivocal evidence for their existence has recently been obtained (Jankowska and Lindström, 1971), as shown in *Fig.* 1.

The potentials shown in the upper records on the right of *Fig.* 1 were recorded intracellularly from a Renshaw cell which responded repetitively to stimulation of a ventral root. The cell was then stained by injecting the dye, Procion Yellow. The

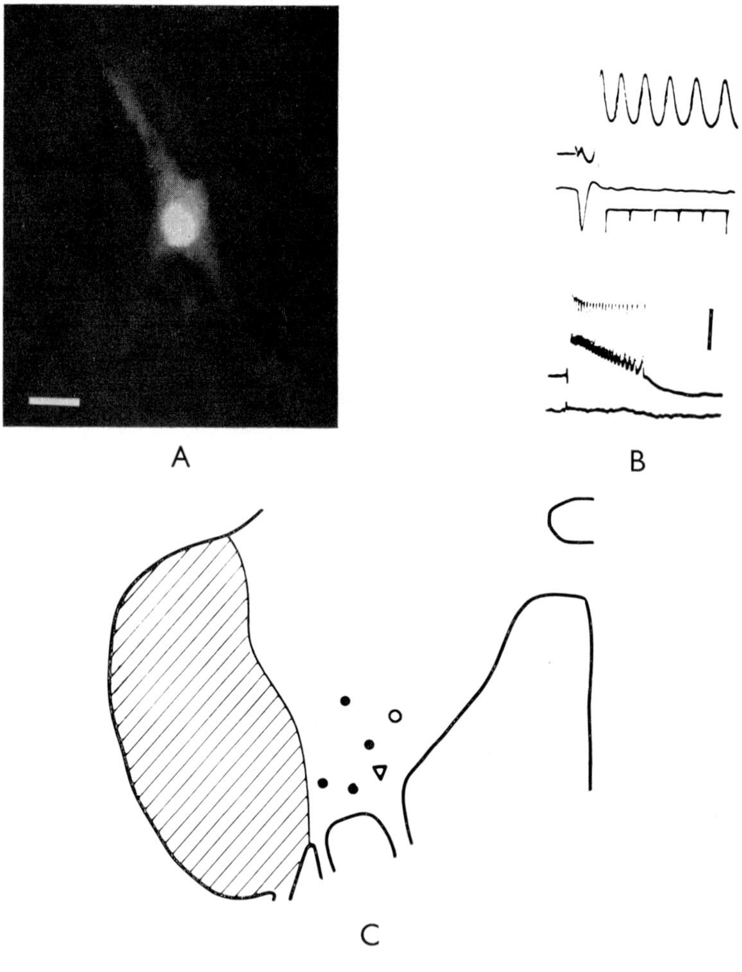

Fig. 1. A, Photomicrograph of a Renshaw cell (the bar corresponds to 10 µm.) lightly stained with Procion Yellow, its approximate location in the spinal cord being represented by the open circle in the diagram (C). B, The records (upper traces intracellular potentials, lower traces cord surface potentials) were obtained just before the intracellular injection of dye, using two different sweep speeds (time calibration; 1 + 5 msec. for upper records, 10 + 50 msec. for lower records; voltage calibration for intracellular potentials; 10 mV). The filled circles and the triangle in the diagram (C) indicate the positions of other Renshaw cells identified in this way. The shaded area represents the extent of motor nuclei. (Jankowska and Lindström, 1971.)

photomicrograph shows the nucleus and the contours of the soma and proximal dendrites of the Renshaw cell, and its approximate position in the spinal cord is indicated by the open circle in the diagram.

Renshaw cells inhibit the firing of motoneurons and some interneurons, including other Renshaw cells (Renshaw, 1941; Eccles and others, 1954; Frank and Fuortes, 1956; Hunt and Kuno, 1956; Brooks and Wilson, 1959; Willis and Willis, 1966; Ryall, 1970; Ryall and Piercey, 1971). The relatively small size of Renshaw cells (*Fig.* 1) is probably the reason why locally administered chemical substances readily influence the firing of these cells. This property, as well as the ease with which they can be identified electrophysiologically, have provided a basis for a number of elegant neuropharmacological studies of Renshaw cells.

Dorsal horn spinal interneurons are of various types which cannot readily be separated into physiologically distinct populations. Both the effects of stimulating sensory inputs and the actions of electrophoretically administered substances on spinal interneurons have been variable. For this reason the pharmacology of spinal interneurons will be mentioned only briefly in this chapter, although the actions of possible transmitters on these cells are clearly as important as those on motoneurons and Renshaw cells. Cells identified as spinocervical tract neurons were excited by glutamate and depressed by both glycine and GABA (Zieglgänsberger and Herz, 1971), but no other pharmacological studies have been made of these neurons. It is possible that pharmacological studies will soon be extended to include certain interneurons in the ventral horn which are monosynaptically excited by impulses in Ia afferents, monosynaptically inhibited by Renshaw cells, and which monosynaptically inhibit motoneurons innervating antagonistic muscles (Jankowska and Roberts, 1972). The recent review by Ryall (1972) provides a comprehensive discussion of the pharmacology of various types of spinal neurons, and deals with a number of important aspects of chemical transmission which are not elaborated here.

Renshaw Cells

Acetylcholine

The best-known example of an identified neuron which has been studied pharmacologically, and for which at least one excitatory transmitter has been identified, is the Renshaw cell. It is now generally accepted that Renshaw cells are excited monosynaptically by ACh released from motor axon collaterals. The first attempt to demonstrate this (Eccles and others, 1954) stemmed from the knowledge that motor nerve-fibres activate skeletal muscle by releasing ACh, and that the same chemical transmitter is probably released from all the synaptic terminals of a neuron (Dale's principle, 1934).

Eccles and others (1954) and Eccles, Eccles, and Fatt (1956) showed that the characteristic high-frequency Renshaw cell discharge evoked by antidromic stimulation of ventral roots was prolonged by systemically administered cholinesterase inhibitors (physostigmine, for example), and depressed by antagonists of ACh (mainly by dihydro-β-erythroidine, but also by atropine to some extent). Intra-arterial injections of ACh had variable effects, but often increased the frequency and duration of Renshaw cell discharges. Nicotine was more potent than ACh, and the effects of ACh, but not those of nicotine, were potentiated by physostigmine.

More direct pharmacological evidence that these compounds act on Renshaw cells was first obtained by Curtis and Eccles (1958a, b), who administered ACh and related substances from multibarrel micropipettes into the immediate extracellular environment of Renshaw cells. Electrophoretically administered ACh, nicotine, and various anticholinesterases consistently increased the discharge rate of Renshaw cells, and these effects, as well as the synaptic responses of Renshaw cells, were antagonized by nicotinic blocking agents.

It has since been shown that the synaptic responses of Renshaw cells evoked by maximal antidromic stimulation of ventral roots are mediated by at least two different types of cholinoceptive receptors (Curtis and Ryall, 1966a, b, c). Nicotinic compounds (including nicotine) were more powerful excitants of Renshaw cells than ACh, whereas muscarinic agents (including muscarine and acetyl-β-methylcholine) were less effective (Curtis and Ryall, 1966a). The differential antagonism by dihydro-β-erythroidine and atropine of the excitation of Renshaw cells by various cholinomimetics indicated that this excitation was mediated by both nicotinic and muscarinic receptors on the same neuron (Curtis and Ryall, 1966b).

An example of this is illustrated in *Fig.* 2, which shows that dihydro-β-erythroidine greatly reduced the rapid excitant action of ACh, but had little effect on the slow excitation produced by acetyl-β-methylcholine (*Fig.* 2A, B, C). Atropine, in a concentration which completely blocked the excitant action of acetyl-β-methylcholine, only reduced that of ACh (*Fig.* 2D, E, F). These and similar observations suggested that the excitation of Renshaw

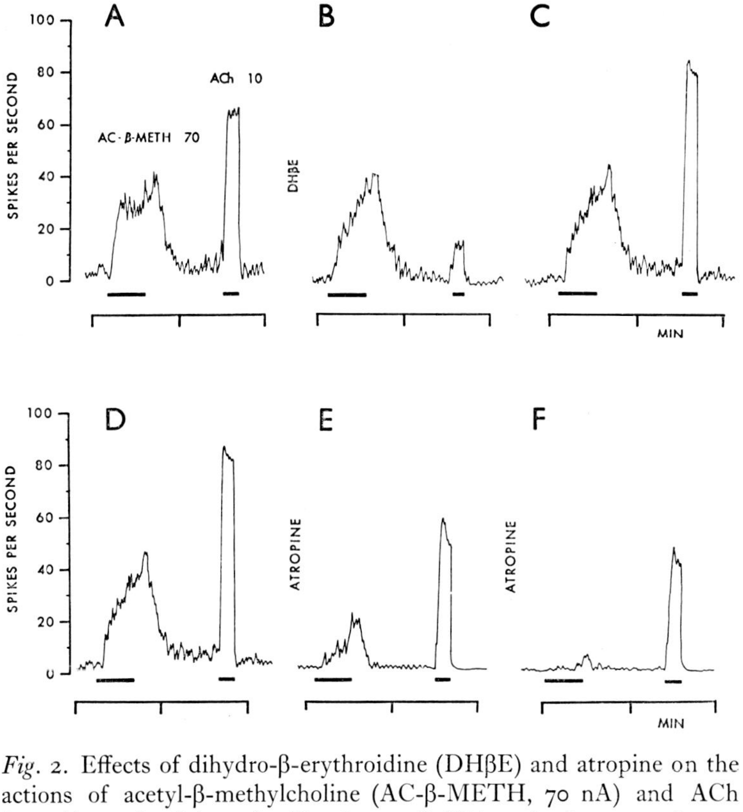

Fig. 2. Effects of dihydro-β-erythroidine (DHβE) and atropine on the actions of acetyl-β-methylcholine (AC-β-METH, 70 nA) and ACh (10 nA) on the firing of a Renshaw cell. This and subsequent figures of this type show chart recordings of the firing frequency (spikes per sec., ordinate) of the cell plotted against time (min., abscissa). Periods of drug ejection are indicated by horizontal bars above or below the tracings. A, Control responses; B, 3 min. after the beginning of an ejection of DHβE (10 nA) from a pipette containing a solution of 50 m*M* DHβE hydrobromide in 165 m*M* NaCl; C,D, 2 min. after the DHβE ejection was terminated; E, 5 min. after atropine sulphate (0·1 mg./kg.) had been injected intravenously; F, 11 min. after a further dose of atropine sulphate (0·5 mg./kg.). (Curtis and Ryall, 1966b.)

cells by ACh (of rapid onset and brief duration) was due predominantly to interaction with nicotinic receptors. When nicotinic receptors were blocked by dihydro-β-erythroidine, ACh often had a weak excitant action which was of slower onset and longer duration. This excitation was due to muscarinic receptors, as it could

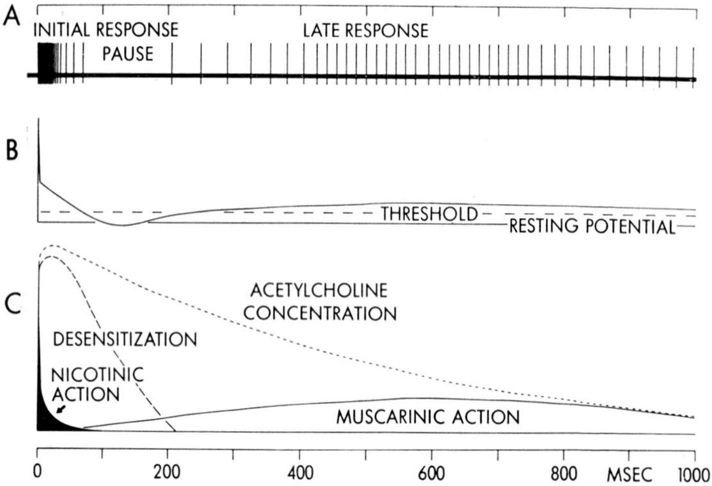

Fig. 3. Diagrammatic representation of the synaptic response of a Renshaw cell to a maximal ventral root volley. A, Firing pattern; B, Intracellularly recorded excitatory postsynaptic potential; C, Characteristics of the nicotinic and muscarinic actions of ACh (*see* text). (Curtis and Ryall, 1966c.)

be blocked by atropine. In the presence of both dihydro-β-erythroidine and atropine, ACh was as effective as acetyl-β-methylcholine in depressing Renshaw cells, suggesting the presence of a third type of cholinoceptive receptor (Curtis and Ryall, 1966b, c).

The relationship between cholinoceptive receptors on Renshaw cells and the Renshaw cell discharge evoked by ventral root stimulation is summarized diagrammatically in *Fig.* 3. A maximal ventral root volley usually evokes an initial high-frequency discharge, followed by a pause which, in turn, is followed by another, lower-frequency discharge (*Fig.* 3A). Dihydro-β-erythroidine markedly reduced the early discharge without affecting the late response. Atropine, however, blocked the late response, but had no action on the early discharge. These observations suggested that the early discharge is mediated by nicotinic receptors, whereas the late discharge is mediated by muscarinic receptors (Curtis and Ryall, 1966c).

The mechanism underlying the 'intermediate pause' is not clear. It has been suggested that either the nicotinic excitant action of synaptically released ACh is terminated by non-specific desensitization of the postsynaptic membrane, which persists for several hundred msec., or that the reduction in sensitivity to DLH and ACh (electrophoretically and synaptically released) is associated with the hyperpolarization which follows the initial depolarization (cf. *Fig.* 3A, and B; Curtis and Ryall 1966c). More recently Ryall (1970) has provided evidence which suggests that the pause may be due to inhibitory synaptic bombardment from the terminals of other Renshaw cells. Thus, Renshaw cells receive direct monosynaptic excitation from motor axon collaterals, and may also receive disynaptic inhibition from other Renshaw cells. It should be stressed, however, that the existence of such a 'mutual' inhibition presents other problems of interpretation, especially the observation that this inhibition is resistant to strychnine as well as bicuculline and picrotoxin (Ryall, 1971; Ryall, Piercey, and Polosa, 1972). Further studies are required to elucidate this problem.

Other pharmacological evidence supports the postulate that ACh is released synaptically from motor axon collaterals on to Renshaw cells. Hemicholinium-3 (which, amongst other actions, is known to block the synthesis of ACh in presynaptic terminals), administered electrophoretically in concentrations which neither excited Renshaw cells nor reduced their sensitivity to ACh, progressively increased the latency of the first and subsequent spikes, and decreased the number of spikes of the Renshaw cell discharge evoked by ventral root stimulation (Quastel and Curtis, 1965).

Studies of the release of ACh from the spinal cord are also consistent with the existence of cholinergic transmission in this structure. Stimulation of peripheral nerves increased the liberation of ACh from the perfused spinal cord of the dog (Bülbring and Burn, 1941) and cat (Kuno and Rudomin, 1966). Similarly, perfusion of the entire subarachnoid space, epidural space, and various segments (Edery and Levinger, 1971), or perfusion of the central canal (Jordan and Webster, 1971) of the cat spinal cord revealed an increase in the efflux of ACh following stimulation of the sciatic and femoral nerves. Since the increase in release of ACh was also apparent when the dorsal roots were cut (Kuno and Rudomin, 1966; Jordan and Webster, 1971), the release of ACh was interpreted as resulting from impulses in motor axon collaterals, especially as the amount of ACh released correlated strongly with the number of Renshaw spikes evoked (Kuno and Rudomin, 1966). It has long been known that ventral roots contain relatively high quantities of ACh and choline acetyltransferase compared with dorsal roots and some other areas of the C.N.S. (Loewi and Hellauer, 1938; MacIntosh, 1941; Wolfgram, 1954).

These findings may be compared with studies of the amphibian spinal cord, the physiology and pharmacology of which resemble that of the mammalian cord. Perfusion of the vertebral canal or the isolated spinal cord of the frog revealed an efflux of ACh which increased following stimulation of the limbs or ventral roots (Angelucci, 1956; Mitchell and Phillis, 1962; Ramwell, Shaw, and Jessup, 1966). Dorsal root stimulation did not evoke such a release of ACh. Although there is no physiological or anatomical evidence for Renshaw cells in the amphibian spinal cord, pharmacological studies suggest the existence of such cells (Kiraly and Phillis, 1961; cf. Esplin, Čapek, and Esplin, 1972).

Renshaw cells may be associated with AChE (*see* Söderholm, 1965; Erulkar and others, 1968; Navaratnam and Lewis, 1970; Odutola, 1972), but there are insufficient histochemical data to make definite conclusions.

Another interesting observation related to cholinergic transmission in the cord is that atropine (injected intravenously) depressed the spontaneous firing of Renshaw cells, suggesting that muscarinic receptors mediate the tonic excitatory bombardment on to these neurons (Curtis and Ryall, 1966c). However, the origin of these cholinergic fibres is unknown.

In conclusion, there is extremely good pharmacological evidence, supported by studies of ACh release and biochemical analyses, that ACh is liberated from the presynaptic terminals of motor axon collaterals on to Renshaw cells. Nothing, however, is known about the ionic mechanisms underlying the excitation of Renshaw cells by ACh, and it is not clear how closely the excitation by electrophoretically administered ACh resembles that of synaptic excitation of Renshaw cells. Such studies have been impeded by the difficulty in recording intracellularly from these cells. However, the recent success with intracellular dye injection (Jankowska and

Lindström, 1971) may rekindle interest in pharmacological experiments involving intracellular recording from Renshaw cells.

Excitant Amino-acids

The synaptic discharge of Renshaw cells evoked by *dorsal* root stimulation was unaffected by dihydro-β-erythroidine and atropine, indicating that a non-cholinergic excitatory transmitter also acts on Renshaw cells (Curtis, Phillis, and Watkins, 1961; Curtis and Ryall, 1966c). The identity of this transmitter is unknown, although it is possibly an excitant amino-acid.

Acidic amino-acids structurally related to glutamic acid have been shown to excite Renshaw cells (Curtis, Phillis, and Watkins, 1960; Curtis and Watkins, 1960, 1963). On the basis of regional distribution studies it has been suggested that both aspartate and glutamate are interneuronal excitatory transmitters in the spinal cord (Graham, Shank, Werman, and Aprison, 1967). This view is supported by the observation that occlusion of the thoracic aorta led to a loss of interneurons and a decrease in the levels of aspartate and glutamate in the grey matter of the lumbar spinal cord (Davidoff, Graham, Shank, Werman, and Aprison, 1967). However, since the loss of interneurons correlated directly only with the decrease in aspartate, aspartate was considered more likely to be the major transmitter of excitatory interneurons. Distribution studies suggest that glutamic acid may be the excitatory transmitter released from primary afferent fibres (Graham and others, 1967; Duggan and Johnston, 1970; Johnson and Aprison, 1970).

Although aortic occlusion resulted in some loss of small neurons in both dorsal and ventral grey matter, the most severe loss of neurons occurred in the central portions of the spinal grey matter (Davidoff and others, 1967). It is not clear, therefore, to what extent these results may be related to Renshaw cells, which are localized predominantly in the ventromedial part of the ventral horn. It is possible that both aspartate and glutamate are excitatory transmitters on Renshaw cells. This problem will be difficult to solve because the course of the non-cholinergic excitatory pathways terminating on Renshaw cells have not been elucidated, and no specific antagonists of amino-acid excitation are available. Since the synaptic responses of Renshaw cells following dorsal root stimulation are extremely short in rare cases, this activation must occur from direct terminations of primary afferent fibres (Frank and Fuortes, 1956). The biochemical evidence for the presence of glutamate in primary afferent fibres (Graham and others, 1967; Duggan and Johnston, 1970; Johnson and Aprison, 1970) therefore suggests that glutamate may be a transmitter acting on at least some Renshaw cells.

Additional evidence for glutamate and aspartate as excitatory transmitters in the spinal cord is reviewed later, in the section on motoneurons. It is perhaps necessary to stress at the beginning, however, that although the results of numerous different types of studies support the view that glutamate and aspartate are excitatory

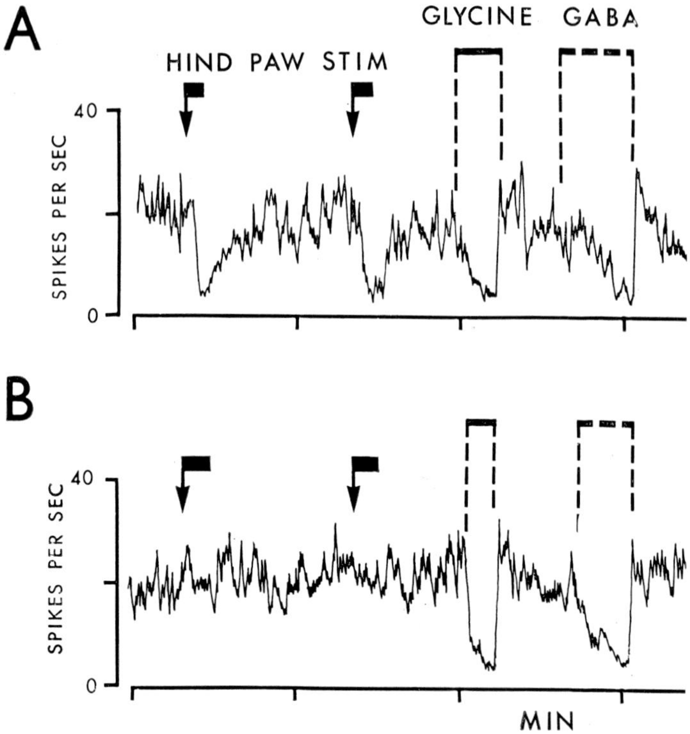

Fig. 4. Firing frequency of a Renshaw cell before (A) and 49 min. after (B) the local administration of tetanus toxin (by pressure from a pipette approximately 90 μm. above the tip of the recording electrode). A, Firing was maintained by ACh, when the anionic retaining current was reduced from —10 nA to —5 nA. The horizontal bars and arrows indicate stimulation of the hind-paw. Glycine (2 nA) and GABA (5 nA) were administered electrophoretically as indicated by the solid and broken horizontal bars. B, Spontaneous firing of the cell. (Curtis and De Groat, 1968.)

transmitters in the mammalian C.N.S. (Curtis and Johnston, 1970; Krnjević, 1970; Curtis, Duggan, Felix, Johnston, Tebēcis, and Watkins, 1972); Johnson, 1972, direct pharmacological evidence has yet to be obtained. The available evidence is certainly not as convincing as that for depressant amino-acids.

Depressant Amino-acids

Many neutral amino-acids structurally related to glycine and GABA have been shown to depress Renshaw cells, glycine being approximately equipotent to GABA (Curtis, Phillis, and Watkins, 1959; Curtis and Watkins, 1960, 1965; Curtis, Hösli, and Johnston, 1968; Larson, 1969). The inhibition of Renshaw cells by hindlimb stimulation, and the depressant action of glycine (but not that of GABA) were blocked by strychnine (Curtis, Hösli, and Johnston, 1968; Ryall and others, 1972), which blocks several other types of postsynaptic inhibition in the spinal cord (Curtis, 1963, 1969). These and similar studies suggest that glycine, or a glycine-like amino-acid, is an inhibitory transmitter released on to Renshaw cells (Curtis, Hösli, and Johnston, 1968), as is probably true for other spinal neurons (Curtis, 1969; Werman and Aprison, 1970).

Tetanus toxin blocks the same types of spinal postsynaptic inhibition as strychnine (Brooks, Curtis, and Eccles, 1957; Curtis, 1963). However, in contrast to strychnine, which blocks both postsynaptic inhibition and the depressant action of glycine, tetanus toxin has been shown to suppress the synaptic inhibition without affecting the depression by either glycine or GABA (Curtis and De Groat, 1968), as illustrated in *Fig.* 4. Stimulation of the ipsilateral hind paw inhibited the firing of a Renshaw cell, and this firing was depressed by glycine and GABA (*Fig.* 4A). Approximately 50 minutes after tetanus toxin had been administered (by pressure from an adjacent pipette) some distance from the cell, the inhibition evoked by squeezing the hind paw was blocked, whereas the depressant actions of the amino-acids remained unaffected (*Fig.* 4B).

The most likely explanation for these results is that tetanus toxin blocks the release of the inhibitory transmitter from presynaptic terminals, although other explanations cannot be excluded (Curtis and De Groat, 1968). Consistent with this hypothesis are the neurochemical findings that tetanus toxin had no action on the content of depressant amino-acids (including glycine and GABA) as well as some excitatory amino-acids in the cat and rat spinal cord, suggesting that the toxin is unlikely to deplete tissue stores of transmitter (Johnston, de Groat, and Curtis, 1969; Fedinec and Shank, 1971).

Other evidence for glycine as an inhibitory transmitter in the spinal cord is reviewed in the section on motoneurons. Although the ionic mechanism underlying the depressant action of glycine on Renshaw cells has not been compared with that of synaptic inhibition of these cells, it is reasonable to assume that it is identical to that of motoneurons.

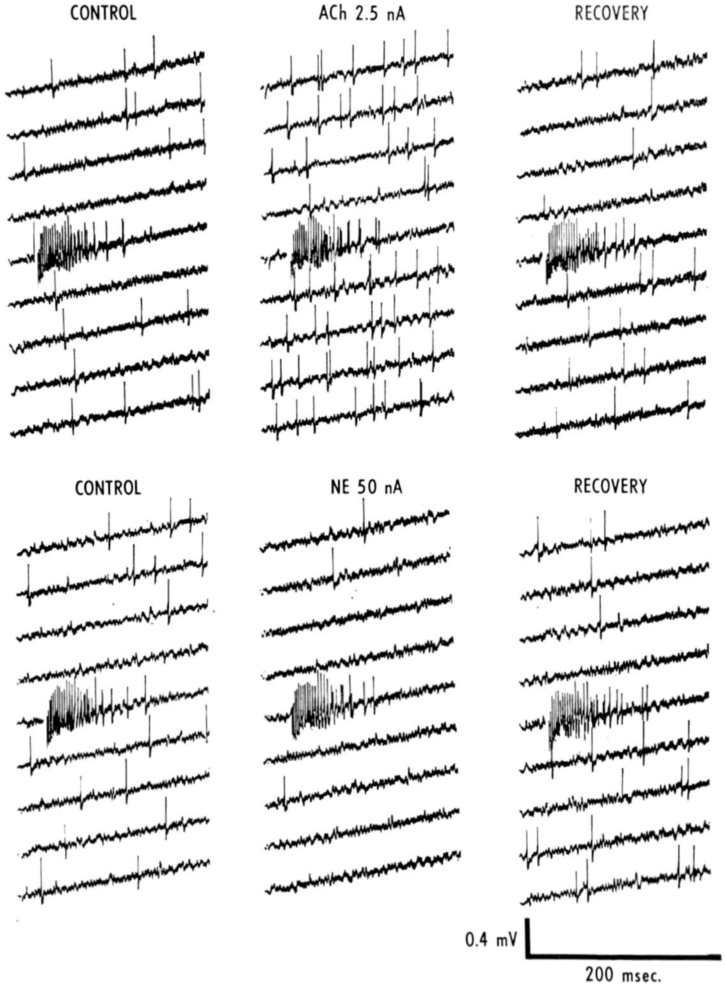

It is well known that antidromic volleys in motor axons generate strychnine-sensitive IPSPs in motoneurons through the mediation of Renshaw cells (Eccles and others, 1954; Curtis, 1963; Larson, 1969). If Renshaw cells inhibit other Renshaw cells (Ryall, 1970), this inhibition must be mediated by the same transmitter released by Renshaw cells in producing recurrent inhibition of motoneurons. In other words, a glycine-like inhibitory transmitter is released both on to the postsynaptic membrane of Renshaw cells and from their axon terminals. 'Mutual inhibition', however, has proved to be remarkably resistant to strychnine (Ryall, 1971; Ryall and others, 1972). Moreover, this inhibition was also unaffected by bicuculline and picrotoxin, indicating that GABA is unlikely to be involved. To explain these results, the presence of both strychnine-sensitive and strychnine-insensitive glycine receptors on Renshaw cells has been postulated (Ryall, 1971; Ryall and others, 1972). Clearly, further studies are required to clarify the mechanism underlying 'mutual inhibition' of Renshaw cells.

MONOAMINES

Microelectrophoretically administered monoamines apparently had no effect on spinal neurons in cats anaesthetized with pentobarbitone sodium (Curtis and others, 1961; but cf. De Groat and Ryall, 1967), but excited or depressed the firing of many spinal neurons in unanaesthetized (decerebrate) cats, or in animals anaesthetized with nitrous oxide and methoxyflurane, or with chloralose (Biscoe and Curtis, 1966; Biscoe, Curtis, and Ryall, 1966; Engberg and Ryall, 1966; Weight and Salmoiraghi, 1966a, b;

Fig. 5. Actions of ACh and NA on the firing of a Renshaw cell. Extracellular action potentials of the cell were recorded on film moving parallel to the *y* axis of the oscilloscope such that successive sweeps progress from bottom to top of each column. The fifth sweep in each column shows the evoked discharge produced by single maximal antidromic (alpha) volleys in the ipsilateral ventral root. In the upper records, ACh was administered for 18 sec. with an electrophoretic current of 2·5 nA, the centre records being taken 2 sec. after the beginning of the ACh ejection, and the recovery records 12 sec. after the end of the ejection. In the bottom records, noradrenaline (NE) was administered for 16 sec. with a current of 50 nA, the centre records being obtained 8 sec. after the beginning of the ejection, and the recovery records 15 sec. after the ejection. (Weight and Salmoiraghi, 1966b.)

Phillis, Tebēcis, and York, 1968a; Ryall and De Groat, 1972). Extensive studies of the effects of monoamines on Renshaw cells have not been made, but the available data indicate that proportions of these cells are either excited or depressed by NA and 5-HT. *Fig.* 5 illustrates the depressant action of NA on the firing of a Renshaw cell which could be readily excited by ACh.

NA depressed the firing of 11 and excited 5 of 16 Renshaw cells tested by Weight and Salmoiraghi (1966b). It depressed the firing by ACh and DLH of 8, and had no action on 5 of 13 Renshaw cells tested by Engberg and Ryall (1966), and, in addition, depressed the spontaneous firing rate of 1 and increased that of 2 other Renshaw cells tested. Biscoe and Curtis (1966) also described depressant actions of NA on Renshaw cells, but the number of neurons tested was not stated. 5-HT depressed the firing of ACh and DLH of 6 Renshaw cells, and increased the spontaneous firing of 2 others (Engberg and Ryall, 1966). DA has not been tested on Renshaw cells. These figures have little quantitative meaning in the present discussion, but serve to illustrate how little is known about the sensitivity of Renshaw cells to monoamines, especially 5-HT, which is just as likely to be a transmitter in the spinal cord as NA.

The studies of Renshaw cells indicate that both NA and 5-HT can have opposite actions on a particular electrophysiologically identified population of neurons, an observation made in various areas of the C.N.S. Studies of the depressant action of NA on spinal motoneurons (this chapter) and cerebellar Purkinje cells (Chapter 3) have revealed some properties of the mechanism of depression by NA, but the mechanism underlying excitation by NA is entirely unknown. Indeed, it has been suggested that in some cases, at least, the excitation of central neurons by NA (which is of extremely slow time course) may be indirect, resulting from localized vascular or *p*H changes (Frederickson, Jordan, and Phillis, 1971; Stone, 1971). At present it is not possible to know what proportion of the monoamine excitations observed have been artefacts. To be concise, therefore, both the excitant and depressant actions of monoamines will be described in terms of transmitter-like actions in this monograph.

Strychnine blocked two types of postsynaptic inhibition of Renshaw cells without affecting the depressant action of NA, indicating that NA is not an inhibitory transmitter at strychnine-sensitive synapses on Renshaw cells (Biscoe and Curtis, 1966). Renshaw cells may, however, be affected by an adrenergic bulbospinal pathway (Weight and Salmoiraghi, 1966b). The biochemical and histochemical evidence for this is similar to that for descending 5-hydroxytryptaminergic pathways, and applies equally well to other neurons in the spinal cord. This evidence is discussed in more detail in the section on motoneurons. In conclusion, it is possible that both NA and 5-HT are transmitters (excitatory or inhibitory) released on to Renshaw cells from fibres which originate in the brain-stem, but direct pharmacological evidence for this is lacking.

Motoneurons

Acetylcholine

It is well established that ACh is the excitatory transmitter released from motoneuron terminals which form neuromuscular junctions with skeletal muscle (*see* Katz, 1966), as well as from recurrent axon collaterals which synapse with Renshaw cells (previous section). Whether ACh affects the motoneuron soma was first investigated by Curtis and others (1961), who found no action of ACh on spinal neurons other than Renshaw cells in cats anaesthetized with pentobarbitone sodium. ACh had both excitant and depressant actions on various types of spinal interneurons in unanaesthetized cats (Curtis, Ryall, and Watkins, 1966; Engberg and Ryall, 1966; cf. Weight and Salmoiraghi, 1966a), and it is possible that ACh is a transmitter (either excitatory or inhibitory) on certain spinal interneurons other than Renshaw cells (cf. Iwata and Sakai, 1971). However, an action of ACh on motoneurons in unanaesthetized animals has yet to be reported. In any case, it is clear that ACh is not the excitatory transmitter mediating the monosynaptic pathway from primary afferent fibres to motoneurons, as little ACh and choline acetyltransferase are present in dorsal roots (Loewi and Hellauer, 1938; MacIntosh, 1941; Wolfgram, 1954), and cholinergic antagonists do not affect the monosynaptic excitatory responses of motoneurons evoked by impulses in primary afferent fibres (*see* Discussion by Curtis and others, 1966; cf. Bülbring and Burn, 1941; Phillis and Tebēcis, 1967a). This transmitter is possibly an excitant amino-acid.

Excitatory Amino-acids

Extensive investigations have been made of the sensitivity of spinal motoneurons to electrophoretically administered amino-acids. Motoneurons have been shown to be excited by glutamate, aspartate, and a number of other structurally related amino-acids, some of which occur naturally in the C.N.S. (Curtis and others,

1960; Curtis and Watkins, 1960, 1963). Intracellular recording from motoneurons revealed that extracellular administration of these amino-acids depolarized the membrane and increased its conductance (Curtis and others, 1960; Bernardi, Zieglgänsberger, Herz, and Puil, 1972), and that changes in the intracellular ion

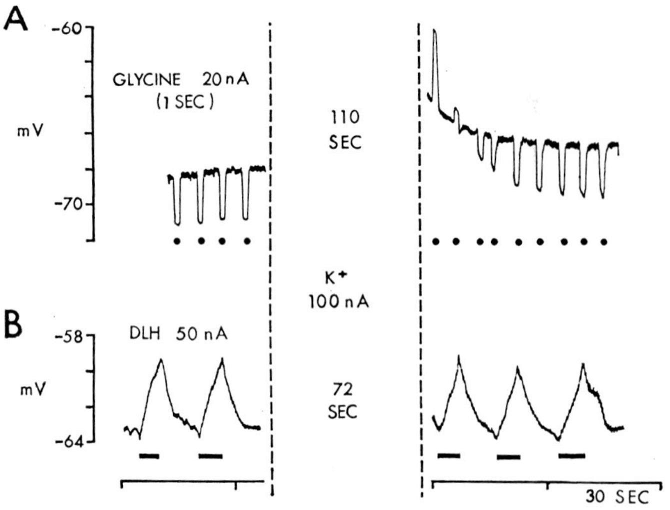

Fig. 6. Effects of potassium ions injected intracellularly on the hyperpolarizing action of glycine and the depolarizing action of DLH on a peroneal motoneuron, recorded intracellularly with a potassium citrate electrode. A, Extracellular administration of glycine (20 nA for 1 sec., indicated by the dot beneath the record). The vertical broken lines represent the period in which potassium ions were injected intracellularly with a current of 100 nA for 110 sec. B, Extracellular administration of DLH (50 nA for 6 sec., black bar beneath the record). Potassium was injected with a current of 100 nA for 72 sec. *Ordinates*: membrane potential, in mV. *Abscissae*: time, in 30-sec. intervals. (Curtis and others, 1972.)

content, which did not alter the amplitude of EPSPs, were also without effect on the depolarizing action of DLH (Curtis and Johnston, 1970; Curtis and others, 1972). *Fig*. 6 illustrates intracellular records from a motoneuron. Immediately after potassium ions had been injected into the cell, the hyperpolarizing action of glycine was reversed to a depolarizing action, and recovery of the hyperpolarization took up to 30 sec. (*Fig*. 6A). However, the intracellular injection of potassium ions had no action on the depolarization produced by DLH (*Fig*. 6B).

Similarly, intracellular injections of chloride ions sufficient markedly to alter IPSPs and the hyperpolarizing action of glycine were without effect on the depolarizing action of DLH on other motoneurons (Curtis and others, 1972). Previous studies have established that these ions injected into motoneurons have no effect on EPSPs (Coombs, Eccles, and Fatt, 1955). In addition, extracellular concentrations of tetrodotoxin sufficient to block the generation of action potentials had no effect on EPSPs and the depolarization of motoneurons by DLH (Curtis and others, 1972). Thus, the ionic mechanism of the depolarization of spinal motoneurons by excitant amino-acids appears to resemble that of synaptic excitation, and is consistent with the hypothesis that glutamate or a related amino-acid is a transmitter on motoneurons. Ideally, it is necessary to demonstrate that the reversal potential for the depolarizing action of glutamate is similar to the equilibrium potential for monosynaptic excitation, in order to show 'identity of action' (*see* Werman, 1966). Such an attempt on spinal motoneurons revealed that the reversal potential for glutamate was at a less depolarized level than that for synaptic excitation (Curtis, 1965). This apparent discrepancy, however, can be explained if it is assumed that the synaptic receptors for glutamate are present on the dendrites of motoneurons, and that motoneurons are likely to be impaled in the soma. At present, therefore, the data on the ionic mechanism underlying the depolarization of motoneurons by amino-acids such as glutamate are not inconsistent with the view that glutamate or a related amino-acid is an excitatory transmitter on motoneurons (Curtis and others, 1972).

Studies of the regional distribution of amino-acids in the spinal cord and roots of the cat have led to the proposal that glutamate is the excitatory transmitter released from the terminals of primary afferent fibres (Graham and others, 1967; Duggan and Johnston, 1970; Johnson and Aprison, 1970). In other words, it is likely that glutamate is the excitatory transmitter which mediates the monosynaptic excitation of motoneurons from dorsal roots. Unfortunately, a specific and effective antagonist of glutamate is not yet available for testing on such reflex responses, although some compounds have recently been shown to reduce excitation by amino-acids (L-glutamate, L-aspartate, DLH) in concentrations which have little or no effect on that by ACh on central neurons (McLennan, Marshall, and Huffman, 1971; Curtis and others, 1972; Davies and Watkins, 1972; Haldeman, Huffman, Marshall, and

McLennan, 1972). The discovery of an amino-acid antagonist which can distinguish between glutamate and aspartate will, no doubt, be a major breakthrough in the identification of excitatory transmitters.

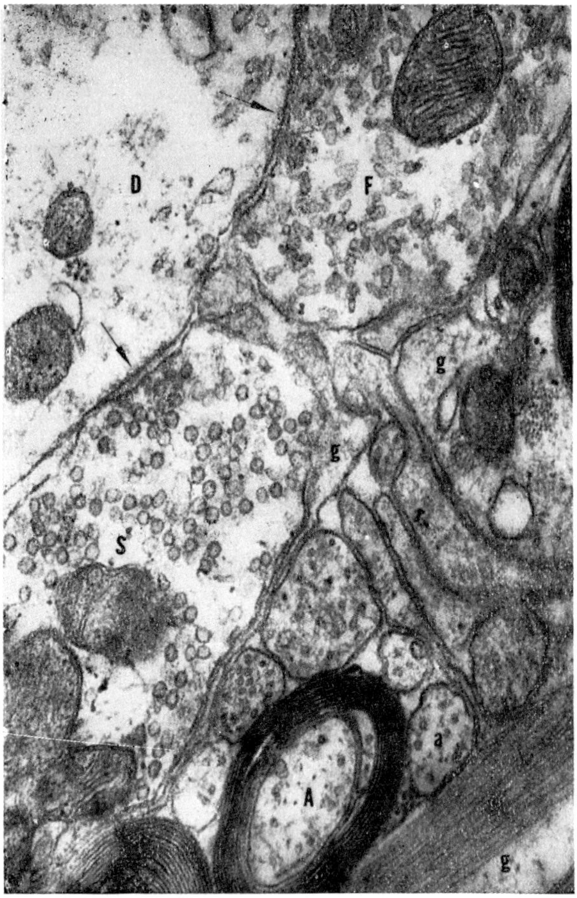

Fig. 7. Electron micrograph from a monkey spinal cord, showing a large dendrite (D), identified by the presence of a ribosomal cluster, and bearing two types of synaptic bulbs (S and F) side by side. The Type S synaptic bulb contains spherical agranular vesicles, whereas the Type F bulb contains mostly flattened agranular vesicles. The arrows indicate synaptolemmal attachment sites. A, Myelinated axon; a, Unmyelinated axon or axon telodendrion; g, Glial process. (× 52,000) (Bodian, 1967.)

The efflux of labelled glutamate (as well as glycine and GABA) from slices of the rat spinal cord was increased by electrical stimulation, which had no effect on the efflux of labelled alanine or urea (Hopkin and Neal, 1971), supporting the hypothesis that glutamate is a transmitter in the mammalian spinal cord.

Another approach which is proving useful in the identification of synaptic transmitters is light and electron microscopy combined with autoradiography. It is becoming increasingly clear that nerve terminals containing spherical synaptic vesicles are probably excitatory, whereas terminals with flattened vesicles are probably inhibitory. This was first suggested by Uchizono (1965) from studies of the cerebellum (cf. Walberg, 1965; inferior olive), but confirmatory evidence for this has been subsequently obtained from studies of many structures, including the spinal cord (Uchizono, 1966; Bodian, 1967; Conradi, 1969). *Fig.* 7 is an electron micrograph of the monkey spinal cord, showing a terminal containing spherical vesicles (S) and an adjacent one with flattened vesicles (F) synapsing with a dendrite (D).

Motoneurons receive synaptic bulbs containing either spherical or flattened vesicles, and some containing large granular (adrenergic ?) vesicles (Uchizono, 1966; Bodian, 1967; Conradi, 1969). Large axodendritic synaptic bulbs of proven dorsal root origin contain irregularly spheroid agranular vesicles, and occasionally receive axo-axonic endings containing flattened vesicles. Autoradiographic studies of the distribution of amino-acids in the spinal cord have so far been concerned only with depressant amino-acids, but will, no doubt, soon be extended to include excitatory amino-acids as well.

Depressant Amino-acids

a. Glycine

Neutral amino-acids related to glycine and GABA have been shown to depress motoneurons (Curtis and others, 1959; Curtis and Watkins, 1960, 1965; Larson, 1969), glycine usually being more effective than GABA (Curtis, Hösli, Johnston, and Johnston, 1968; Werman, Davidoff, and Aprison, 1968). Regional distribution studies of amino-acids in the spinal cord have indicated that the expected distribution for a postsynaptic inhibitory transmitter is satisfied by concentration profiles for glycine and GABA (Graham and others, 1967), whereas other naturally occurring depressant amino-acids (cystathionine, L-serine, taurine,

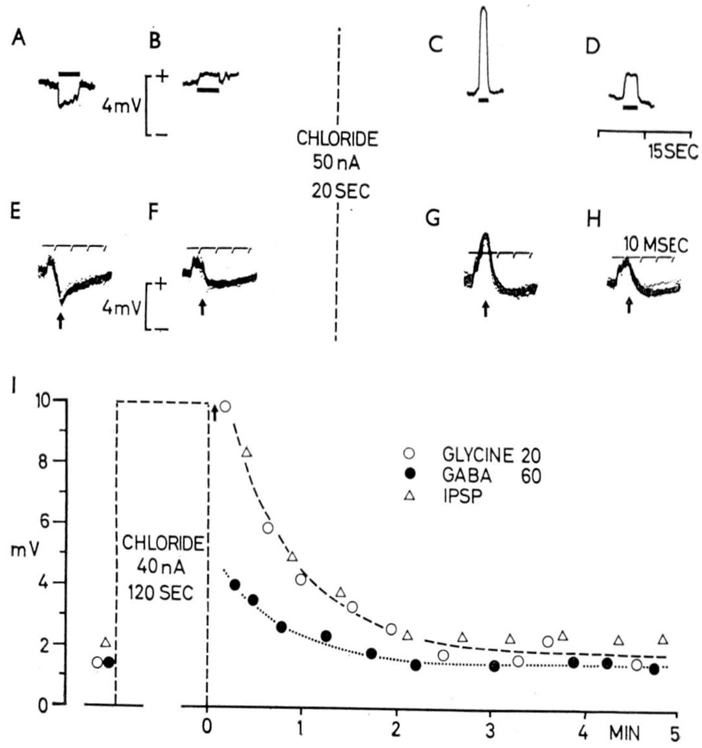

Fig. 8. Influence of intracellular chloride injection on the action of glycine and an IPSP. Unidentified motoneuron, resting potential −65 mV, KCl recording electrode. A–D, Alterations of membrane potential by glycine (30 nA, indicated by black bar). E–H, Postsynaptic potentials evoked by stimulating the sural nerve; the arrow indicates the peak of the IPSP in E. A, E, 4–5 min. after impalement. B, F, 90 sec. later. C, G, Immediately after the intracellular injection of chloride ions (50 nA for 20 sec.). D, H, 3 min. later. I, Another series from the same neuron showing the effect of an intracellular chloride injection (40 nA for 120 sec.) on depolarizing potentials (mV) produced by glycine (o, 20 nA for 3 sec.) and GABA (●, 60 nA for 4 sec.) and the depolarizing component (Δ) of the postsynaptic potential corresponding to the original hyperpolarization of *Fig.* 9E by Curtis, Hösli, Johnston, and Johnston (1968). The arrow indicates that this depolarization exceeded threshold for spike production. *Ordinate*: change in resting potential, mV. *Abscissa*: time in minutes. The time of chloride injection is not to scale. (Curtis, Hösli, Johnston, and Johnston, 1968.)

L-α-alanine) do not satisfy this criterion (Johnston, 1968). The high levels of glycine in the ventral horn, together with the low content in ventral roots, and the decrease in both glycine content and the number of interneurons after temporary aortic occlusion, suggest that glycine is an inhibitory transmitter associated with spinal interneurons (Aprison and Werman, 1965; Graham and others, 1967; Davidoff and others, 1967). Studies involving intracellular recording support the hypothesis that glycine is an inhibitory transmitter in the spinal cord (Bruggencate and Engberg, 1968; two papers by Curtis and others, 1968; Werman and others, 1968; Larson, 1969). The depression of motoneurons and interneurons by glycine and GABA was accompanied by a hyperpolarization of the membrane and an increase in its conductance; the reversal potentials for synaptic inhibition and the amino-acid-induced hyperpolarizations were similar, and alterations in intracellular ion concentrations affected both types of hyperpolarization in the same manner.

Fig. 8 illustrates the effects of injecting chloride ions into a motoneuron. Shortly after impalement of the cell with a KCl-containing electrode, glycine hyperpolarized the membrane (*Fig.* 8A) and stimulation of the sural nerve evoked a small EPSP followed by a larger IPSP (*Fig.* 8E). Shortly afterwards, glycine produced a small depolarization (*Fig.* 8B) and sural nerve stimulation evoked a smaller IPSP (*Fig.* 8F), presumably as a result of the diffusion of chloride ions into the cell from the electrode. Immediately after chloride ions were injected, both glycine and sural nerve stimulation produced a marked depolarization (*Fig.* 8C, G), which decayed with time (*Fig.* 8D, H). The time course of the changes in the IPSP, glycine, and GABA depolarizations after another intracellular injection of chloride into the same motoneuron is plotted in *Fig.* 8I.

Strychnine blocks various types of postsynaptic inhibition in the spinal cord (Curtis, 1963, 1969). The alkaloid consistently blocked the depressant action of glycine, but not that of GABA, on motoneurons and other spinal neurons (Curtis and others, 1968, two papers; Larson, 1969). These observations are illustrated in *Fig.* 9. On one motoneuron, intravenously administered strychnine blocked the IPSP evoked by peroneal nerve stimulation (*Fig.* 9A, C) but had no action on the EPSP evoked by stimulation of the gastrocnemius nerve (*Fig.* 9B, D). On another motoneuron, electrophoretically administered strychnine reversibly blocked the hyperpolarizing action of glycine (*Fig.* 9E, G, I), but not that of GABA (*Fig.* 9F, H, J).

Although the evidence for glycine being an inhibitory transmitter in the spinal cord has recently been reviewed (Aprison, Davidoff, and Werman, 1970; Curtis and Johnston, 1970; Werman, 1972b), more recent findings provide even stronger evidence. The termination of the depression by glycine is thought

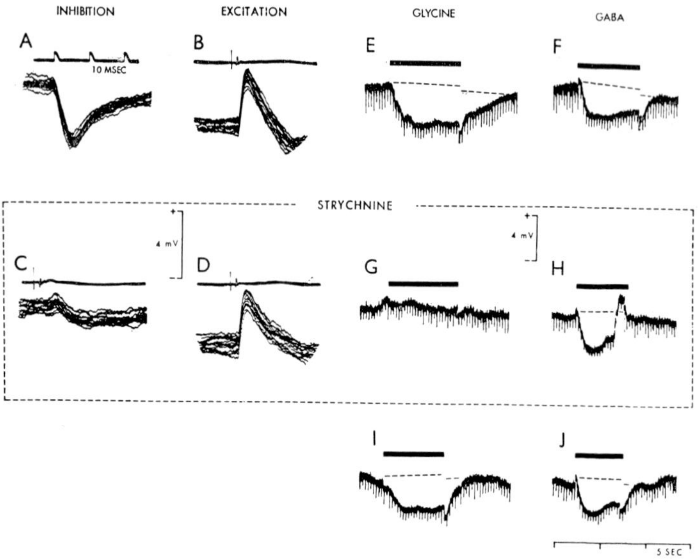

Fig. 9. Intracellular potentials recorded from two different motoneurons (A–D and E–J) demonstrating the action of strychnine on postsynaptic potentials. A–D, Gastrocnemius motoneuron, recorded with the central barrel of a co-axial electrode. E–J, Tibial motoneuron, recorded with a parallel electrode. A, IPSP evoked by stimulating the peroneal nerve. B, EPSP evoked by stimulating the gastrocnemius nerve. C, D, The same responses after strychnine was allowed to diffuse from the outer barrel of a co-axial electrode for 3 min. E, Hyperpolarization produced by extracellularly administered glycine (30 nA, black bar). The fast vertical deflections are IPSPs evoked by stimulating the sural nerve. The broken lines provide a baseline from which the amount of hyperpolarization can be measured. F, Hyperpolarization produced by GABA (60 nA). G, H, The same responses during the extracellular administration of strychnine from one barrel of a parallel electrode. I, J, Several minutes after terminating the strychnine ejection. (Curtis, 1969.)

to be the result of transport processes, rather than of enzymic inactivation (Neal and Pickles, 1969; Curtis, Duggan, and Johnston, 1970; Shank and Aprison, 1970; Johnston and Iversen, 1971; Neal, 1971). Although glycine can be taken up by a relatively non-specific system in slices and homogenates of various areas of the C.N.S., a specific high-affinity uptake of glycine was observed only in tissue of the spinal cord, medulla, and pons (Neal and Pickles, 1969; Johnston and Iversen, 1971; Neal, 1971), which also contain high concentrations of glycine compared with other structures of the C.N.S. (Shaw and Heine, 1965; Kandera, Levi, and Lajtha, 1968; Rizzoli, 1968; Aprison, Shank, and Davidoff, 1969; Shank and Aprison, 1970). A proportion of the glycine content of the rat spinal cord appeared to be associated with synaptosomes (Johnston and Iversen, 1971). In addition, although both glycine and GABA were taken up by synaptosomes of spinal cord homogenates and nerve-endings of spinal cord slices, there was some evidence that these two amino-acids were taken up by different populations of nerve terminals (Iversen and Johnston, 1971; Iversen and Bloom, 1972).

Autoradiographic studies of slices of the rat spinal cord incubated with labelled glycine revealed that this amino-acid was distributed to a greater extent in the grey matter than in the white matter in the ventral horn, and that it was associated with terminals containing flattened vesicles, synapsing with motoneurons and a proportion of small neurons in the ventral horn (Hökfelt and Ljungdahl, 1971a; Matus and Dennison, 1971).

Perfusion of the central canal of the cat spinal cord with a solution containing a blocker of glycine uptake revealed a spontaneous efflux of glycine which was increased 2–3 times by stimulation of the femoral and sciatic nerves (Jordan and Webster, 1971). Moreover, depolarization of slices of the rat spinal cord by electrical stimulation or by solutions containing a high concentration of potassium greatly increased the efflux of glycine (Hopkin and Neal, 1971). Evoked release of glycine has also been demonstrated from the amphibian spinal cord (Aprison, 1970; Mitchell and Roberts, 1972).

b. GABA

Various studies suggest that GABA is also an inhibitory transmitter in the mammalian spinal cord. The levels of GABA and its synthesizing enzyme, glutamic acid decarboxylase, are higher in the dorsal than in the ventral grey matter of the spinal cord (Albers and Brady, 1959; Graham and others, 1967; Rizzoli, 1968; Graham and Aprison, 1969), and since temporary aortic occlusion did not affect the spinal GABA levels or the majority of neurons in the dorsal horn (Davidoff and others, 1967), most GABA-releasing spinal neurons are likely to be located in the dorsal horn. Histochemical studies of an enzyme which destroys GABA revealed strong activity in the neuropile of the substantia gelatinosa (Sims, Weitsen, and Bloom, 1972). Regional uptake studies have shown that GABA is taken up to a different degree in different regions of the C.N.S., and that this uptake closely parallels endogenous GABA levels (Hökfelt, Jonsson, and Ljungdahl, 1970; Iversen and Johnston, 1971). Thus, GABA uptake occurs in slices and homogenates of the rat spinal cord, although the density of uptake sites in this region is only about one-tenth of that in the cerebral cortex (Iversen and Johnston, 1971). A proportion of the spinal GABA content is associated with synaptosomes, which form a different population from the glycine-containing synaptosomes (Iversen and Johnston, 1971; Iversen and Bloom, 1972).

The depressant action of GABA on motoneurons and other spinal neurons is essentially identical with that of glycine (Bruggencate and Engberg, 1968; Curtis and others, 1968, two papers; Werman and others, 1968; Larson, 1969), except that GABA is resistant to strychnine (Curtis and others, 1968, both papers; Larson, 1969). Although most types of spinal postsynaptic inhibition can be blocked by strychnine (Curtis, 1963, 1969), some postsynaptic inhibitions of motoneurons are unaffected, or even potentiated by this convulsant (Kellerth and Szumski, 1966a; Wagner and Klee, 1968; Curtis, Duggan, Felix, and Johnston, 1971). In particular, picrotoxin and bicuculline, two relatively specific antagonists of GABA in various areas of the C.N.S., including the spinal cord (Engberg and Thaller, 1970a; Curtis, Duggan, Felix, and Johnston, 1971), had no effect on strychnine-sensitive spinal inhibition, but blocked certain inhibitions which were resistant to strychnine; these were typically of long duration (several tens or hundreds of msec.) (Eccles, Schmidt, and Willis, 1963; Kellerth and Szumski, 1966b; Kellerth, 1968; Wagner and Klee, 1968; Schmidt, 1971; Curtis, Duggan, Felix, and Johnston, 1971; Levy, Repkin, and Anderson, 1971; Huffman and McFadin, 1972; Tanji and Kato, 1972).

Fig. 10 illustrates an example of a strychnine-insensitive, picrotoxin-sensitive postsynaptic inhibition in the spinal cord. The intracellular records of the motoneuron were obtained from a cat into which strychnine (0·2 mg./kg.) had been injected intravenously. The maintained firing produced by injected current

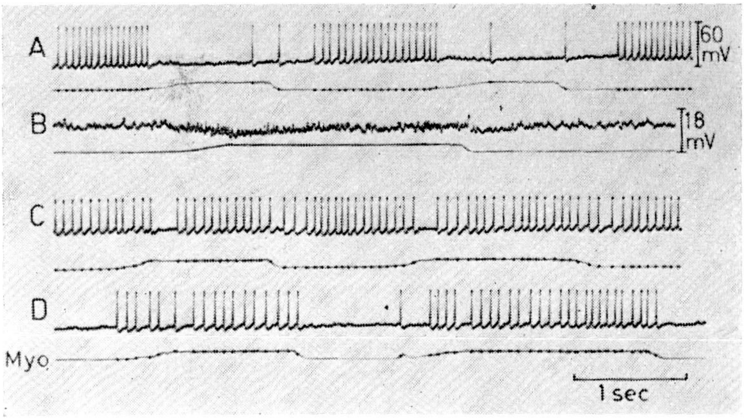

Fig. 10. Intracellular records from a spinal motoneuron of a cat which had received strychnine (0·2 mg./kg.) intravenously. Peroneal motoneuron of 70 mV spike height. A, Maintained firing was caused by injected current, and stretches of triceps surae (500 g., bottom trace), inhibited the discharge. B, Higher d.c. gain showing a hyperpolarization (3·5 mV maximum) of the average membrane potential during muscle stretch (500 g.). C, 1·5 min. after picrotoxin (2·0 mg./kg.) had been injected intravenously. D, 5 min. after the injection of picrotoxin. During stimulation of the neuron with just subthreshold injected depolarizing current, stretching of triceps surae now fired the neuron. (Kellerth and Szumski, 1966b.)

was inhibited by stretching the triceps surae muscle (*Fig.* 10A), the inhibition resulting from a membrane hyperpolarization (*Fig.* 10B). Shortly after picrotoxin (2 mg./kg.) had been administered intravenously, the inhibition was largely blocked (*Fig.* 10C), and a few minutes later muscle stretch resulted in firing of the motoneuron (*Fig.* 10D).

The mechanism of the picrotoxin- and bicuculline-sensitive prolonged inhibition of motoneurons by impulses in cutaneous and muscle afferents is not clear. The possible explanations for 'presynaptic inhibition', 'presynaptic depolarization', or 'remote inhibition' (Frank, 1959; Eccles, 1964; Kellerth, 1968) are too complex and controversial to be discussed here, and the reader should consult the excellent review by Schmidt (1971). Just two of the problems may be mentioned in passing. Firstly, there are opposing opinions as to whether GABA depolarizes presynaptic terminals (Schmidt, 1963; Curtis and Ryall, 1966d; Tebēcis and Phillis, 1967, 1969; Galindo, 1969; Davidson and Southwick, 1971; Barker and Nicoll, 1972; Davidoff, 1972a, b). Secondly, the connexions of presynaptic terminals with interneurons, and the proportions of round-vesicle (excitatory) and flattened-vesicle (inhibitory) axo-axonic synapses involved are not entirely clear (*see* references by Conradi, 1969). Recently, Barker and Nicoll (1972) demonstrated that Na^+ ions are necessary for the depolarizing actions of both GABA and glutamate on dorsal root terminals in the amphibian spinal cord. However, it is nevertheless possible that both presynaptic and postsynaptic mechanisms are responsible for the different 'remote inhibitions' studied (*see* Curtis and Felix, 1971a; Curtis, Duggan, Felix, and Johnston, 1971). Clearly, many more studies are required to clarify these problems.

MONOAMINES

Effects of monoamines on spinal motoneurons have been observed only in unanaesthetized cats and in animals anaesthetized with nitrous oxide and methoxyflurane. Noradrenaline has been shown to depress motoneurons sensitive to it (Engberg and Ryall, 1966; Weight and Salmoiraghi, 1967; Curtis, 1968; Phillis and others, 1968a; Engberg and Thaller, 1970b), although the samples of motoneurons tested are small compared with similar studies using amino-acids. Studies involving intracellular recording from motoneurons have revealed that the depression by NA results from a hyperpolarization of the membrane (Curtis, 1968; Phillis and others, 1968a; Engberg and Thaller, 1970b). Although no clear changes in membrane conductance were observed initially, Engberg and Marshall (1971) recently found that the hyperpolarization of motoneurons by NA is accompanied by a small *decrease* in membrane conductance. This is in contrast to glycine and GABA, which produce an increase in conductance.

Fig. 11 illustrates the effects of NA on intracellularly recorded potentials from a lumbar motoneuron. Sodium ions had no action on the resting membrane potential and on the membrane potential shifts resulting from current pulses passed across the membrane (*Fig.* 11A), indicating that positive current flow alone did not affect this motoneuron. However, increasing doses of NA produced increasing hyperpolarizations and increased the amplitude of the membrane potential shifts resulting from the current injections, indicative of an increase in membrane resistance (*Fig.* 11B, C).

The hyperpolarization by NA was decreased and sometimes reversed to a depolarization when the membrane was depolarized by current injection, and increased when the membrane was hyperpolarized (Engberg and Marshall, 1971). Similar observations have been made previously on cerebellar Purkinje cells (Siggins, Oliver, Hoffer, and Bloom, 1971) and may indicate that the hyperpolarization by NA in various areas of the C.N.S. is mediated by comparable mechanisms.

In addition, a number of pharmacological investigations involving systemic injections of the precursor of NA and compounds which interfere with the action, metabolism, storage, and release of NA suggest that this catecholamine (and possibly DA)

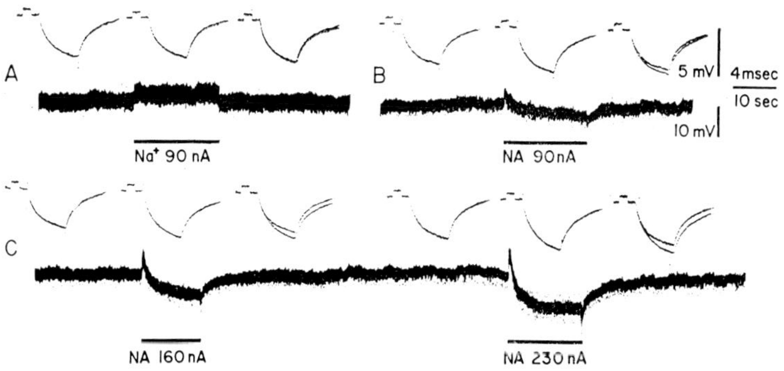

Fig. 11. Intracellular records from a lumbar motoneuron, showing the actions of increasing concentrations of NA administered extracellularly. *Lower* (*thick*) *traces*: membrane potential recorded at a slow speed (10 sec. time calibration). A, An injection of Na^+ ions (90 nA) only produced a coupling artefact. B, C, NA (90, 160, and 230 nA) produced hyperpolarizations. *Upper traces*: membrane potential shifts due to the injection of current pulses (5 nA) through the recording electrode, showing the membrane conductance. The voltage drop over the electrode resistance has been subtracted and 16 pulses averaged electronically in each record. The first record in each set of three was obtained immediately before the corresponding drug ejection, the second record during the maximal drug action, and the third record consists of the other two superimposed. *Calibration*: 5 mV for upper records, 10 mV for lower records. (Engberg and Marshall, 1971.)

functions as an inhibitory transmitter in the feline spinal cord (Cranmer, Bram, and Bach, 1959; Andén, Jukes, and Lundberg, 1966; Andén, Jukes, Lundberg, and Vyklický, 1966a, b; Jankowska, Lund, and Lundberg, 1966; Grillner, Hongo, and Lundberg, 1967; Jankowska, Jukes, Lund, and Lundberg, 1967a, b; Engberg, Lundberg, and Ryall, 1968a, b; Baker and Anderson, 1970a, b; Barnes and Pompeiano, 1971b; Fedina, Lundberg, and Vyklický, 1971). The results from all these studies are too detailed to discuss adequately here, but have been interpreted in terms of a noradrenergic projection from the brain-stem to the spinal cord. The evidence for such a projection is similar to that for a descending 5-hydroxytryptaminergic pathway.

Histochemical studies of rats have demonstrated that NA- and 5-HT-containing neurons of the caudal ventromedial area of the medulla oblongata give rise to axons that descend in the ventral and lateral funiculi of the spinal cord (Carlsson, Falck, Fuxe, and Hillarp, 1964; Dahlström and Fuxe, 1965a, b; Fuxe, Hökfelt, and Ungerstedt, 1970; Ungerstedt, 1971). NA- and 5-HT-containing nerve terminals have been observed in the dorsal and ventral horns of the spinal cord, and some of these terminals are localized close to the cell bodies of motoneurons (Dahlström and Fuxe, 1965b). Transection of the spinal cord led to an almost complete loss of NA and 5-HT below the level of transection (Andén, Häggendal, Magnusson, and Rosengren, 1964) and a complete disappearance of terminals containing NA and 5-HT (Dahlström and Fuxe, 1965b). Moreover, stimulation of descending fibres resulted in a release of NA and 5-HT from the spinal cord (Andén, Carlsson, Hillarp, and Magnusson, 1964, 1965), and a reduction in the fluorescence intensity of monoamine-containing terminals (Dahlström, Fuxe, Kernell, and Sedvall, 1965). Both NA and 5-HT are localized in the synaptosomal fraction of homogenates of the rat and guinea-pig spinal cord (Fuxe, Grobecker, Hökfelt, and Jonsson, 1967; Ross, Andreoli, and Marchbanks, 1971).

Monoamines other than NA have not been tested on motoneurons to any great extent. Engberg and Ryall (1966) and Weight and Salmoiraghi (1967) observed no action of 5-HT on the few motoneurons studied, but Phillis and others (1968a) found that 5-HT often depressed motoneurons, the depression resulting from a membrane hyperpolarization. Pharmacological investigations involving intravenous administration of the precursor of 5-HT (5-HTP) and related compounds, as well as biochemical findings, also suggest that 5-HT is a transmitter in the spinal cord, possibly involved in motor regulation (Anderson and Holgerson, 1966; Anderson and Shibuya, 1966; Marley and Vane, 1967; Anderson, Baker, and Banna, 1967; Banna and Anderson, 1968; Clineschmidt and Anderson, 1970; Clineschmidt, Pierce, and Lovenberg, 1971; Ahlman, Grillner, and Udo, 1971). In particular, stimulation of the raphe nuclei and ventromedial reticular formation either facilitated or inhibited spinal reflexes; and the bulbospinal inhibition, but not the facilitation, was blocked by substances which also blocked the action of 5-HT (Clineschmidt and Anderson, 1970). Although histochemical and neurochemical evidence suggests the presence of a 5-hydroxytryptaminergic projection from the raphe nuclei and neighbouring areas of the brain-stem to the spinal cord, various observations led Clineschmidt and Anderson (1970) to interpret their results in terms of a 5-HT-containing interneuron within the spinal cord. The antagonists of 5-HT were more effective when injected into the arterial supply of the spinal cord than when injected into the arterial supply of the brain-stem, and the minimal conduction velocity (10 m./sec.) observed for the bulbospinal inhibitory pathway was considered too rapid for the small-diameter unmyelinated 5-HT-containing axons which have been described. Furthermore, an earlier investigation revealed that approximately 25 per cent of the spinal 5-HT content remained several weeks after transection of the spinal cord (Shibuya and Anderson, 1968).

These results are difficult to explain, as there is no histochemical evidence for monoamine-containing cell bodies in the rat spinal cord. Moreover, it is perhaps too ambitious to estimate likely conduction velocities of the monoamine-containing descending axons (Dahlström and Fuxe, 1965b) if the diameters of the axons have been calculated from the width of fluorescence. The actual width of these fibres may be greater, allowing conduction velocities of the order of 10 m./sec. However, it is also possible that the histochemical studies of rats do not apply to the results obtained by Anderson and his colleagues, who performed their experiments on cats. At present, therefore, it is not possible to make definite conclusions about spinal pathways which operate with 5-HT as a transmitter, but the bulk of the evidence suggests that at least some 5-hydroxytryptaminergic fibres in the spinal cord descend from the brain-stem. The 'status' of 5-HT in the spinal cord has recently been reviewed (Anderson, 1972).

Histamine had depressant actions on spinal motoneurons (Phillis and others, 1968a), but the significance of this is not clear. An action of DA on motoneurons has not been reported.

In conclusion, it is important to remember that although the course of the monoamine-containing fibres is reasonably well-defined (Ungerstedt, 1971), their precise origins and terminations have not been elucidated by fluorescence histochemistry. Moreover, almost all the fluorescence studies have been done of rats, which may well differ from those of cats (*see* Pin, Jones, and Jouvet, 1968; Sladek, 1971). In addition, the recent histochemical evidence for a descending pathway containing an indoleamine other than 5-HT (Björklund, Falck, and Stenevi, 1970) has complicated the situation even more. It is possible that some results interpreted in terms of a 5-hydroxytryptaminergic pathway may have involved this other tryptamine-containing pathway.

Autonomic Spinal Neurons

Few pharmacological studies have been made of spinal neurons which project to autonomic ganglia, but the available data are consistent with the view that 5-HT and NA are transmitters on some of these cells.

Sympathetic Preganglionic Neurons

Sympathetic preganglionic neurons are located in the intermediolateral cell column of the thoracic spinal cord (Hongo and Ryall, 1966), a region particularly rich in 5-HT- and NA-containing nerve terminals (Dahlström and Fuxe, 1965b). Sympathetic preganglionic neurons may be identified electrophysiologically by the antidromic activation that follows stimulation of the splanchnic nerves or the sympathetic rami. These neurons either do not fire spontaneously, or fire at low rates (of the order of 2 spikes/sec.), and the maximum sustained average frequency of firing produced by DLH and other excitant substances is approximately 15/sec., which corresponds extremely well with the firing rate of preganglionic nerve-fibres that produces a maximal physiological effect in the animal (De Groat and Ryall, 1967; Ryall, 1967).

Electrophoretically administered 5-HT excited the majority of sympathetic preganglionic neurons tested, whereas NA depressed the firing of less than half of these cells; ACh had no effect (Hongo and Ryall, 1966; De Groat and Ryall, 1967; Ryall, 1967). The sample of neurons studied is relatively small, probably because the experiments are difficult to perform. Nevertheless, the results are consistent, and together with the histochemical data (Dahlström and Fuxe, 1965b) raise the possibility that 5-HT is an excitatory, whereas NA is an inhibitory transmitter released from 'descending' fibres which terminate on sympathetic preganglionic neurons (De Groat and Ryall, 1967; Ryall, 1967).

Parasympathetic Preganglionic Neurons

Parasympathetic preganglionic neurons are located in the spinal grey matter (lateral region) between the dorsal and ventral horns in the second and third sacral segments, and may be identified antidromically by stimulating the pelvic nerve or the second or third ventral roots (De Groat and Ryall, 1968a). Repetitive stimulation of parasympathetic axons produced a recurrent inhibition of pathways to the urinary bladder, and in most cases this inhibition was reduced by intravenous strychnine, but unaffected by dihydro-β-erythroidine (De Groat and Ryall, 1968b). Both glycine and GABA depressed the firing of parasympathetic preganglionic neurons (their relative potencies being similar), and the depressant action of glycine was blocked by strychnine in concentrations which had no effect on the depression by GABA (De Groat, 1970; Ryall and De Groat, 1972). It seems likely, therefore, that glycine is the transmitter mediating recurrent inhibition of parasympathetic neurons.

In contrast to *sympathetic* preganglionic neurons (most of which were excited by 5-HT), *parasympathetic* neurons were usually depressed by 5-HT, no excitant actions being observed (Ryall and De Groat, 1972; cf. De Groat and Ryall, 1968a). Moreover, the action of ACh on parasympathetic neurons was similar to that of 5-HT (Ryall and De Groat, 1972), whereas sympathetic neurons were unaffected by ACh (*see above*). NA depressed the firing of the majority of parasympathetic neurons, although some such cells were excited (Ryall and De Groat, 1972). Thus, the pharmacology of parasympathetic neurons differs from that of sympathetic neurons, but resembles that of various types of spinal interneurons. The particularly intense monoamine fluorescence in the parasympathetic 'nucleus' of the spinal cord (Dahlström and Fuxe, 1965b), together with the pharmacological results of Ryall and De Groat (1972) suggest that 5-HT and NA are inhibitory transmitters released on to some parasympathetic neurons, but direct pharmacological evidence for this has yet to be obtained.

Descending Pathways

The literature on the possible transmitters released from excitatory and inhibitory pathways which 'descend' to the spinal cord from the brain is confused and will not be reviewed here. However, it is at least clear that different pathways releasing different transmitters must be involved. Since the monoamine-containing axons are of relatively small diameter and unmyelinated (Dahlström and Fuxe, 1965b) they must have slow conduction velocities (of the order of a few m./sec.) and therefore cannot account for the short-latency mono- and di-synaptic EPSPs and IPSPs generated in motoneurons by stimulation of vestibulo- and reticulospinal pathways (Llinás and Terzuolo, 1964, 1965; Shapovalov, 1966; Jankowska, Lund, Lundberg, and Pompeiano, 1968; Lund and Pompeiano, 1968; Wilson and Yoshida, 1969a, b; Wilson, Yoshida, and Schor, 1970). These rapidly-conducting pathways possibly involve amino-acids. For example, studies with convulsants suggest that some descending inhibitions may be mediated by GABA (Bremer, 1953; Llinás, 1964; *see also* Curtis and Felix, 1971a), whereas others may be mediated by glycine (Llinás, 1964; Felpel, 1972; Tebēcis and Di Maria, 1972a; cf. Huffman and McFadin, 1972; *see also* Rizzoli, 1968).

SUMMARY

The excitatory and inhibitory transmitters, which are thought to act on spinal motoneurons and Renshaw cells, may be represented diagrammatically, as in *Fig.* 12. This, of course, is a highly simplified scheme of what actually occurs in the spinal cord, but will serve as a basis for summarizing the main points.

Acetylcholine

There is no doubt that ACh is the excitatory transmitter released from motor axons on to skeletal muscle, as well as from recurrent collaterals on to Renshaw cells. The cholinergic monosynaptic excitation of Renshaw cells by impulses in motor axons is mediated predominantly by nicotinic receptors.

Although it is possible that ACh is a transmitter (excitatory or inhibitory) on other types of spinal interneurons, there is no evidence to suggest that it is a transmitter on motoneurons. In fact, it is not clear whether ACh has an action on mammalian motoneurons. In any case, it is certainly not the excitatory transmitter released from primary afferent fibres on to these cells.

AMINO-ACIDS

The two naturally occurring excitant amino-acids, glutamate and aspartate, are likely to be transmitters in the spinal cord,

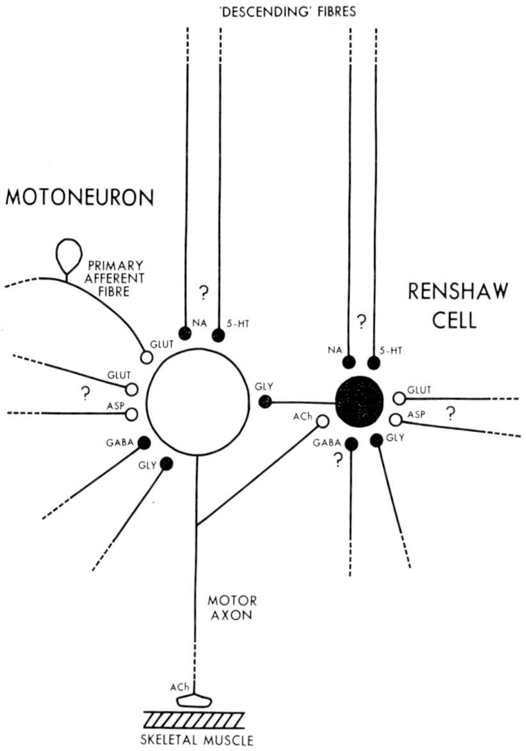

Fig. 12. Schematic diagram showing the main transmitters which have been either established or which are thought to act on spinal motoneurons and Renshaw cells. Excitatory neurons and terminals are white, inhibitory ones are black. The question marks indicate doubt as to the existence of the connexions, or if the evidence is not sufficiently convincing for the presence of such terminals. *See* text for details.

mediating at least some of the synaptic excitations which are resistant to antagonists of ACh and monoamines. Amino-acids structurally related to glutamate depolarize motoneurons, and the ionic mechanism underlying this depolarization resembles that of synaptic excitation. However, there is no direct pharmacological evidence to implicate either glutamate or aspartate in a known spinal pathway. Nevertheless, distribution studies of amino-acids in the spinal cord suggest that glutamate is the excitatory transmitter released from primary afferent fibres on to motoneurons (and possibly on to a small proportion of Renshaw cells). The synaptic excitation of Renshaw cells by volleys in *dorsal* roots is non-cholinergic, and may well be mediated by an amino-acid, possibly glutamate or aspartate. In addition, other excitatory fibres terminating on both motoneurons and Renshaw cells possibly release these amino-acids as transmitters.

Results from a wide variety of investigations (pharmacological, neurochemical, autoradiographic, evoked release, and others) indicate that both glycine and GABA are inhibitory transmitters in the spinal cord. Both amino-acids mimic the actions of inhibitory transmitters, and studies with strychnine (a relatively specific antagonist of glycine) and picrotoxin or bicuculline (relatively specific antagonists of GABA), have established these amino-acids as transmitters at certain synapses in the spinal cord.

Glycine is released from Renshaw axons on to motoneurons, and Renshaw cells themselves receive inhibitory terminals from other sources which release glycine as a transmitter. In addition to the recurrent inhibition (by glycine), motoneurons also receive glycine-releasing inhibitory terminals from other sources. A 'prolonged' inhibition of motoneurons is mediated by GABA, but the origins of these inhibitory fibres are not clear. Whether GABA is also an inhibitory transmitter on Renshaw cells is not known.

Monoamines

The actions of monoamines on spinal motoneurons and Renshaw cells have been poorly studied in comparison with those of amino-acids. However, histochemical and biochemical studies suggest the existence of noradrenergic and 5-hydroxytryptaminergic pathways which descend from the brain-stem to the spinal cord, and it is not unlikely that some of these fibres terminate on motoneurons and Renshaw cells. If so, these fibres are likely to be predominantly inhibitory, as NA and 5-HT have mainly depressant actions on motoneurons and Renshaw cells. The observation that some Renshaw cells are excited by NA and 5-HT, however, raises the possibility that some of these fibres may be excitatory. Similarly, histochemical and pharmacological observations suggest that NA and 5-HT are transmitters released from 'descending' fibres which terminate on sympathetic and parasympathetic preganglionic neurons. The mechanism underlying the hyperpolarization of motoneurons by NA is different from that producing the hyperpolarization by glycine and GABA, suggesting that the mechanism underlying noradrenergic transmission also differs from that of 'classic' synaptic transmission studied in the spinal cord. Despite these studies, it is necessary to investigate further the possible role of NA and 5-HT in the spinal cord, using both histochemical and pharmacological methods in the same animal. The function of DA, if any, in the spinal cord is unknown.

Chapter

2

BRAIN-STEM

THE brain-stem is a complex structure consisting of a large number of different fibre tracts and nuclei which vary considerably in their anatomy, physiology, and pharmacology. Although investigations of the brain-stem have not been as extensive as those of the spinal cord, in the last few years marked progress has

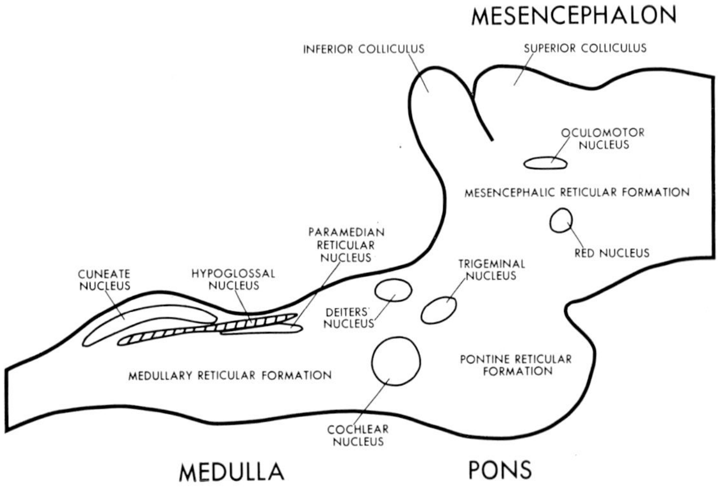

Fig. 13. Schematic diagram of a saggital view of the cat brain-stem, showing the approximate locations of some of the nuclei discussed in this chapter. *See* text for details.

been made in the understanding of the pharmacology of synaptic transmission in some areas of the brain-stem (Deiters' nucleus, for example) by studies involving intracellular recording during extracellular drug administration. This chapter deals with the pharmacology of neurons in the cuneate, hypoglossal, paramedian reticular, raphe, cochlear, Deiters', trigeminal, oculomotor, and red nuclei, the inferior and superior colliculi, and certain neurons in the medullary, pontine, and mesencephalic reticular formation. The approximate locations of most of these areas (of the cat brain) are represented diagrammatically in *Fig.* 13. The actions of putative transmitters on unidentified neurons are described only if the pharmacology of physiologically identified neurons has not been studied in these nuclei. Clearly, this chapter can give no more than a superficial knowledge of the possible identity of transmitters in the brain-stem, as many areas of this structure have not been explored.

Cuneate Neurons

The cuneate nucleus is an important relay station on the primary somatosensory pathway. Primary afferent fibres from cutaneous and joint receptors terminate on cuneate neurons, which project to the contralateral ventrobasal complex of the thalamus via the medial lemniscus. Extremely large synaptic knobs contact cuneate neurons (Rozsos, 1958), and both presynaptic and postsynaptic inhibition are thought to be important in the cuneate nucleus (Andersen, Eccles, Oshima, and Schmidt, 1964; Andersen, Etholm, and Gordon, 1970). Andersen, Eccles, Schmidt, and Yokota (1964) defined cuneothalamic relay neurons as those cuneate neurons which were activated antidromically by stimulating the medial lemniscus, and orthodromically by afferent nerve volleys; they also demonstrated the presence of cuneate interneurons which were excited from the cerebral cortex and/or peripheral nerves.

Amino-acids

The effects of electrophoretically administered substances on particular (identifiable) types of cuneate neurons were first studied by Galindo, Krnjević, and Schwartz (1967), who classified a relatively large number of cells as 'hair cells', which discharged when hairs were moved (fast-adapting); 'touch cells', which discharged when the skin was prodded and when a weight was placed on the limb (non-adapting); 'proprioceptive cells', which discharged according to the position of the limb, but not to stimulation of cutaneous receptors (slow-adapting); and those cells which could not be activated by these forms of peripheral stimulation. This investigation is particularly interesting in that it is one of the very few studies in which the effects of amino-acids on certain types of neurons are correlated with the effects of different forms of 'natural' stimulation. The differential sensitivity to L-glutamate and GABA of these cell types (many of which were also identified as cuneothalamic relay neurons by antidromic stimulation of the medical lemniscus) are indicated in *Table I*. Other physiological and pharmacological properties of these cells (including their sensitivity to glutamate and GABA) were reported by Galindo, Krnjević, and Schwartz (1968).

Table 1. SUMMARY OF SOME DATA ON CUNEATE NEURONS REPORTED BY GALINDO AND OTHERS (1967)

1 CELL TYPE	2 CTR (%)	3 DEPTH (mm.)	4 GLUTAMATE	5 GABA
Hair	72	0·2—0·8	+++	— —
Touch	80	0·4—0·9	++	— —
Proprioceptive	23	0·6—1·3	++	— — —

Column 1, Types of neurons classified according to their sensitivity to various forms of peripheral stimulation; Column 2, The proportions of these neurons which were also identified as cuneo-thalamic relay (CTR) neurons by antidromic stimulation of their axons; Column 3, Their approximate distances from the dorsal surface of the medulla; Columns 4, 5, Their relative sensitivity to the excitant (+) action of glutamate and the depressant (—) action of GABA, respectively.

All the cuneate neurons tested by Galindo and others (1967, 1968) were excited by glutamate, whereas in an earlier study of dorsal column nuclei only about half the neurons appeared to respond to glutamate (Steiner and Meyer, 1966). Hair cells were particularly sensitive to glutamate, an ejecting current of less than 10 nA sometimes being sufficient to produce excitation (Galindo and others, 1967, 1968). Touch and proprioceptive cells were approximately equally sensitive to glutamate, but both were clearly less sensitive than hair cells. No other studies have been made of the differential sensitivity of cuneate neurons to excitant amino-acids, although glutamate has been used to excite cuneate neurons in order to study the effects of depressant amino-acids and convulsants (Galindo, 1969; Kelly and Renaud, 1971). It has been suggested that glutamic acid is an excitatory transmitter in the cuneate nucleus (Galindo and others, 1967), a postulate supported by the large quantities of glutamic acid in dorsal roots, dorsal columns, and dorsal column nuclei (Graham and others, 1967; Duggan and Johnston, 1970; Johnson and Aprison, 1970, 1971).

Electrophoretically administered GABA readily depressed the spontaneous, orthodromic, and chemically evoked firing of cuneate neurons, but only rarely prevented antidromic activation of cuneothalamic relay neurons, probably because of the high safety factor in antidromic conduction (Galindo and others, 1967, 1968; Galindo, 1969; Kelly and Renaud, 1971). GABA depressed proprioceptive cells more readily than hair and touch cells (*Table I*). Glycine also depressed cuneate neurons, being approximately equipotent to GABA (Galindo and others, 1967; Galindo, 1969; Kelly and Renaud, 1971), but its relative depressant action on different types of cuneate neurons has not been studied to the same extent as that of GABA.

Although it has been suggested that both GABA and glycine may be inhibitory transmitters in the cuneate nucleus (Galindo and others, 1967; Galindo, 1969), direct pharmacological evidence for this has been obtained only for GABA (Kelly and Renaud, 1971). The two antagonists of GABA, picrotoxin and bicuculline, blocked the depressant action of GABA, but not that of glycine, on cuneate neurons (Galindo, 1969; Kelly and Renaud, 1971). In particular, bicuculline almost abolished the postsynaptic inhibitions evoked by stimulation of the pyramidal tract, contralateral median nerve, and ipsilateral forepaw (Kelly and Renaud, 1971). The post-stimulus latency histograms illustrated in *Fig.* 14 show the inhibitory effects of stimulation of the pyramidal tract and median nerve on the firing produced by glutamate of a proprioceptive cell before (A, D), during (B, E), and after (C, F) the administration of bicuculline. Bicuculline reversibly reduced both inhibitions.

Strychnine reversibly blocked the depression caused by glycine without affecting that by GABA, and it has been suggested that glycine may be a postsynaptic inhibitory transmitter, whereas GABA may be both a post- and a presynaptic inhibitory transmitter in the cuneate nucleus (Galindo and others, 1967; Galindo, 1969). This is supported by the observations that intravenous picrotoxin and bicuculline antagonized 'presynaptic inhibition' (presumably mediated by GABA), whereas strychnine antagonized postsynaptic inhibition (presumably mediated by glycine) in this

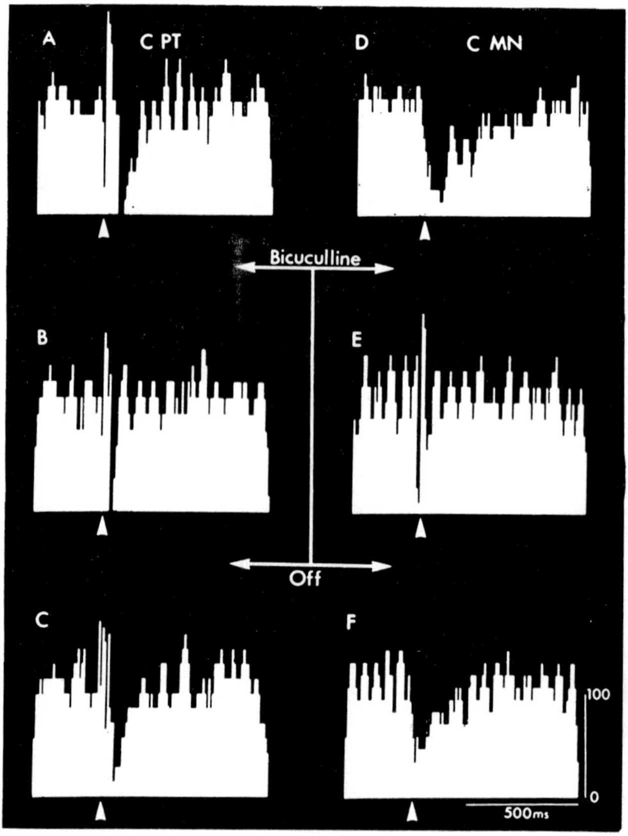

Fig. 14. Post-stimulus histograms of the firing of a cuneate neuron, photographed from an oscilloscope display. A, B, C, Stimulation (2 V, a train of three stimuli at 400 Hz, 1/sec.) of the contralateral pyramidal tract (CPT). D, E, F, Stimulation (1 V, single stimuli, 1/sec.) of the contralateral median nerve (CMN). A, D, Controls. B, E, 4 and 3 min. after the beginning of a bicuculline application with a current of 112 nA. C, F, 6 and 3 min. after the bicuculline current had been terminated. Each histogram was computed from 64 sweeps; arrows mark the stimulus artefact. The ordinate value of each bin of the histograms displayed was derived by averaging the contents of four adjacent bins which were then normalized as a percentage of the mean content of the initial 27 bins of the histogram (15–19 spikes/sec.). (Kelly and Renaud, 1971.)

nucleus (Boyd, Meritt, and Gardner, 1966; Banna and Jabbur, 1969; Banna, Naccache, and Jabbur, 1972). Subsequent studies involving topical application of amino-acids to a restricted area of the rat cuneate nucleus suggested that GABA depresses the postsynaptic excitability of cuneate neurons and depolarizes primary afferent fibres, whereas glycine reduces the excitability of primary afferents (Davidson and Southwick, 1971). The actions of GABA on primary afferent fibre excitability and 'presynaptic inhibition' in the cuneate nucleus were blocked by picrotoxin and bicuculline (Davidson and Southwick, 1971; Davidson and Reisine, 1971). However, the mechanisms underlying 'presynaptic inhibition' in the cuneate nucleus are, no doubt, analogous to those in the spinal cord (Curtis, Duggan, Felix, and Johnston, 1971; Schmidt, 1971) and must therefore await clarification. It is interesting in this regard that the neuropile of the dorsal column nuclei stained heavily for an enzyme which destroys GABA, as in the substantia gelatinosa of the spinal cord (Sims and others, 1972).

Acetylcholine and Monoamines

Little is known about the actions of ACh and the monoamines on cuneate neurons. ACh, 5-HT, catecholamines, and histamine usually had weak depressant or no actions on the few cuneate neurons tested (Steiner and Meyer, 1966; Galindo and others, 1967, 1968). However, it is possible that barbiturate anaesthesia masked the actions of these substances, and that cholinergic and monoaminergic mechanisms are, in fact, important in this nucleus, considering the biochemical and histochemical data for the presence of some of these substances and the enzymes which metabolize them in the brain-stem (*see* later). The actions of ACh and monoamines on cuneate neurons in unanaesthetized animals require investigation before conclusions may be made concerning their possible transmitter function.

Hypoglossal Motoneurons

The hypoglossal nucleus, the XIIth cranial nerve nucleus, has been subdivided into several morphologically discrete areas (Barnard, 1940), but from a physiological point of view may be broadly divided into at least two major regions—one related to tongue protrusion, the other for tongue retraction (Morimoto Takata, and Kawamura, 1968; *see also* Porter, 1965). The antidromic responses of tongue protrusive motoneurons were inhibited by conditioning stimulation of the lingual nerve, but those of tongue retractive motoneurons were initially facilitated and subsequently inhibited (except in a few cases in which only facilitation was observed). Similarly, stimulation of the lingual nerve evoked IPSPs in tongue protrusive motoneurons, but in tongue retractive motoneurons EPSP-IPSP sequences or only EPSPs were observed (Morimoto and others, 1968). Of particular interest in the present context is the finding that intravenous strychnine diminished this inhibition of hypoglossal motoneurons. An example of this is illustrated in *Fig.* 15.

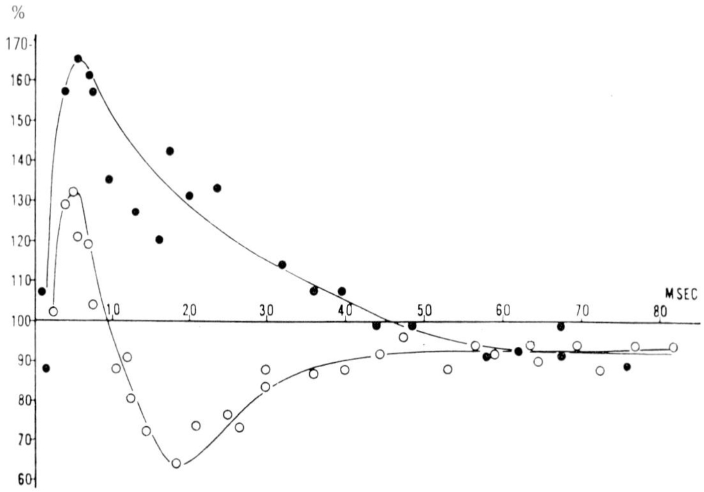

Fig. 15. Effects of strychnine on the activity of tongue retractive motoneurons in the hypoglossal nucleus. The plotted points indicate the relative changes in the amplitude of antidromic responses evoked by stimulation of the hypoglossal nerve at different intervals following conditioning volleys to the lingual nerve. Open circles: before strychnine. Solid circles: 5 min. after strychnine (0·08 mg./kg.) had been injected intravenously. (Morimoto, Takata, and Kawamura, 1968.)

The points plotted in the graphs of *Fig.* 15 represent the percentage change in the amplitude of tongue retractive motoneuron responses evoked by stimulation of the hypoglossal nerve ('retractive' branch) at varying intervals following stimulation of the lingual nerve. Lingual nerve stimulation initially facilitated and subsequently inhibited the antidromic responses (open circles). Five minutes after strychnine (0·08 mg./kg.) had been injected intravenously, the facilitation was increased and the inhibition was almost blocked (solid circles).

Results from subsequent pharmacological studies of hypoglossal motoneurons involving both intra- and extracellular recording during extracellular administration of amino-acids, suggest that glycine is likely to be the strychnine-sensitive inhibitory transmitter in the hypoglossal nucleus (Bruggencate and Sonnhof,

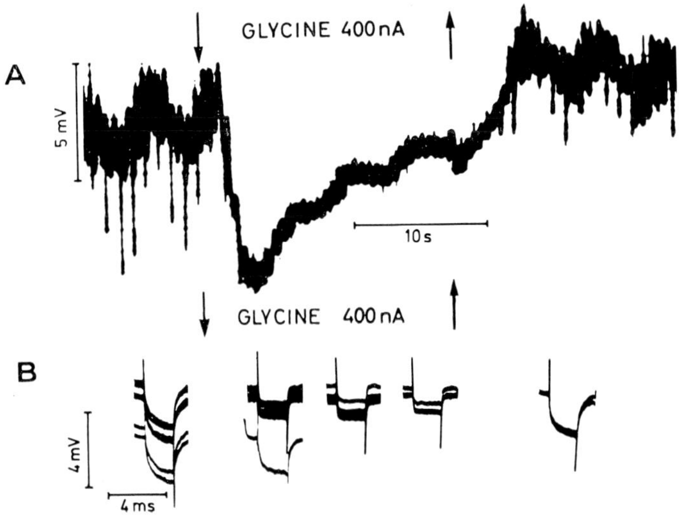

Fig. 16. Intracellular recordings showing the action of glycine on the membrane potential and membrane conductance of a hypoglossal motoneuron (50 mV resting membrane potential). A, Slow-speed (time calibration, 10 sec.) record of the membrane potential (positivity upwards). The rapid deflections are polysynaptic IPSPs mixed with EPSPs evoked by stimulating the lingual nerve. Glycine (400 nA) was administered extracellularly between the arrows. B, Measurements of membrane conductance obtained by the injection of short hyperpolarizing current pulses (2 nA, 2 per sec.) through the recording electrode (the voltage drop across the electrode resistance being compensated). Superimposed oscilloscope traces were obtained before, during (between the arrows) and after the glycine application. The first of the pulses after the beginning of the glycine ejection was unaffected, whereas the subsequent pulses were reduced, indicating a delay in the increase in conductance. Direct current amplification in A, alternating current in B. (Bruggencate and Sonnhof, 1972.)

1971, 1972). Glycine and GABA depressed hypoglossal motoneurons, the depression resulting from a membrane hyperpolarization and an increase in membrane conductance. *Fig.* 16A shows the hyperpolarizing action of glycine administered extracellularly to a hypoglossal motoneuron from which the membrane potential was recorded intracellularly. The decreased amplitude of the potential changes produced by injected pulses of intracellular current indicate an increase in membrane conductance during the hyperpolarization produced by glycine (*Fig.* 16B).

Although glycine was usually a more effective depressant than GABA of hypoglossal motoneurons in the more central areas of the nucleus, the reverse was true in the dorsal dendritic region, and the effects of these amino-acids on different components of the antidromic field potentials suggested that GABA acts preferentially on the dendrites (Altmann, Bruggencate, and Sonnhof, 1972). In particular, the depressant action of glycine was blocked by strychnine, which had no effect on that of GABA (Bruggencate and Sonnhof, 1971, 1972). It is likely, therefore, that the strychnine-sensitive inhibition observed by Morimoto and others (1968) is mediated by glycine, especially since the pharmacology of hypoglossal motoneurons (as regards depressant amino-acids and convulsants) is similar to that of spinal motoneurons, which are inhibited by synaptically released glycine (Chapter 1). Nevertheless, antagonism by strychnine of both synaptic inhibition and the depressant action of glycine on the same hypoglossal neuron has yet to be demonstrated.

GABA may also be an inhibitory transmitter in the hypoglossal nucleus, as this nucleus shows a good histochemical reaction for GABA metabolism (van Gelder, 1965b; Sims and others, 1972), and contains medium levels of GABA compared with various areas of the C.N.S. (Okada, Nitsch-Hassler, Kim, Bak, and Hassler, 1971). The depressant action of GABA on hypoglossal motoneurons was blocked by picrotoxin in concentrations which had little or no effect on that of glycine (Bruggencate and Sonnhof, 1971, 1972), but an action of this convulsant on synaptic inhibition in the hypoglossal nucleus has not been shown.

Otherwise, virtually nothing is known about the possible identity of other transmitters operating in the hypoglossal nucleus. The recent studies mentioned here provide the first suggestions as to what at least one inhibitory transmitter may be, and should stimulate further pharmacological experiments on hypoglossal neurons.

Paramedian Reticular Neurons

The paramedian reticular nucleus consists of three well-defined groups of neurons, most, if not all, of which send their axons to the cerebellum (Brodal, 1953; Brodal and Torvik, 1954). Neurons in the area of the paramedian reticular nucleus which were activated antidromically by stimulating the ipsilateral cerebellar peduncle were relatively homogeneous in their responses to electrophoretically administered ACh and monoamines (Avanzino, Bradley, and Wolstencroft, 1966). ACh and 5-HT excited, whereas NA depressed, the firing of almost all the paramedian reticular neurons tested. These observations are interesting because all paramedian reticular neurons are thought to be of a single type. Unfortunately, no other pharmacological studies have been made of these neurons, so it is not possible at this stage to make firm conclusions about the significance of these observations.

Reticulospinal Neurons

More than half of the neurons in the medullary and pontine reticular formation are reticulospinal neurons (Torvik and Brodal, 1957; *see also* Magni and Willis, 1963; Wolstencroft, 1964; Ito, Udo, and Mano, 1970), and reticulospinal tracts may be either excitatory or inhibitory (Lundberg, 1966; Ito, Udo, and Mano, 1970). Inhibitory reticulospinal neurons are distributed mainly in the ventrocaudal area of the medullary reticular formation (nucleus reticularis gigantocellularis) and mediate inhibition monosynaptically on to themselves and on to a small proportion of Deiters' neurons via collaterals of their descending axons (Ito, Udo, and Mano, 1970). Excitatory reticulospinal neurons are located predominantly in the dorsocranial area of the pontomedullary reticular formation, and mediate excitation monosynaptically to and around cells of their origin via recurrent axon collaterals (Ito, Udo, and Mano, 1970). Some reticulospinal neurons have axons that dichotomize in both ascending and descending directions (Torvik and Brodal 1957; Scheibel and Scheibel, 1958; Magni and Willis, 1963; Ito, Udo, and Mano, 1970; Tebēcis and Di Maria, 1972a).

Identification of particular inputs to the reticular formation is complicated because afferent projections largely overlap with the origins of reticulospinal fibres (Rossi and Brodal, 1957; Torvik and Brodal, 1957). However, Udo and Mano (1970) demonstrated that the ascending spinoreticular tract and probably all the tracts descending from or through the regions of the medial and lateral vestibular nuclei, superior colliculi, and interstitial nuclei of Cajal (*see* Busch, 1964) exert a direct excitatory action on reticular neurons, independently of the collateral systems of the reticulospinal tracts.

Amino-acids

Glutamate, aspartate, and DLH excited medullary reticular neurons (including reticulospinal neurons), and some evidence was obtained for a differential sensitivity of these neurons to the excitant amino-acids (Hösli and Tebēcis, 1970; Tebēcis and Di Maria, 1972a). Other than these observations, there is no evidence to suggest that glutamate or a related amino-acid is an excitatory transmitter in the medullary reticular formation. In fact, this applies to the whole brain-stem.

Glycine and GABA depressed the firing of reticulospinal and unidentified reticular neurons, glycine being more effective than GABA (Hösli, Tebēcis, and Filias, 1969; Hösli and Tebēcis, 1970; Boakes, Bradley, Briggs, and Dray, 1970; Tebēcis and Di Maria, 1972a). Studies involving intracellular recording during extracellular drug administration to medullary reticular neurons have revealed that the depression by glycine results from a hyperpolarization of the membrane (which is reversed to a depolarization when chloride ions are injected intracellularly) and an increase in membrane conductance and time constant (Hösli, Tebēcis, and Haas, 1971; Tebēcis, 1972b). Strychnine blocked the depressant action of glycine, but not that of GABA, on medullary reticular neurons, whereas bicuculline blocked the depression by GABA in concentrations which had little or no effect on that by glycine (Hösli and Tebēcis, 1970; Tebēcis, Hösli, and Haas, 1971; Tebēcis and Di Maria, 1972a). In particular, stimulation of the cervical spinal cord (ventrolateral area) often inhibited the firing of neurons in the medullary reticular formation, and strychnine reversibly reduced or blocked this inhibition, as well as the depression of these cells by glycine, but did not affect the depression by GABA (Tebēcis and Di Maria, 1972a). An example of such a strychnine-sensitive inhibition in the medullary reticular formation is illustrated in *Fig.* 17.

Stimulation of the pontine reticular formation (PRF) in the area of nucleus reticularis pontis caudalis evoked a synaptic discharge of 12–17 action potentials from a medullary reticular neuron (*Fig.* 17A). When PRF stimulation was preceded by a volley to the ipsilateral cervical spinal cord, PRF stimulation evoked a discharge of only 2–5 spikes (*Fig.* 17B). However, after strychnine (20 nA) had been administered, the inhibitory effect of spinal stimulation was reduced (*Fig.* 17D), and after strychnine (80 nA), inhibition was virtually blocked (*Fig.* 17F). Strychnine had no action on the firing evoked by PRF stimulation alone (*Fig.* 17A, C, E, G). In addition, the depressant action of glycine, but not that of GABA, on this neuron was reversibly blocked by strychnine (not illustrated).

These and similar observations provide direct pharmacological evidence that glycine is an inhibitory transmitter in the medullary reticular formation (Tebēcis and Di Maria, 1972a). Since the recordings were made in the region where inhibitory reticulospinal

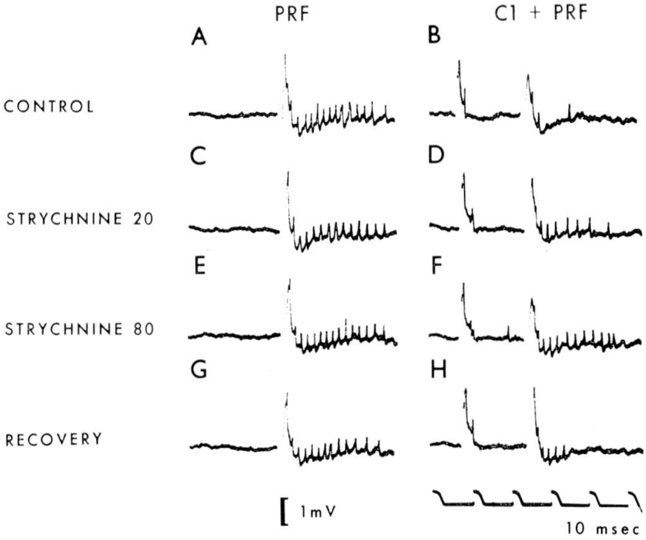

Fig. 17. Antagonism by strychnine of synaptic inhibition in the medullary reticular formation of an unanaesthetized (decerebrate) cat. A, C, E, G, Repetitive discharges of a medullary reticular neuron evoked by stimulating the pontine reticular formation (PRF; 3 V, 0·5/sec.). B, D, F, H, PRF stimulation was preceded by a single volley to the cervical spinal cord (C.1, ventrolateral area, 6·5 V). A, B, Recorded immediately before strychnine. C, D, Strychnine (20 m*M* in 165 m*M* NaCl) was administered with a current of 10 nA for 1 min., followed by 20 nA for about 3 min., and the records were obtained at the end of this period. E, F, At the end of a further ejection of strychnine (80 nA) for 1·5 min. G, H, Approximately 7 min. after the current ejecting strychnine had been terminated. (Tebēcis, unpublished observations.)

neurons predominate, and since the spinal stimulating electrode was positioned in the vicinity of their axons (cf. Ito, Udo, and Mano, 1970), the inhibition was interpreted as a result of impulses in recurrent collaterals of inhibitory reticulospinal axons. This is shown diagrammatically in *Fig.* 18.

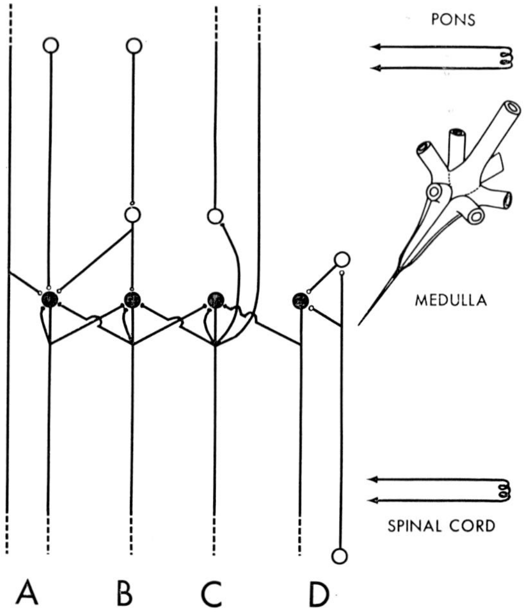

Fig. 18. Schematic diagram illustrating possible arrangements of neurons to account for the inhibition evoked by stimulating the spinal cord. Inhibitory neurons and terminals are black; excitatory are white. The broken vertical lines represent long ascending or descending axons. By means of recurrent axon collaterals, inhibitory reticulospinal neurons can monosynaptically inhibit the monosynaptic (A), disynaptic (B), or antidromic (C) excitation evoked by stimulation of the PRF. Polysynaptic inhibition from the spinal cord can arise from the activation of spinoreticular fibres (D). *See* text for details (Tebēcis and Di Maria, 1972a.)

Recurrent collaterals of inhibitory reticulospinal axons can monosynaptically inhibit the monosynaptic (*Fig.* 18A), di- or polysynaptic (*Fig.* 18B) or antidromic (*Fig.* 18C) excitation of medullary reticular neurons evoked by stimulating the pontine reticular formation. In the case of polysynaptic inhibition from the spinal cord, spinal stimulation presumably activates spinoreticular fibres which excite reticulospinal neurons directly or indirectly (*Fig.* 18D). However, the possibility that other pathways are involved cannot be excluded. Another interesting implication is that if glycine is the inhibitory transmitter released from recurrent collaterals of inhibitory reticulospinal axons, it follows that glycine is also released synaptically from reticular axon terminations on spinal and possibly some Deiters' neurons (Tebēcis and Di Maria, 1972a).

Stimulation of nucleus reticularis pontis caudalis also inhibited the firing of medullary reticular neurons by a mechanism which was blocked by strychnine, but the origin of this inhibitory pathway is not clear (Tebēcis and Di Maria, 1972a).

Various other findings are compatible with the view that glycine is an inhibitory transmitter in the medullary reticular formation. Relatively high concentrations of glycine are present in the mammalian medulla oblongata, the levels being not only 2–3 times those of GABA, but much higher than in any of the more rostral brain areas, and comparable with those in the spinal cord (Shaw and Heine, 1965; Kandera and others, 1968; Aprison and others, 1969). Studies of glycine uptake in different areas of the C.N.S. have indicated that a high affinity uptake mechanism for this amino-acid occurs only in the medulla, pons, and spinal cord, which are also the structures containing the highest glycine levels (Neal, 1971; Johnston and Iversen, 1971). Thus, a comparison of the data on the brain-stem with those on the spinal cord, where glycine is known to be an inhibitory transmitter (Chapter 1), strengthens the view that glycine is also involved in synaptic inhibition in the medulla, and possibly the pons.

Monoamines

Electrophoretically administered monoamines had relatively consistent actions on reticulospinal neurons of the medullary reticular formation, in the area of the nucleus reticularis gigantocellularis (Hösli, Tebēcis, and Schönwetter, 1971). Although it was not determined whether the reticulospinal neurons studied were excitatory or inhibitory, the majority were recorded in the area where inhibitory reticulospinal neurons predominate (Ito, Udo, and Mano, 1970). 5-HT and NA excited almost all the reticulospinal neurons tested, whereas DA usually had a slight depressant action or no action at all. One example of this is illustrated in *Fig.* 19. It is possible, however, that a study of the effects of 5-HT, NA, and DA on a larger sample of reticulospinal neurons would reveal a more varied pharmacology, since more than half of the neurons in the medullary reticular formation are reticulospinal neurons (Torvik and Brodal, 1957), and these monoamines have been shown to have a variety of effects on unidentified reticular

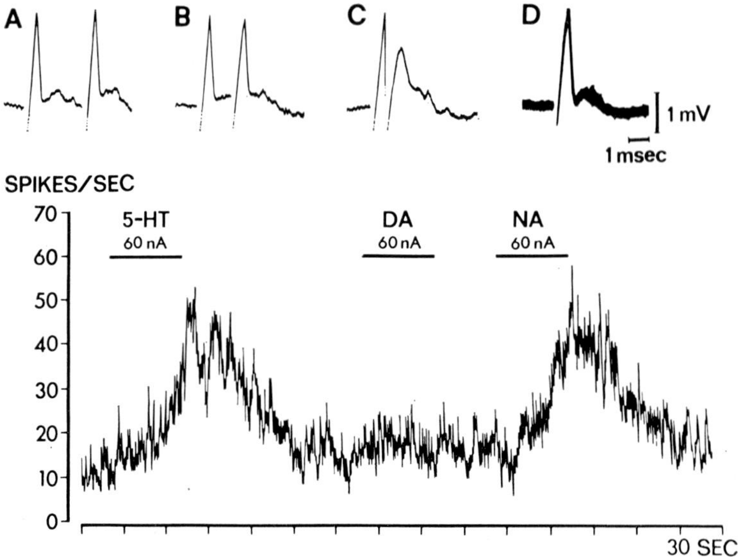

Fig. 19. Actions of 5-HT, NA, and DA, each administered with a current of 60 nA, on the spontaneous firing of a reticulospinal neuron. A–D, Action potentials evoked antidromically by stimulating the cervical spinal cord (C.1, ventrolateral portion). A–C, The interval between the two stimuli was progressively reduced, until the second action potential failed (C). D, Superimposed records of the spike following stimulation at 110/sec. (Hösli, Tebēcis, and Schönwetter, 1971.)

neurons (Bradley and Wolstencroft, 1965; Bradley, 1968; Boakes and others, 1970; Boakes, Bradley, Brookes, Candy, and Wolstencroft, 1971; Hösli and others, 1971).

The actions of monoamines on brain-stem neurons have not been related to synaptic activity in this structure. However, biochemical evidence for the presence of monoamines and their metabolizing enzymes in the brain-stem, as well as fluorescence histochemical evidence for the presence of catecholamine- and 5-HT-containing nerve terminals in various brain-stem areas (Dahlström and Fuxe, 1965a, b; Fuxe, 1965; Pin and others,

1968; Fuxe and others, 1970; Björklund and others, 1971; Sladek, 1971; Ungerstedt, 1971) suggest that at least 5-HT and NA are transmitters acting at different sites in this structure. Of especial interest is the finding that of all the regions of the C.N.S. studied, the brain-stem is the only structure (with the exception of the retina and 'reticular region' of the diencephalon) in which monoamine-containing cell-bodies are located. That is, virtually all the monoamine-containing nerve terminals found in the brain and spinal cord arise from a few discrete sites in the brain-stem (Andén, Dahlström, Fuxe, Larsson, Olson, and Ungerstedt, 1966; Björklund and others, 1970, 1971; Fuxe and others, 1970; Ungerstedt, 1971). One serious problem, however, is that most of the histochemical data have been obtained from studies of rats, and recent histochemical studies of both rats and cats have revealed marked differences between these species in the distribution of fluorescence in the brain-stem (Pin and others, 1968; Sladek, 1971). Clearly, histochemical studies of the C.N.S. of various groups of mammals are required before generalizations may be made about the monoaminergic systems of the C.N.S.

Histamine usually depressed the firing of medullary reticular neurons (Bradley, 1968; Hösli and Haas, 1971), but the significance of this is not clear, as this area of the brain-stem contains low levels of histamine (Taylor, Gfeller, and Snyder, 1972).

Acetylcholine

ACh had little or no action on the few reticulospinal neurons tested in unanaesthetized cats, although low concentrations of glutamate excited the same neurons (Tebēcis, unpublished observations). Otherwise, an effect of ACh on reticulospinal neurons has not been reported, although ACh had excitant or depressant actions on different proportions of unidentified reticular neurons (Salmoiraghi and Steiner, 1963; Bradley, Dhawan, and Wolstencroft, 1966). An interesting finding was that ACh had no action on almost all the respiratory neurons tested in the medulla, although it excited or depressed the firing of many neurons which were not obviously related to respiratory activity (Salmoiraghi and Steiner, 1963). If respiratory neurons have an intrinsic pacemaker, they may require few or no inputs from other types of neurons. It would therefore be intriguing to know whether other suspected transmitters (amino-acids, monoamines) have actions on respiratory neurons.

Rostrally Projecting Reticular Neurons

More than half of the giant neurons in the pontine reticular formation have ascending axons, whereas only a few small cells of the medullary reticular formation project rostrally (Brodal and Rossi, 1955; Torvik and Brodal, 1957; Ito, Udo, and Mano, 1970). Electrophoretically administered NA depressed a little over one-half, and excited almost all the rest of the pontomedullary reticular neurons (nucleus reticularis gigantocellularis), which could be activated antidromically by stimulating the central tegmental tract in the mesencephalon (Bradley, Wolstencroft, Hösli, and Avanzino, 1966). These neurons may not necessarily belong to one particular type, as the axons stimulated may have been of many different projection systems. Both excitant and depressant actions of NA, 5-HT, and ACh have been observed on unidentified neurons in the pontine reticular formation of the rat (Schmidt, Krug, Maier, Pohle, and Matthies, 1967; Krug, Schmidt, and Matthies, 1970). The comments on the possible significance of the actions of the monoamines and ACh on reticulospinal neurons apply equally to this section.

Glycine and GABA depressed the firing (glycine being more effective than GABA) of the rostrally projecting neurons tested in the medullary reticular formation (Tebēcis and Di Maria, 1972a). Similarly, the possible role of glycine on reticulospinal neurons discussed in the previous section is relevant here. Indeed, some reticulospinal neurons tested with the amino-acids also belong to the present section, as they also had a rostrally projecting axon collateral (Tebēcis and Di Maria, 1972a).

Raphe Neurons

The raphe nuclei of the cat constitute a complex, ill-defined series of eight nuclei in the midline of the brain-stem, ranging from the lower medulla to the rostral mesencephalon (Taber, Brodal, and Walberg, 1960). The majority of efferent connexions make up long ascending fibres which originate in all the raphe nuclei and project to many rostral areas of the brain (Brodal, Taber, and Walberg, 1960). Descending efferents are less numerous and originate mainly in the nucleus raphe magnus, whereas efferents to the cerebellum are even less numerous, originating mainly in the nucleus raphe pontis. Afferent projections to some raphe nuclei have been shown to originate in the cerebral cortex, the spinal cord, and the fastigial nucleus of the cerebellum (Brodal,

Walberg, and Taber, 1960). Since these are relatively few, however, most afferent projections are thought to originate in other structures, presumably from other areas in the brain-stem.

Most studies concerned with the functions of the raphe nuclei have been made of rats. Virtually all the 5-HT-containing neurons in the rat brain are localized in the raphe nuclei, and these neurons are the main origins of ascending and descending 5-HT-containing pathways which terminate in many areas of the C.N.S. (Dahlström and Fuxe, 1965a, b; Andén, Dahlström and others, 1966; Aghajanian, Bloom, and Sheard, 1969; Fuxe and others, 1970; Ungerstedt, 1971; Bloom, Hoffer, Siggins, Barker, and Nicoll, 1972). However, the histochemical and biochemical evidence for raphe neurons, which contain a tryptamine derivative other than 5-HT (Björklund and others, 1970, 1971; Aghajanian and Asher, 1971; Aghajanian, 1972; *see also* Saavedra and Axelrod, 1972), raises the possibility that this compound has been confused with 5-HT in some previous investigations. Various pharmacological and biochemical studies have revealed relationships between 5-HT and the activity of raphe neurons (Aghajanian, Rosecrans, and Sheard, 1967; Aghajanian, Foote, and Sheard, 1968; Jouvet, 1969; Padjen and Randić, 1970; Sheard and Zolovick, 1971; Kuhar, Roth, and Aghajanian, 1972). Both the raphe nuclei and 5-HT are thought to be important in various brain functions, including temperature regulation (Sheard and Aghajanian, 1967; Myers, 1969), sensory habituation (Sheard and Aghajanian, 1968), sleep (Jouvet, 1969; Koella, 1969), and various patterns of behaviour (Aprison and Hingtgen, 1972).

Monoamines and Acetylcholine

The only comprehensive microelectrophoretic study of raphe neurons has been done of rats anaesthetized with pentobarbitone sodium (Couch, 1970; cf. Schmidt and others, 1967). A comparison of the effects of 5-HT, NA, and ACh on neurons of the pontine and midbrain raphe nuclei revealed that these neurons are not pharmacologically homogenous. Virtually all the raphe neurons excited by stimulation of the nucleus paragigantocellularis lateralis (NPL) were also excited by 5-HT. NA excited about one-half and depressed the other half of these cells, whereas ACh had even more variable effects. *Fig.* 20 illustrates an example of such a neuron, this one being excited by all three substances. The oscilloscope traces show the typically long-latency, small-amplitude synaptic discharges evoked from these neurons, which were located mainly in the dorsal half of the pontine raphe nuclei.

In contrast, most raphe neurons which were inhibited or unaffected by NPL stimulation were depressed by 5-HT, whereas NA and ACh had variable actions. It is also of interest that 5-HT resembled ACh in its effects on most of the neurons excited or depressed by NPL stimulation.

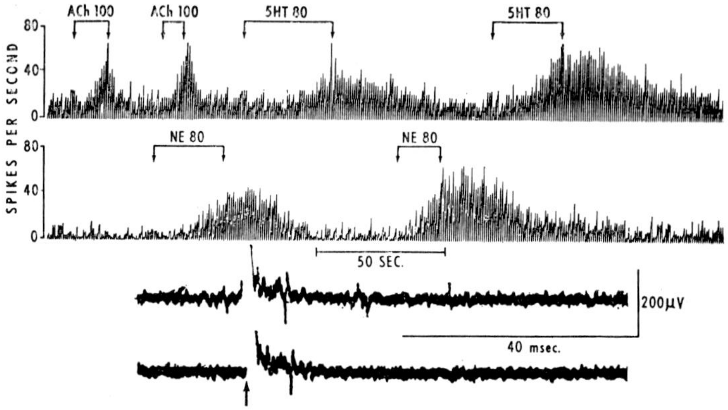

Fig. 20. Responses of a spontaneously firing raphe neuron to electrophoretically administered ACh (100 nA), 5-HT (80 nA), and noradrenaline (NE, 80 nA). The top records were obtained with a chart recorder, and show the change in firing frequency of the neuron with time. The lower two rows are oscillograph records showing the responses of the cell to two consecutive stimuli (0·9 V, 0·5 msec. duration, 0·5/sec.; onset of stimulus indicated by an upward arrow) delivered to the NPL area. (Couch, 1970.)

Couch (1970) suggested the presence of a 5-hydroxytryptaminergic pathway from the NPL to the raphe nuclei, which could be either monosynaptic (slowly conducting, 'C-fibre type') or polysynaptic with a 5-hydroxytryptaminergic neuron as the last neuron in the pathway. The excitation by both NPL stimulation and 5-HT on some raphe neurons was apparently blocked by LSD, but the significance of this is not clear because LSD is not a specific antagonist of 5-HT. Although Boakes and others (1970) reported that LSD specifically blocked the excitant action of 5-HT (as well as that of glutamate on 5-HT-excited neurons) in the medullary reticular formation, it has been shown also that LSD blocked the excitant and depressant actions of both 5-HT and NA in this area (Hösli, Tebēcis, and Schönwetter, 1970) and in the rabbit olfactory bulb (Bloom, Costa, and Salmoiraghi, 1964). Methiothepin may prove to be a more specific antagonist of 5-HT in the C.N.S. (Monachon, Burkard, Jalfre, and Haefely, 1972; Tebēcis, 1972a), and should be tested on the synaptic excitation evoked by NPL stimulation.

Although some 5-HT-containing neurons of the rat brain are localized in the area corresponding to NPL (Dahlström and Fuxe, 1965b), 5-HT-containing nerve terminals have not been described in the raphe nuclei. However, such terminals would be difficult to demonstrate by fluorescence microscopy when fluorescence is already present within the soma. Recent autoradiographic studies have, in fact, revealed the presence of degenerating nerve terminals that can accumulate 5-HT, following lesions in NPL (Bloom, Hoffer, Siggins and others, 1972). It seems possible, therefore, that 5-HT may be released synaptically on to certain raphe neurons which themselves release 5-HT on to neurons elsewhere in the C.N.S. Nevertheless, this problem requires further investigation in view of the histochemical evidence for another unidentified monoamine projection which originates in the raphe nuclei (Björklund and others, 1971).

Since practically all the raphe neurons tested responded to NA with either an excitation or a depression, it was suggested that this catecholamine may also be a transmitter in the raphe nuclei, possibly acting on the same neurons as 5-HT (Couch, 1970). Histochemical evidence for a NA-containing fibre projection from the locus coeruleus to the raphe region has been described (Loizou, 1969; Olson and Fuxe, 1971). Moreover, catecholamine-containing nerve terminals on neurons of the dorsal raphe nucleus degenerated following treatment of the animals with 6-hydroxydopamine (Bloom, Algeri, Groppetti, Revuelta, and Costa, 1969).

Similarly, the observation that a high proportion of raphe neurons responded to ACh raises the possibility of cholinergic inputs to the raphe nuclei (Couch, 1970), but there is no other evidence for this at present.

Cochlear Neurons

The cochlear nucleus is the region where second-order neurons of the auditory pathway terminate. Most pharmacological studies of cochlear neurons have been made by means of the micro-tap (Comis, Evans, and Whitfield, 1964). Thus, locally administered

ACh decreased the threshold of many cochlear neurons to sound stimuli, and this effect was blocked by intravenous atropine or dihydro-β-erythroidine (Comis and Whitfield, 1966). NA depressed the firing of some cochlear neurons (Comis and Whitfield, 1967). Although the micro-tap is useful for the administration of non-ionized substances, a major disadvantage is that only one drug can be administered at a time, thus limiting an evaluation of the pharmacological specificity of the effect observed. Nevertheless, investigations with the micro-tap have led to the suggestion that ACh is an excitatory transmitter in the cochlear nucleus (Comis and Whitfield, 1968).

Acetylcholine

The cochlear nucleus receives a centrifugal projection from the superior olive, some fibres of which contain AChE (*see* references by McDonald and Rasmussen, 1971). Stimulation of the S segment of the superior olive and local administration of ACh often fired the same neuron and lowered its threshold to sound stimuli in the ventral cochlear nucleus; the effects of electrical stimulation were reduced by locally administered atropine, dihydro-β-erythroidine, and gallamine triethiodide (Comis and Whitfield, 1968). The significance of these results is difficult to evaluate, however. The latencies of the cochlear responses evoked by olivary stimulation ranged from 25 to 500 msec., suggesting the involvement of several neurons in the 'pathway', particularly as the cholinergic antagonists sometimes also reduced the firing evoked by sound stimuli (Comis and Whitfield, 1968, *Fig.* 6). Moreover, the pharmacological specificity of the actions of the administered substances was not determined.

More recent evidence, however, supports the postulate that ACh is a transmitter in a centrifugal pathway to the cochlear nucleus (Comis and Davies, 1969). Hemicholinium applied by the micro-tap caused a slow and progressive decrease in the responses of cochlear neurons to sounds and stimulation of the superior olive. Moreover, after hemicholinium had been injected into the lateral ventricles, stimulation of the superior olive and sound stimulation led to a decrease in the concentration of ACh in the ipsilateral cochlear nucleus. Thus, ACh may, in fact, be a transmitter in the cochlear nucleus, but the available evidence in itself is not sufficiently convincing to make a definite conclusion.

Depressant Substances

Locally administered NA depressed the firing of all cochlear neurons which were synaptically inhibited by stimulation of the lateral part of the superior olive (Comis, 1970), but there is no other reason to suppose that this inhibitory transmitter on cochlear neurons is a monoamine. Although glycine and GABA depressed the firing of cochlear neurons (Comis, unpublished observations), this inhibitory transmitter is unlikely to be glycine, as the synaptic inhibition in the cochlear nucleus was resistant to locally administered strychnine (Comis, 1970). Such inhibition was, however, blocked by application of strychnine to the round window (*Fig.* 21), indicating that the olivo-cochlear bundle is involved (*see below*).

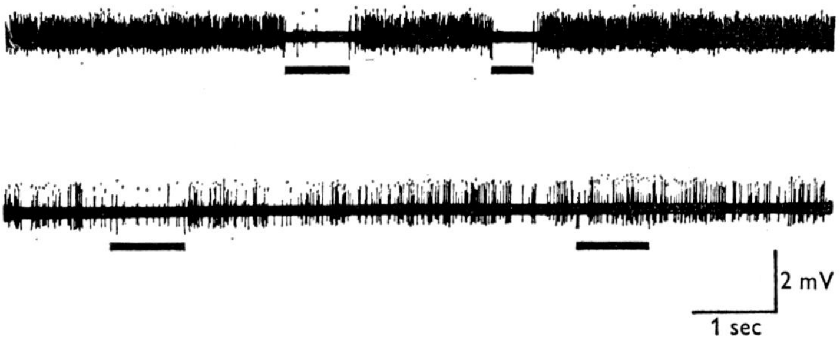

Fig. 21. An effect of strychnine on synaptic inhibition in the cochlear nucleus. The top trace shows the normal inhibitory effect of ipsilateral olivary stimulation (duration indicated by the horizontal bars) on the firing of a neuron in the cochlear nucleus. The bottom trace shows the abolition of the inhibitory effect following the topical application of strychnine (1:400) on the round window. (Comis, 1970.)

Recent evidence has suggested that GABA may be an inhibitory transmitter in the cochlear nucleus, although its site of action is unknown (Watanabe, 1971). The inhibitory effect of one tone on the excitation evoked in the cochlear nucleus by another tone ('two-tone inhibition') was reversibly reduced by picrotoxin (Watanabe, 1971). Such inhibition has previously been shown to be insensitive to strychnine (Whitfield and Allanson, 1958). It is now necessary to locate the site of action of this synaptic inhibition, and to compare its properties with those of GABA on the same neuron, both before and during the application of picrotoxin and/or bicuculline.

Cochlear Hair Cells

ACh has also been implicated in a 'synaptic' inhibition of cochlear hair cells, but interpretation of the relevant studies is confused (*see* references by Guth and Bobbin, 1971). Stimulation of the crossed tract of Rasmussen, or olivo-cochlear bundle, inhibits VIII nerve action potentials evoked by sound, and since olivo-cochlear bundle fibres contain AChE, and their endings (which terminate on hair cells) contain synaptic vesicles, it has been thought that the inhibition may be mediated by a neuro-humoral mechanism involving ACh (Guth and Bobbin, 1971). This is supported by the recent observation that stimulation of the olivo-cochlear bundle resulted in a release of ACh (Norris, Guth, and Stockwell, 1972). Of various substances (including ACh, neostigmine, GABA, strychnine, atropine, adrenaline, and in organic ions) administered electrophoretically in the vicinity of cochlear hair cells, only ACh and neostigmine depressed cochlear responses (Tanaka and Katsuki, 1966). However, atropine, hyoscine, dihydro-β-erythroidine, *d*-tubocurarine, and mecamylamine did not antagonize the inhibition from the olivo-cochlear bundle (Desmedt and Monaco, 1961; Tanaka and Katsuki, 1966; Brown, Daigneault, and Pruett, 1969; Guth and Amaro, 1969), although hemicholinium slowly reduced this inhibition (Guth and Amaro, 1969, Brown and others, 1969). Olivo-cochlear inhibition was blocked by strychnine (Desmedt and Monaco, 1961; Desmedt and La Grutta, 1963; Fex, 1967; Brown and others, 1969; Comis, 1970), which also blocked the depressant action of ACh on cochlear responses (Brown and others, 1969). Moreover, most cochlear nerve-endings which exhibit AChE activity contain flattened vesicles (McDonald and Rasmussen, 1971).

All these findings are difficult to evaluate. One interpretation is that ACh is an inhibitory transmitter released on to cochlear hair cells, and that the inhibition is blocked by strychnine. If this is true, why is the inhibition resistant to antagonists of ACh, which are effective in other areas of the C.N.S. ? One answer may be that the junction between nerve terminals and hair cells do not form true synapses in the classical sense, and the ACh receptors on the post-synaptic membrane may well differ from those on central neurons. Although strychnine has been found to be a relatively specific antagonist of the depressant action of glycine in various areas of the C.N.S., in higher concentrations it can also block depression by ACh (Phillis and York, 1967a). However, although it is feasible that strychnine blocks ACh on the postsynaptic membrane of cochlear hair cells, the possibility that glycine plays a role in olivo-cochlear inhibition should not be overlooked. Alternatively, the situation may be more comparable to the peripheral nervous system (superior cervical ganglion), where strychnine has been shown to block the release of ACh from nerve terminals (McKinstry and Koelle, 1967; *see also* Guth and Bobbin, 1971). Olivo-cochlear inhibition was unaffected by picrotoxin (Desmedt and La Grutta, 1963; Brown and others, 1969), and studies with GABA, thiosemicarbazide, and AOAA indicate further that the inhibitory transmitter is unlikely to be GABA (Bobbin and Guth, 1970). However, the action of glycine has not been studied.

In conclusion, it is possible that both ACh and glycine are involved in olivo-cochlear inhibition, but this is far from clear. Many have attempted to unravel this extremely complex problem, but interpretation remains confused.

Deiters' Neurons

The vestibular nuclei are involved in maintaining posture and equilibrium, and the production of patterned movements of the head, eyes, trunk, and limbs. Deiters' (lateral vestibular) nucleus is composed almost exclusively of neurons which have long axons projecting to the spinal cord (Brodal, Pompeiano, and Walberg, 1962), and Deiters' neurons can be readily identified by antidromic responses evoked by stimulating the ventral surface of the cervical spinal cord (Ito, Hongo, Yoshida, Okada, and Obata, 1964). Cerebellar Purkinje cells project to the dorsal portion of Deiters' nucleus (Brodal and others, 1962) and mediate monosynaptic inhibition on to these neurons (Ito and Yoshida, 1966; Ito, Obata, and Ochi, 1966; Ito, Kawai, and Udo, 1968). There is now extremely good evidence that the transmitter which mediates this inhibition is GABA. Indeed, this is the best example of an identified neuron for which an inhibitory transmitter is identified in the mammalian C.N.S.

Amino-acids

The first observations to suggest this were made by Obata, Ito, Ochi, and Sato (1967), who showed that GABA, administered electrophoretically to Deiters' neurons, hyperpolarized the postsynaptic membrane and increased its conductance. Thus, GABA decreased the potential changes across the cell membrane caused by intracellular injection of rectangular current pulses; depressed EPSPs and IPSPs; initially increased the amplitude of action potentials and then blocked them; increased the amplitude of the after-depolarization, but decreased that of the after-hyperpolarization; and depressed antidromic field potentials evoked by stimulation of the cervical spinal cord. In addition, hydroxylamine, which

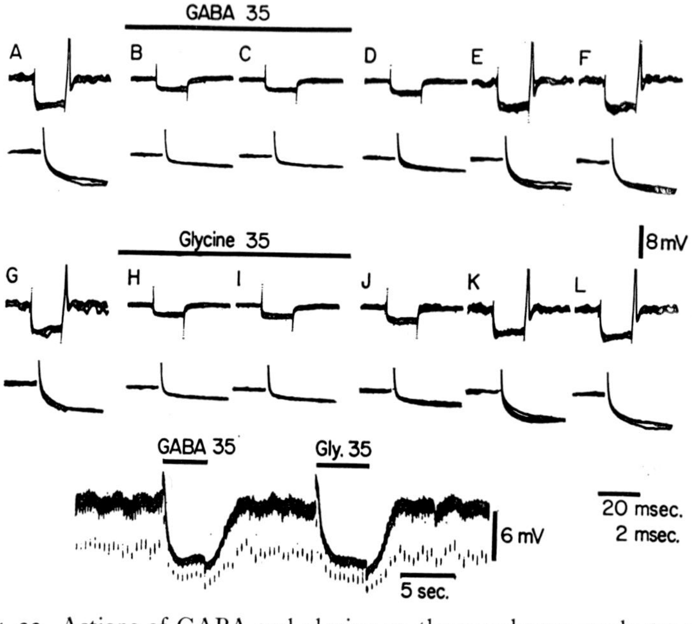

Fig. 22. Actions of GABA and glycine on the membrane conductance and membrane potential of a Deiters' neuron. Records A–L (each with two different sweep speeds) show measurements of membrane conductance by the injection of short hyperpolarizing current pulses (7·5 nA) through the recording electrode. Sequences A–F and G–L were recorded before, during, and after the applications of GABA (35 nA) and glycine (35 nA) respectively, indicated above the records. The total duration of each sequence was about 9 sec. Lower traces show only the first parts of the pulses. A marked reduction in amplitude and time constant is seen during and shortly after the drug applications. The slow-speed membrane potential record below shows the hyperpolarizing effects of the same drug applications. There is no shift in the baseline level between the oscilloscope traces, since A.C. amplification was used in these recordings. (Bruggencate and Engberg, 1971.)

blocks the action of GABA transaminase (amongst other actions; *see* Baxter and Roberts, 1958), sometimes increased the amplitude of IPSPs and positive field potentials (which represent inhibitory postsynaptic currents).

These observations have been confirmed and extended (Curtis, Duggan, and Felix, 1970; Obata, Takeda, and Shinozaki, 1970; Bruggencate and Engberg, 1971; Curtis, Duggan, Felix, Johnston, and McLennan, 1971). The action of glycine on Deiters' neurons was similar to that of GABA, the two amino-acids usually being either equipotent or GABA being slightly more effective. *Fig.* 22 illustrates the hyperpolarization and increase in conductance of the membrane of a Deiters' neuron produced by GABA and glycine (Bruggencate and Engberg, 1971).

During alterations of the membrane potential, the changes in amplitude and direction of the polarization by GABA and glycine correlated well with those of the IPSP evoked by stimulation of Purkinje axons (Obata and others, 1970; Bruggencate and Engberg, 1971). Moreover, intracellular injection of chloride ions affected the hyperpolarization by the amino-acids and the IPSP in the same manner. *Fig.* 23 shows that the reversal potential for the hyperpolarization of the Deiters' cell membrane by GABA was at the same level as that for the IPSP.

Current pulses of different intensity and polarity were passed through the intracellular recording electrode before (*Fig.* 23A) and during (*Fig.* 23B) the ejection of GABA on to the external surface of a Deiters' neuron. During the administration of GABA, the amplitude of the potential changes evoked by the current pulses and that of the IPSPs evoked by cerebellar cortical stimulation was reduced, indicating an increase in membrane conductance. A plot of the current–voltage relationship before and during the application of GABA indicated that the reversal potentials of the IPSPs and the GABA induced hyperpolarizations were −73 mV for this neuron (*Fig.* 23D).

Strychnine did not affect the Purkinje cell inhibition of Deiters' neurons, or the action of GABA, although it blocked the depression by glycine (Obata and others, 1967, 1970; Curtis, Duggan, and Felix, 1970; Bruggencate and Engberg, 1971; Curtis, Duggan, Felix, Johnston, and McLennan, 1971). However, Purkinje cell inhibition and the depressant action of GABA on Deiters' neurons were blocked by picrotoxin and bicuculline, which had no effect on that of glycine (Obata and others, 1970; Curtis, Duggan, and

Felix, 1970; Bruggencate and Engberg, 1971; Curtis, Duggan, Felix, Johnston, and McLennan 1971; cf. Huffman and McFadin, 1972). The blocking action of bicuculline on

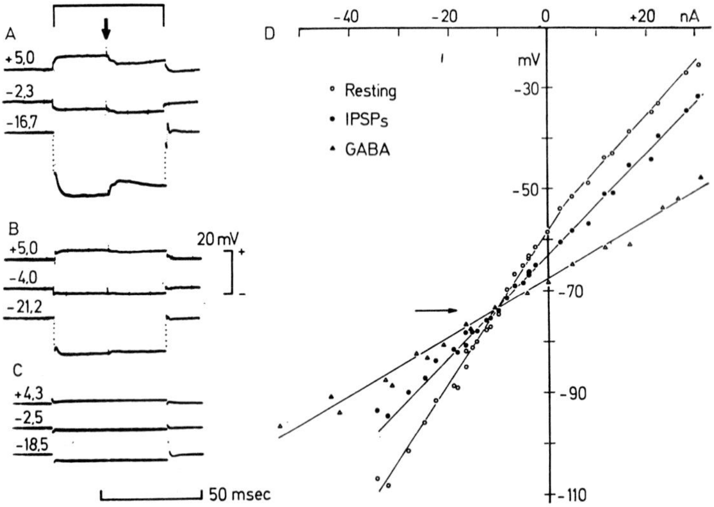

Fig. 23. Current–voltage relationship of a Deiters' neuron membrane during the resting state, cerebellar IPSPs, and administration of GABA, obtained with a parallel electrode. A, Specimen records obtained with an intracellular double-barrel electrode. During the period shown by the uppermost horizontal bar (56 msec. duration), polarizing currents (intensity indicated on each record in nA) were passed through one barrel. Plus sign (+) represents outward currents through the membrane, and minus (—) inward currents. The cerebellar cortex was stimulated at the time indicated by the downward arrow. B, During the electrophoretic administration of GABA (270 nA) from a 1 *M* solution (*p*H 7) contained in the single extracellular pipette which was glued to the recording electrode. C, Artefacts due to coupling resistance between the two barrels, recorded extracellularly by passing currents similar to those in A and B. D, Current–voltage relationship derived from the series of records partly shown in A–C. *Ordinate*: membrane potential level. *Abscissa*: depolarizing (+) and hyperpolarizing (—) currents (nA) passed across the membrane. O, Measured just before cerebellar stimulation; ●, at the peak of the IPSPs; △, similar to O, but during GABA (270 nA) administration. Coupling artefacts (as shown in C) were corrected in plotting each point in D. The artefact produced by GABA ejecting currents was so small (0·6 mV) that it was neglected in plotting D. Horizontal arrow marks the crossing point of the three curves. (Obata and others, 1970.)

Purkinje cell inhibition of a Deiters' neuron is illustrated in *Fig.* 24. *Fig.* 24A, C, E are action potentials superimposed on fields evoked antidromically by stimulating the spinal cord. In *Fig.* 24B, D, F, spinal stimulation was preceded by cerebellar stimulation. Immediately after bicuculline had been administered, the inhibition of the antidromic spike by cerebellar stimulation was blocked (*Fig.* 24D), followed by recovery 2·5 min. later (*Fig.* 24F).

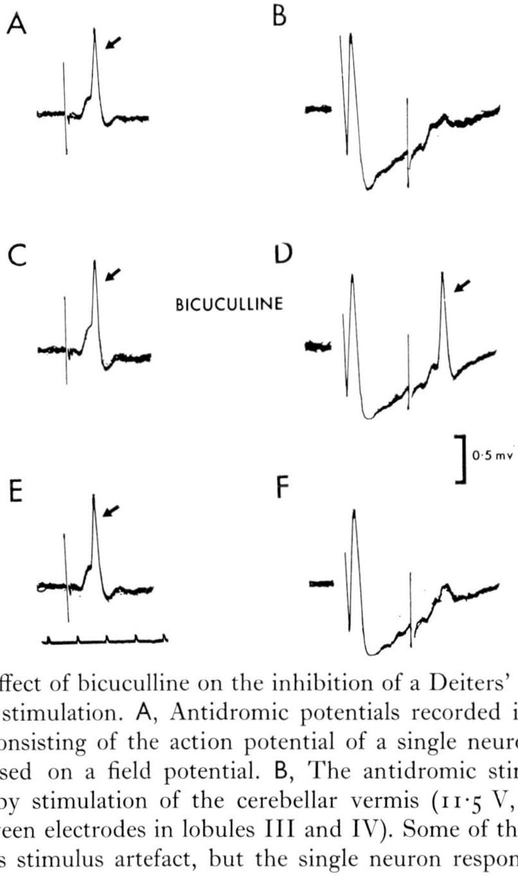

Fig. 24. Effect of bicuculline on the inhibition of a Deiters' neuron by cerebellar stimulation. A, Antidromic potentials recorded in Deiters' nucleus, consisting of the action potential of a single neuron (arrow) superimposed on a field potential. B, The antidromic stimulus was preceded by stimulation of the cerebellar vermis (11·5 V, 0·2 msec. pulse between electrodes in lobules III and IV). Some of the complex potential is stimulus artefact, but the single neuron response is consistently absent, and the field potential is reduced in amplitude. C, D, As for A, B, but at the end of an electrophoretic ejection of bicuculline (5 m*M* in 165 m*M* NaCl, 150 nA for 2 min.). E, F, 2·5 min. after the bicuculline ejection had been terminated. *Time calibration*: 1 msec. (Curtis, Duggan, and Felix, 1970.)

Thus, GABA mimics the action of the inhibitory transmitter released from Purkinje axon terminals on to Deiters' neurons, and both effects are influenced by convulsants in the same manner. Several other observations support the postulate that GABA is the inhibitory transmitter released from Purkinje axons. Relatively high concentrations of GABA are localized in or on cerebellar Purkinje and Deiters' neurons, compared with spinal ganglion cells, spinal motoneurons and cerebral Betz cells, and the GABA content of *dorsal* Deiters' neurons is 2–3 times that of *ventral* Deiters' neurons (Otsuka, Obata, Miyata and Tanaka, 1971; *see also* Sims and others, 1972). Moreover, the GABA levels in dorsal Deiters' neurons were greatly reduced after removal of the cerebellar vermis, whereas those of the ventral nucleus were unaffected (Otsuka and others, 1971). Similarly, the levels of glutamate decarboxylase, the enzyme that synthesizes GABA (Roberts and Eidelberg, 1960), are higher in the dorsal than in the ventral Deiters' nucleus, and destruction of the anterior cerebellar vermis resulted in a loss of glutamate decarboxylase activity only in the dorsal Deiters' nucleus (Fonnum, Storm-Mathisen, and Walberg, 1970). Subcellular fractionation studies of the dorsal nucleus suggested that glutamate decarboxylase (and therefore GABA) is localized in synaptosomes (Fonnum and others, 1970). Furthermore, stimulation of the cerebellar cortex (where Purkinje cells are located) increased the release of GABA into the fourth ventricle, the floor of which contains the Deiters' nuclei (Obata and Takeda, 1969). Similarly, studies involving perfusion of Deiters' nucleus with a 'push-pull' cannula revealed an increase in the release of GABA following cerebellar stimulation (Obata and Shinozaki, unpublished observations).

The similarities between the actions of GABA and glycine on Deiters' neurons suggested that there may be some synaptic connexions to Deiters' neurons operating with glycine as the transmitter, yet involving the same ionic mechanisms as Purkinje cell inhibition (Obata and others, 1970; Bruggencate and Engberg, 1971). If glycine is released synaptically from recurrent collaterals of inhibitory reticulospinal axons, it is probably an inhibitory transmitter acting on some Deiters' neurons (Tebēcis and Di Maria, 1972a). However, this particular source of glycine-releasing afferents to Deiters' nucleus cannot be large, as only a small proportion of monosynaptic IPSPs were evoked in Deiters' neurons following stimulation of the spinal cord (Ito, Udo, and Mano, 1970).

High doses of picrotoxin (10 mg./kg.), but not of strychnine (0·3 mg./kg.), blocked the inhibition of secondary vestibular neurons by stimulation of the cerebellar flocculus (Ito, Highstein, and Fukuda, 1970). Although the pharmacology of this inhibition has not been elucidated, the IPSPs evoked from the flocculus are thought to be mediated by Purkinje cell axons of the flocculovestibular projection (Ito, Highstein, and Tsuchiya, 1970).

Deiters' neurons have been shown to be excited by amino-acids structurally similar to glutamate (Obata and others, 1970; Curtis, Duggan, Felix, Johnston, and McLennan, 1971), but this excitation has not been related to a transmitter function.

Acetylcholine and Monoamines

Little is known about other transmitters in Deiters' nucleus. From one-half to three-quarters of the Deiters' neurons tested were excited by ACh, the remainder being unaffected (Steiner and Weber, 1965; Yamamoto, 1967). Physostigmine potentiated the excitant action of ACh on Deiters' neurons which responded to caloric and galvanic stimulation of the labyrinth (Steiner and Weber, 1965), but the action of this anticholinesterase on synaptic excitation in Deiters' nucleus has not been reported. The excitant action of ACh on Deiters' neurons was blocked by atropine (and unaffected by dihydro-β-erythroidine), but atropine had no effect on the synaptic responses of Deiters' neurons evoked by vestibular or sciatic nerve volleys (Yamamoto, 1967). However, there is some recent pharmacological evidence for cholinergic links in pathways to vestibular neurons (Barnes and Pompeiano, 1971a; Magherini, Pompeiano, and Thoden, 1971), although the details of these pathways are not clear.

In cats in which the cerebellum was removed, NA excited one-half of the Deiters' neurons tested, the remainder being unaffected (Yamamoto, 1967). *n*-(2-Dichloroethyl)-dibenzylamine (DCI) blocked the excitant action of NA, but not the synaptic activation of Deiters' neurons evoked by vestibular stimulation. Otherwise, no studies of the effects of monoamines on Deiters' neurons have been made.

Trigeminal Motoneurons

Motoneurons of the trigeminal nucleus are involved in the opening and closing of the jaws. Although the effects of possible transmitters on trigeminal motoneurons have not been investigated, studies involving systemic injections of agents known to block or potentiate the effects of transmitters elsewhere in the C.N.S. have provided indications of what some transmitters may be in the trigeminal nucleus.

Amino-acids

Stimulation of the inferior dental afferent nerve inhibited the monosynaptic reflex response of masseter motoneurons in two

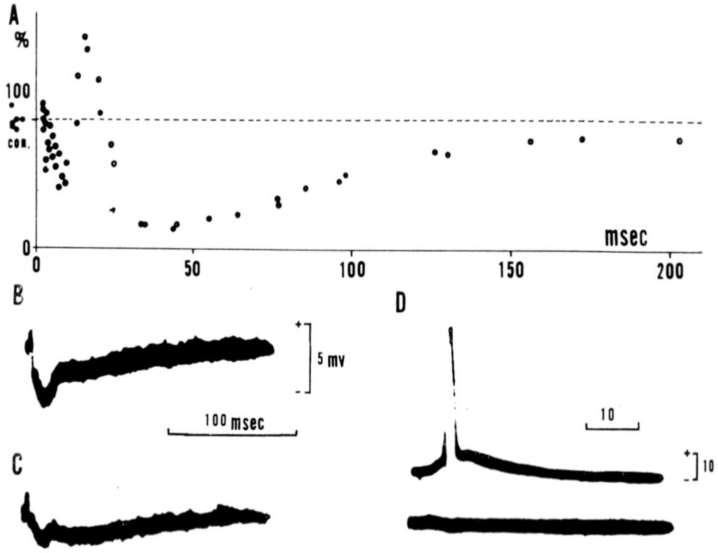

Fig. 25. Effects of strychnine on the inhibition of the masseteric monosynaptic reflex (MSR) and IPSPs of masseter motoneurons produced by stimulation of the inferior dental nerve. The MSR recorded in the masseter nerve following stimulation of the mesencephalic tract was conditioned with a supramaximal inferior dental shock. Before strychnine was administered, no facilitation was observed, and the early inhibition lasted approximately 15 msec. (*Fig*. 2 by Kidokoro, Kubota, Shuto, and Sumino, 1968). A, Following an intravenous injection of strychnine (0·1 mg./kg.). B, IPSPs recorded from a jaw-closing motoneuron before, and C, after the administration of strychnine (0·1 mg./kg.). D, After strychnine (0·1 mg./kg.), showing a depolarization with a spike (upper record) and its extracellular potential recorded outside the cell (lower record) produced by stimulation of the inferior dental nerve. (Kidokoro and others, 1968.)

phases—an early strong inhibition which reached a maximum at approximately 5-15 msec. after the conditioning shock, and a later less pronounced inhibition with a maximum at 40–50 msec. after the shock (Kidokoro, Kubota, Shuto, and Sumino, 1968). These two phases of inhibition were the result of two different types of IPSP. The early IPSP was reversed to a depolarization following intracellular injection of chloride ions, whereas the late IPSP was unaffected by chloride. Moreover, the early IPSP was associated with a greater increase in membrane conductance than the late IPSP, and the equilibrium potentials for the two IPSPs were usually different. In particular, the early IPSP was reduced by intravenous strychnine, which had little or no effect on the late IPSP. Results from some trigeminal motoneurons are illustrated in *Fig.* 25.

Fig. 25A shows the effects of stimulating the inferior dental nerve at various intervals preceding the masseteric monosynaptic reflex responses, 2–10 minutes after strychnine (0·1 mg./kg.) had been given intravenously. The early phase of the inhibition was reduced by strychnine, the monosynaptic reflex response being about 60 per cent of the unconditioned value. Facilitatory effects became evident during the transition from early to late phases of inhibition. Similarly, the early, but not the late IPSP was reduced by strychnine (*Fig.* 25B, C). After strychnine had been administered intravenously, stimulation of the inferior dental nerve sometimes induced a depolarization which could elicit an action potential (*Fig.* 25D).

Similarly, in a subsequent study (Sauerland and Mizuno, 1969), conditioning stimulation of the cerebral cortex evoked two phases of inhibition of the monosynaptic masseteric reflex (recorded from the masseter nerve following a single shock to the nucleus of the trigeminal mesencephalic tract)—an early, relatively intense inhibition and a later, less intense, prolonged inhibition. Intravenous strychnine (0·1 mg./kg.) reduced the early, but not the late inhibition, whereas picrotoxin (0·44 mg./kg.) reduced the late, but not the early inhibition. These results may be compared with those of an earlier study (Shende and King, 1967), in which the primary afferent depolarization evoked in the trigeminal nucleus by stimulating the cerebral cortex was reduced by picrotoxin (1 mg./kg.), but not by strychnine (0·1 mg./kg.). More recently it has been shown that the masseteric reflex was also inhibited in two phases (an early and a late phase) following single stimuli to the lingual nerve, and that intravenous picrotoxin reduced only the late inhibition (Goldberg, 1972).

Several explanations are possible for the mechanisms underlying the two types of inhibition, but the evidence favours the hypothesis that they are mediated by two different transmitters. The early, short-duration inhibition which is sensitive to strychnine is possibly mediated by glycine, whereas the late, prolonged, picrotoxin-sensitive inhibition is possibly mediated by GABA. It is interesting to note that the trigeminal nerve and mesencephalic trigeminal nuclei have been shown to stain particularly strongly for sites of GABA metabolism (van Gelder, 1965b; Sims and others, 1972). However, the actions of glycine, GABA, and the convulsants on single trigeminal motoneurons have yet to be determined.

Acetylcholine

Studies of trigeminal motoneurons have suggested that cholinergic transmission may also be important in the activity of these cells (Iwata, Sakai, and Deguchi, 1971). The monosynaptic reflex response of trigeminal motoneurons (recorded from the masseter nerve) evoked by stimulation of the mesencephalic trigeminal tract was depressed by intravenous physostigmine (0·1–0·3 mg./kg.), and this action was blocked by atropine (1·0 mg./kg.), but not by dihydro-β-erythroidine (0·5 mg./kg). In addition, the inhibition of this monosynaptic reflex evoked by prior stimulation of the superficial radial nerve (Kubota, Kidokoro, and Suzuki, 1968) was potentiated by physostigmine, and this action was also blocked by atropine, but not by dihydro-β-erythroidine. Similarly, physostigmine potentiated the IPSPs evoked in masseter and jaw-closing motoneurons by stimulation of the superficial radial nerve, and this was antagonized by atropine. Examples of this are illustrated in *Fig.* 26.

Physostigmine (0·15 mg./kg.) increased the amplitude of IPSPs evoked by stimulating the superficial radial nerve, and also produced a late membrane hyperpolarization in a masseter motoneuron (*Fig.* 26A, B). Both effects were blocked by atropine (*Fig.* 26C). The intracellular records in *Fig.* 26E were obtained from another masseter motoneuron, after physostigmine had been injected intravenously into the animal.

In addition, after the administration of physostigmine, small IPSPs (less than 2 mV in amplitude and of 5–10 msec. duration) occurred 'spontaneously' in about half the trigeminal motoneurons investigated, and these were blocked by atropine (Iwata and others, 1971). In contrast, the late IPSP evoked by stimulation of the inferior dental nerve studied by Kidokoro and others (1968) was not potentiated by physostigmine (Iwata and others, 1971).

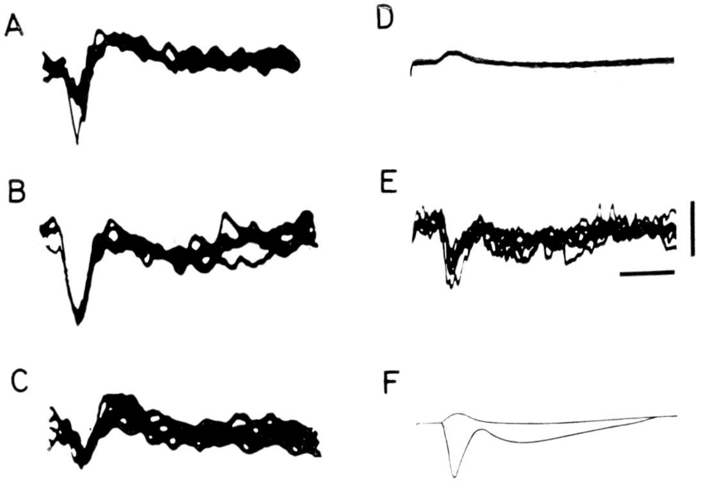

Fig. 26. The effects of physostigmine on IPSPs recorded from masseter motoneurons, and antagonism by atropine. A–C and E, IPSPs evoked by supramaximal stimulation of the superficial radial nerve. A–C were obtained from one, and D–F from another, masseter motoneuron of different preparations. A, Before, B, 8 min. after the intravenous injection of physostigmine (0·15 mg./kg.). C, About 15 min. after physostigmine and 4 min. after atropine (1·0 mg./kg.). E, About 10 min. after physostigmine (0·15 mg./kg.). D, Extracellular field potentials evoked by stimulation of the same nerve as in E and recorded just outside the motoneuron from which the records in E were obtained. F, Traced records from D and E. *Calibration*: 20 msec. and 2 mV. (Iwata and others, 1971.)

The results of these investigations suggest the presence of a cholinergic link in the inhibition of trigeminal motoneurons (Iwata and others, 1971; cf. spinal cord, Iwata and Sakai, 1971). However, details of the cholinergic pathway are not known. It is possible that inhibitory interneurons are synaptically excited by ACh and, in turn, exert a 'tonic' inhibitory bombardment on to trigeminal motoneurons, but this cannot be a result of a Renshaw cell-type of recurrent inhibition because cranial motoneurons have no axon collaterals (Lorente de Nó, 1947; Iwata and others, 1971). Another possibility is that there are inhibitory cholinergic synapses on trigeminal motoneurons, the receptors being muscarinic. In this regard it is interesting that the depressant actions of ACh on brain-stem neurons are mediated by muscarinic receptors, whereas the excitant actions possess both muscarinic and nicotinic properties (Bradley and others, 1966). Little more can be said at this stage. Experiments are required involving the electrophoretic administration of ACh and its antagonists and potentiators on to trigeminal motoneurons, as well as a comparison of the effects of these agents with synaptic inhibition.

Inferior Collicular Neurons

The inferior colliculus is the mesencephalic relay nucleus of the auditory pathway. Few pharmacological studies have been made of inferior collicular neurons, but recent evidence suggests the possibility that ACh and GABA are transmitters in the inferior colliculus.

Acetylcholine

ACh excited only 10 per cent of the inferior collicular neurons studied (in unanaesthetized cats) by Curtis and Koizumi (1961). All these ACh-responsive cells were located within 1–3 mm. of the surface of the inferior colliculus and 2–3 mm. lateral to the midline, and most of them reponded to auditory stimulation with a multiple spike discharge of 10–25 msec. latency. Neostigmine sometimes prolonged the action of ACh, but did not potentiate the auditory responses or increase the rate of spontaneous firing. In a subsequent study, ACh excited approximately one-half of the inferior collicular neurons studied at depths of up to 3 mm. from the dorsal surface of the inferior colliculus, whereas it often depressed the firing of collicular neurons located deeper (Tebēcis, unpublished observations).

Only recently, however, has direct pharmacological evidence been obtained for cholinergic transmission in the inferior colliculus (Watanabe and Simada, 1970). Tone and click stimuli evoked small-amplitude, long-duration burst discharges from some inferior collicular neurons, and both physiological and pharmacological studies suggested that these cells may be auditory interneurons which resemble Renshaw cells (Watanabe and

Simada, 1970). Electrophoretically administered dihydro-β-erythroidine reversibly blocked the synaptic excitation of these neurons to sound stimulation, as illustrated in *Fig.* 27.

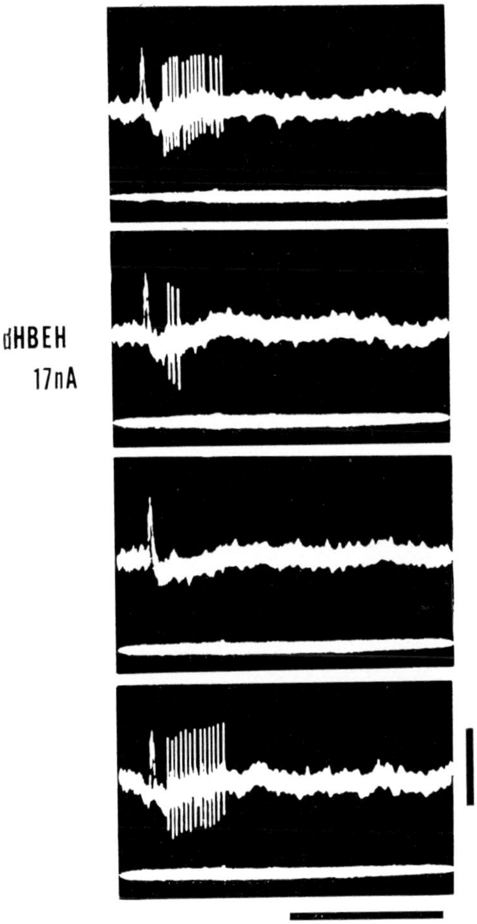

Fig. 27. Effect of dihydro-β-erythroidine on the synaptic discharge of an auditory interneuron in the inferior colliculus. The top record is a control response of the neuron to a click stimulus (—40 dB). The second and third records were obtained 13·5 and 21 min. after the beginning of an electrophoretic ejection of dihydro-β-erythroidine (17 nA). The bottom record shows recovery of the synaptic response after the ejecting current had been terminated. The characteristic frequency of this neuron was 7·5 kHz. *Calibration*: 0·5 mV and 100 msec. (Watanabe and Simada, 1970.)

Although these results point to the probable importance of cholinergic transmission in the inferior colliculus, further studies are required to establish this with more certainty. The specificity of antagonism by dihydro-β-erythroidine must be determined by comparing its action on more than one type of synaptic response and/or its action on ACh and unrelated excitant substances. It would also be useful to know whether muscarinic antagonists of ACh affect synaptic excitation in the inferior colliculus, and the proportion of auditory interneurons sensitive to these agents.

Amino-acids

Glutamate excited and GABA depressed the firing of all the inferior collicular neurons tested (Curtis and Koizumi, 1961; Tebēcis, unpublished observations). Locally administered picrotoxin, but not strychnine, reversibly prolonged the synaptic discharge of phasic ('on'-type) inferior collicular neurons to tones, the prolongation being interpreted as a block of inhibition (Watanabe and Simada, 1971). It is therefore likely that GABA is an inhibitory transmitter in the inferior colliculus, although its site of action has to be determined. It is noteworthy that the inferior colliculus contains relatively high concentrations of GABA and the enzymes for its metabolism (Salvador and Albers, 1959; van Gelder, 1965b; Fahn and Côté, 1968a; Okada and others, 1971).

Monoamines

The monoamines, 5-HT, NA, DA, and histamine usually had a depressant action or no action at all on the few inferior collicular neurons tested (Curtis and Koizumi, 1961; Tebēcis, unpublished observations), but the data are too limited to draw conclusions about the possible significance of these actions.

Superior Collicular Neurons

The superior colliculus is involved in mechanisms responsible for visually guided behaviour and visual attention. Superior collicular neurons respond especially to moving visual stimuli, and the majority are sensitive to the direction of stimulus movement, particularly in the deeper layers of the nucleus (*see* references by Straschill and Hoffmann, 1969).

Acetylcholine

Of a relatively large number of tectal neurons tested with cholinomimetics, none was affected by ACh, and only 7 per cent were affected by carbamylcholine, which is usually far more effective than ACh in different areas of the C.N.S. (Straschill and Perwein, 1971). Since general anaesthesia was not maintained in these experiments, anaesthetics could not have been responsible for the lack of effect of the cholinomimetics, suggesting that ACh is unlikely to be a transmitter in the superior colliculus. However, it is also possible that cholinergic mechanisms in the superior colliculus are localized in discrete areas (possibly superficial layers) which have not been studied. For example, in the rat the superficial and intermediate layers of the superior colliculus receive a rich innervation of cholinesterase-containing fibres (Shute and Lewis, 1967), and in the rhesus monkey, ACh is released spontaneously from the superficial, but not the deep layers (Beleslin and Myers, 1970).

Amino-acids

L-Glutamate was a powerful excitant of most superior collicular neurons, and, in addition, it restored the responsiveness of units adapted to repetitive visual stimulation and increased the receptive field of adapted neurons (Straschill and Perwein, 1971). GABA readily depressed the firing of superior collicular neurons.

Monoamines

5-HT depressed the firing of approximately half the superior collicular neurons tested (Straschill and Perwein, 1971). Examples of this are illustrated in *Fig.* 28. Although 5-HT occasionally increased the rate of firing of superior collicular neurons, this excitation was always preceded by a depression. In particular, 5-HT affected a higher proportion of neurons sensitive to the direction of moving stimuli than those directionally unselective. Although 5-HT depressed both orthodromic and glutamate-induced firing, the relative sensitivity of these two types of firing to 5-HT has not been determined for the same neuron. It is therefore possible that 5-HT in low concentrations selectively depresses a proportion of the synaptic responses of collicular neurons evoked by optic nerve impulses, as occurs in the lateral geniculate nucleus (Tebēcis and Di Maria, 1972b). On the other hand, since the dominant afferent input to the lateral geniculate nucleus arises from a different population of retinal ganglion cells than the dominant input to the superior colliculus (Bishop,

Clare, and Landau, 1969; Clare, Landau, and Bishop, 1969; Hoffmann, 1972; cf. Stone and Hoffmann, 1972), the predominant action of 5-HT on superior collicular neurons may not necessarily resemble the dose-dependent actions of 5-HT on lateral geniculate neurons. This problem has yet to be resolved. It is particularly important in that the excitatory transmitter released from optic nerve terminals has not been identified, but may be an unusual substance, hitherto not considered as a possible transmitter (*see* Chapter 4 on the lateral geniculate nucleus).

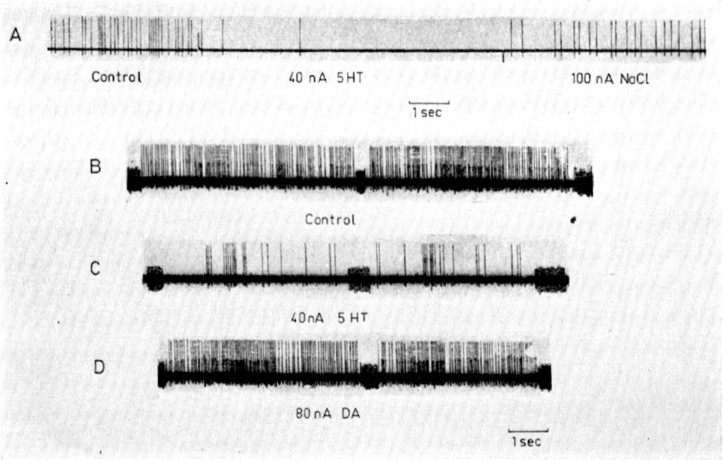

Fig. 28. Actions of 5-HT and DA on the firing of neurons in the superior colliculus. A, 5-HT (40 nA), administered electrophoretically for the period indicated by the upward and downward artefacts, depressed the spontaneous firing of a tectal neuron much more effectively than did Na^{+} (100 nA) ions. B, Discharge of a movement-sensitive neuron during alternating horizontal movements of a bright disk. Reversal of movement direction is indicated by the 50-cycle artefact. C, Discharge of the same neuron as in B, during the administration of 5-HT (40 nA). D, As in B, C, but during an ejection of DA with a current of 80 nA. (Straschill and Perwein, 1971.)

The presence of 5-HT in nerve terminals of the rat superior colliculus (Fuxe, 1965), its spontaneous release from the deeper (but not the superficial) layers of the monkey colliculus (Beleslin and Myers, 1970), and neuropharmacological studies of this region in guinea-pigs and cats (Kawai and Yamamoto, 1969; Straschill and Perwein, 1971) collectively suggest that 5-HT is an inhibitory transmitter in the mammalian superior colliculus. The release of 5-HT from slices of the superior colliculus following stimulation of the optic tract need not necessarily indicate a release from optic tract axons (Kawai, 1970), but may equally well be explained by assuming a release from 5-HT-containing terminals which were stimulated (directly or via interneurons) by optic nerve impulses. Histochemical evidence indicates that 5-HT-containing neurons are located almost exclusively in the raphe nuclei (Dahlström and Fuxe, 1965a, b; Fuxe, 1965; Ungerstedt, 1971), suggesting that 5-hydroxytryptaminergic afferents to the superior colliculus originate in these brain-stem areas.

The catecholamines, NA and DA, depressed almost 60 per cent of the superior collicular neurons tested, no excitant actions being observed (Straschill and Perwein, 1971). In contrast to 5-HT, neither NA nor DA showed a preference for directionally selective neurons. The presence of NA-containing terminals in the superior colliculus, which arise in the locus coeruleus (Fuxe, 1965; Olson and Fuxe, 1971), raises the possibility that NA functions as a transmitter in this nucleus.

Mesencephalic Reticular Neurons

Neurons of the mesencephalic reticular formation are characterized by multimodal sensitivity and habituation to repeated stimuli, and may be involved in arousal and wakefulness (amongst other functions). The pharmacology of mesencephalic reticular neurons at the level of the superior colliculus was similar to that of superior collicular neurons (Curtis and Koizumi, 1961; Straschill and Perwein, 1971). Thus, glutamate excited, GABA depressed, and ACh had no effect on these reticular neurons. Although Curtis and Koizumi (1961) observed no action of 5-HT and NA on mesencephalic reticular neurons, these monoamines and DA depressed the majority of such neurons in another study (Straschill and Perwein, 1971). The effects of these substances have not been related to particular functional types of neurons in the mesencephalic reticular formation.

Oculomotor Neurons

The oculomotor nucleus lies beneath the superior colliculus, and oculomotor neurons may be identified by antidromic activation resulting from stimulation of the third cranial nerve. Although little is known about the pharmacology of oculomotor neurons, the available evidence suggests that GABA is an inhibitory transmitter in this area.

Stimulation of the rostral region of the vestibular nucleus complex in rabbits evoked a monosynaptic inhibition of oculomotor

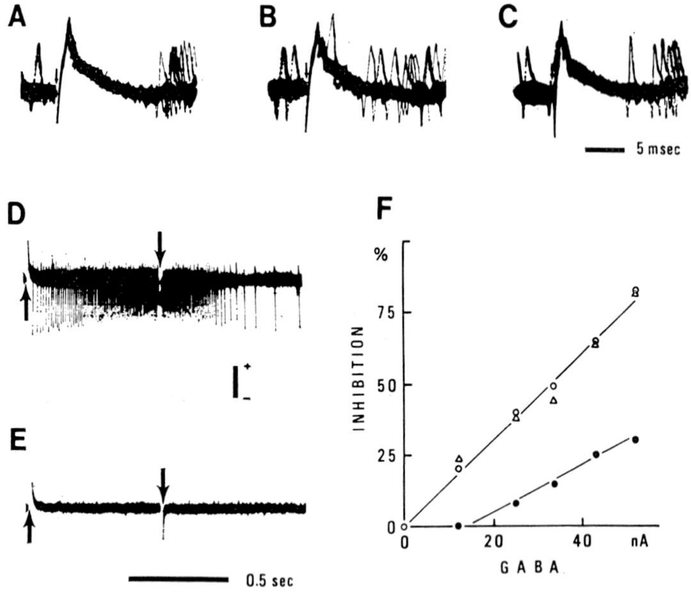

Fig. 29. A–C, Effect of electrophoretically administered picrotoxin (70 nA) on the inhibition of an oculomotor neuron produced by impulses in vestibular fibres. Firing was induced by a continuous ejection of glutamate. The onset of the stimulus is indicated by a downward arrow. A, Before; B, During; C, After the administration of picrotoxin. The records were obtained by superimposing 10 traces. D, E, Spike discharges of an oculomotor neuron produced by glutamate (D) and their suppression by GABA (E). Glutamate (12 nA) was administered between the upward and downward arrows. D, Before; E, During GABA (20 nA). Voltage calibration is 1 mV for A–C, and 2 mV for D and E. F, Effect of picrotoxin on the depression by GABA on the same cell as in A–C. *Ordinate*: percentage depression by GABA of glutamate-induced firing before (O), during (●), and after (△) the administration of picrotoxin (70 nA). *Abscissa*: values of the currents used to eject GABA. (Obata and Highstein, 1970.)

neurons, which was similar to the disynaptic inhibition evoked by vestibular nerve volleys (Ito, Highstein, and Tsuchiya, 1970; Highstein, Ito, and Tsuchiya, 1971; cf. Fukuda, Highstein, and Ito, 1972). This vestibulo-ocular postsynaptic inhibition was reduced or blocked by picrotoxin (3–20 mg./kg.), but not by strychnine (0·2 mg./kg.). Electrophoretically administered GABA and glycine depressed antidromic fields and the spontaneous and glutamate evoked firing of neurons in the oculomotor nucleus, GABA usually being a more effective depressant than glycine (Obata and Highstein, 1970). In particular, picrotoxin reduced the inhibition of oculomotor neurons produced by secondary vestibular neurons, as well as the depressant action of GABA (*Fig.* 29).

Stimulation of the vestibular nerve inhibited the firing of an oculomotor neuron produced by glutamate (*Fig.* 29A). Electrophoretically administered picrotoxin almost blocked this inhibition (*Fig.* 29B), followed by recovery after the current ejecting picrotoxin had been terminated (*Fig.* 29C). Similarly, the depressant action of GABA on the firing by glutamate (*Fig.* 29D, E) was markedly reduced by picrotoxin (plotted in *Fig.* 29F).

These results suggest that GABA is the inhibitory transmitter released from secondary vestibular fibres on to oculomotor neurons (Obata and Highstein, 1970; Highstein and others, 1971). It is noteworthy that the oculomotor nucleus contains relatively high concentrations of GABA and the enzymes for its metabolism (van Gelder, 1965b; Okada and others, 1971; Sims and others, 1972).

The action of glycine on oculomotor neurons was blocked by strychnine, but not by picrotoxin, and Obata and Highstein (1970) suggested that these neurons may also receive a glycine-mediated inhibition from a source yet unknown.

Rubral Neurons

The red nucleus consists of two arbitrarily divided groups of neurons (magno- and parvi-cellular parts) in the mesencephalon (Brodal and Gogstad, 1954; Pompeiano and Brodal, 1957). It receives a monosynaptic excitatory projection from the contralateral intracerebellar nuclei (mainly nucleus interpositus) via the brachium conjunctivum (Courville, 1966; Toyama, Tsukahara, and Udo, 1968; Davis, 1969), and the 'spontaneous' firing of red nucleus neurons is mainly due to the constant discharging of brachium conjunctivum fibres (Massian and Albe-Fessard, 1963; Toyama and others, 1968). Efferent fibres of the red nucleus terminate mainly in the spinal cord (Pompeiano and Brodal, 1957).

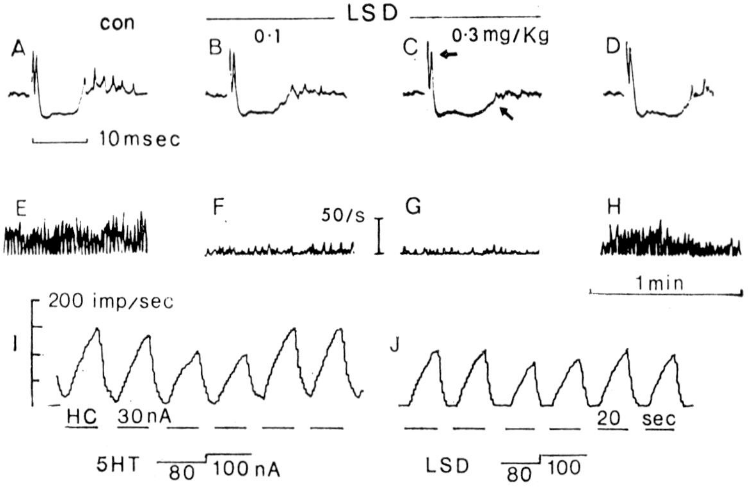

Fig. 30. Actions of LSD on different types of firing of neurons in the red nucleus. A–D, Synaptic responses of a neuron evoked by stimulating the brachium conjunctivum (unanaesthetized cat). E–H, 'Spontaneous' firing frequency of the neuron with the largest spike. A and E are control records. B, F, Within 60 sec. of an intravenous injection of LSD (0·1 mg./kg.). C, G, Following a total dose of 0·3 mg./kg. of LSD approximately 5 min. later. D, H, Partial recovery 10 min. later. I, J, Chart recordings showing the firing evoked by DL-homocysteate (HC) of a rubral neuron in an animal anaesthetized with pentobarbitone sodium. 5-HT and LSD (each administered with currents of 80 nA followed by 100 nA) had slight depressant actions on the firing evoked by DLH. (Davis and Vaughan, 1969.)

Few pharmacological studies have been made of neurons in the red nucleus. Of a number of substances administered electrophoretically to rubral neurons which were firing spontaneously or synaptically excited by stimulating the brachium conjunctivum, only glutamate and DLH consistently excited these neurons; glycine, GABA, 5-HT, NA, DA, and in most cases ACh, depressed the firing of rubral neurons sensitive to them (Davis and Vaughan,

1969; Davis and Huffman, 1969). The synaptic excitation of rubral neurons evoked by impulses in brachium conjunctivum fibres was not affected by either dihydro-β-erythroidine or atropine, indicating further that ACh is not the excitatory transmitter released from brachium fibres on to rubral neurons (Davis and Huffman, 1969; Davis and Vaughan, 1969). Of particular interest, however, is the finding that electrophoretically and intravenously administered LSD readily depressed spontaneous firing and the postsynaptic excitation evoked by stimulating the brachium conjunctivum, but only slightly reduced the excitation produced by amino-acids (Davis and Vaughan, 1969). *Fig.* 30 shows such actions of LSD. This raises the possibility that the excitatory transmitter released from brachium conjunctivum fibres on to rubral neurons is identical to that released from optic nerve-fibres on to lateral geniculate neurons (*see* Chapter 4). This certainly warrants further investigation.

Histochemical studies of the distribution of AChE in the red nucleus have been interpreted in terms of cholinergic afferents and efferents of this nucleus (for example, the rubrospinal tract), but the evidence is so conflicting and variable between species (*see* Discussion by Gwyn, 1971) that no such conclusions may be made.

GABA was usually a more effective depressant of rubral neurons that glycine, and the action of GABA was blocked by picrotoxin in concentrations which had little or no effect on that of glycine (Altmann, Bruggencate, Sonnhof, and Steinberg, personal communication). The actions of amino-acids on rubral neurons, and those of convulsants on synaptic inhibition in the red nucleus therefore also warrant further investigation.

SUMMARY

Cuneate Neurons

Cuneothalamic relay neurons are readily excited by glutamate, hair cells being more sensitive than touch and proprioceptive cells. Glutamic acid is possibly the excitatory transmitter released from primary afferent fibres on to cuneate neurons, but there is yet no direct pharmacological evidence for this.

Glycine and GABA depress the firing of cuneothalamic relay neurons, proprioceptive cells being more sensitive than hair and touch cells. Picrotoxin and bicuculline block the depressant action of GABA without affecting that of glycine, whereas strychnine blocks the depression by glycine without affecting that by GABA. Bicuculline blocks the postsynaptic inhibitions evoked by stimulating the pyramidal tract, median nerve, and forepaw, suggesting that GABA is the inhibitory transmitter acting at these synapses on cuneate neurons. Strychnine also blocks a postsynaptic inhibition in the cuneate nucleus, suggesting that glycine is the transmitter involved, but details of the pathway are not clear. In addition, studies with various amino-acids and convulsants suggest that GABA may also be involved in 'presynaptic inhibition' in the cuneate nucleus, but its precise role is not clear.

The role, if any, of ACh and monoamines in the cuneate nucleus is unknown.

Hypoglossal Motoneurons

The firing of hypoglossal motoneurons is depressed by glycine and GABA, this depresion resulting from a membrane hyperpolarization and an increase in its conductance. The depressant action of glycine is blocked by strychnine, which does not affect the action of GABA. Intravenous strychnine has been shown to block the inhibition evoked in the hypoglossal nucleus by stimulating the lingual nerve, suggesting that glycine is the inhibitory transmitter involved. However, the effect of strychnine on the depression by glycine and synaptic inhibition of the same hypoglossal neuron has yet to be shown.

The depressant action of GABA on hypoglossal motoneurons is blocked by picrotoxin in concentrations which have little or no effect on that of glycine, but an action of picrotoxin on synaptic inhibition in the hypoglossal nucleus has not been shown. Nevertheless, these and neurochemical findings raise the possibility that GABA may also be an inhibitory transmitter in the hypoglossal nucleus.

Paramedian Reticular Neurons

ACh and 5-HT excite, whereas NA depresses, the firing of paramedian reticular neurons, all of which project to the cerebellum. However, the actions of these substances have not been related to synaptic activity in the paramedian reticular nucleus.

Reticulospinal Neurons

Glycine and GABA depress the firing of reticulospinal neurons, glycine being more effective than GABA. Strychnine blocks the action of glycine, but not that of GABA on reticulospinal neurons, whereas bicuculline blocks the action of GABA in concentrations which have little or no action on glycine. Strychnine (but not bicuculline) blocks the inhibitory effects of stimulation of the cervical spinal cord (ventrolateral area), and sometimes of the pontine reticular formation, on the firing of medullary reticular neurons. This suggests that glycine is an inhibitory transmitter in the medullary reticular formation, possibly released from recurrent collaterals of inhibitory reticulospinal axons. The relatively high levels of glycine and the presence of a specific mechanism for its uptake in the medulla support the postulate that this amino-acid is a transmitter in this region.

Glutamate, aspartate, 5-HT, and NA excite, whereas DA depresses or has no action on the firing of medullary reticulospinal neurons, but these actions have not been related to synaptic activity in the medullary reticular formation. An action of ACh on reticulospinal neurons has not been reported.

'Rostrally Projecting' Reticular Neurons

'Rostrally projecting' neurons of the pontomedullary reticular formation have not been extensively tested with possible transmitter substances. A few of these neurons have been shown to be depressed by glycine and GABA (glycine being a more effective depressant than GABA), and either depressed or excited by NA. It is likely that glycine is an inhibitory transmitter acting on some of these neurons.

Raphe Neurons

Raphe neurons are pharmacologically heterogeneous and probably have different functions, some of which involve 5-HT. In the rat, 5-HT excites pontine raphe neurons, which can be synaptically excited by stimulation of the nucleus paragigantocellularis lateralis (NPL), and depresses most of those inhibited by NPL stimulation. 5-HT may be released synaptically on to raphe neurons from a pathway originating in the NPL, but there is yet no direct pharmacological evidence for this. NA and ACh have variable effects on raphe neurons.

Cochlear Neurons and Hair Cells

Locally administered ACh and stimulation of the superior olive excite a proportion of neurons in the ventral cochlear nucleus, and both forms of excitation have been shown to be reduced by antagonists of ACh. Hemicholinium reduces cochlear neuron responses to olivary stimulation and decreases the concentration of ACh in the cochlear nucleus. It has therefore been suggested that ACh is an excitatory transmitter released centrifugally from the superior olive on to ventral cochlear neurons, but the specificity of the actions observed and the details of the excitatory pathway have yet to be elucidated.

Glycine, GABA, and NA have been reported to depress the firing of cochlear neurons, and the observation that picrotoxin blocks 'two-tone inhibition' suggests that GABA is an inhibitory transmitter in the cochlear nucleus. These findings require further investigation to enable a more adequate evaluation of their significance.

The transmitter mediating inhibition of cochlear *hair cells* (via the olivo-cochlear bundle) is unknown. It is blocked by strychnine, and various observations suggest that it may be ACh, but the possibility that glycine is also involved in olivo-cochlear inhibition should not be overlooked.

Deiters' Neurons

There is overwhelming evidence that the transmitter which mediates monosynaptic inhibition from Purkinje axons on to dorsal Deiters' neurons is GABA. Both GABA and glycine depress the firing of Deiters' neurons, this depression resulting from a membrane hyperpolarization and an increase in its conductance. Furthermore, there is a close similarity between the reversal potentials for the IPSP and the polarization caused by GABA and glycine. Picrotoxin and bicuculline block the Purkinje cell inhibition and the depression by GABA, but not that by glycine, on dorsal Deiters' neurons. Strychnine blocks the depression by glycine, but does not affect that by GABA, or the inhibition by Purkinje axons. Relatively high concentrations of GABA are present in or on Purkinje and dorsal Deiters' neurons, and the GABA levels of dorsal, but not ventral Deiters' neurons are reduced following section of Purkinje cell axons. Similarly, the levels of glutamic acid decarboxylase are greater in the dorsal than in the ventral Deiters' nucleus, and destruction of Purkinje cells results in a decrease of glutamic decarboxylase activity only in the dorsal nucleus. Stimulation of the cerebellar cortex increases the release of GABA into the fourth ventricle, suggesting that there is a release of GABA from Purkinje cell terminals in Deiters' nucleus.

Glutamate and structurally related amino-acids, ACh, and NA excite Deiters' neurons sensitive to these substances, but their actions have not been related to synaptic functions in Deiters' nucleus.

Trigeminal Motoneurons

The effects of possible transmitters on trigeminal motoneurons have not been investigated. However, the observation that intravenous strychnine blocks a short-latency synaptic inhibition in the trigeminal nucleus evoked by stimulating the inferior dental nerve, lingual nerve, and cerebral cortex suggests that glycine is the transmitter acting at the inhibitory synapses involved. Picrotoxin (but not strychnine) reduces the primary afferent depolarization evoked in the trigeminal nucleus following stimulation of the cerebral cortex, as well as certain long-latency postsynaptic inhibitions evoked by stimulating the lingual nerve and cerebral cortex, suggesting that GABA also mediates inhibition in this nucleus.

Similarly, studies involving the intravenous administration of physostigmine, atropine, and dihydro-β-erythroidine suggest that a cholinergic mechanism is involved in the inhibition evoked in the trigeminal nucleus following stimulation of the superficial radial nerve.

Inferior Collicular Neurons

Some inferior collicular neurons are excited by ACh, and dihydro-β-erythroidine has been shown to block the synaptic excitation of certain auditory interneurons. This suggests that ACh is an excitatory transmitter in the inferior colliculus, but further studies are required to substantiate this.

GABA depresses the firing of inferior collicular neurons, and picrotoxin (but not strychnine) prolongs the synaptic discharge of phasic ('on'-type) collicular neurons to tones, the prolongation representing a block of inhibition. It is therefore likely that GABA mediates this synaptic inhibition in the inferior colliculus, but the effects of amino-acids and convulsants on the same neuron have yet to be investigated.

The significance of the excitant action of glutamate, or the depressant actions of monoamines on inferior collicular neurons is not known.

Superior Collicular Neurons

An action of ACh on superior collicular neurons has not been observed, but histochemical and release studies nevertheless indicate the possibility that ACh may function as a transmitter in the superficial layers of the tectum.

Glutamate excites, whereas GABA and the monoamines (5-HT, NA, DA) depress the firing of superior collicular neurons sensitive to them, but these actions have not been related to synaptic activity in the superior colliculus. Histochemical, biochemical, and release studies, however, raise the possibility that both 5-HT and NA function as transmitters in the superior colliculus.

Mesencephalic Reticular Neurons

The pharmacology of the few neurons tested in the mesencephalic reticular formation is similar to that of superior collicular neurons.

Oculomotor Neurons

GABA depresses the firing of oculomotor neurons, and this action of GABA, as well as the inhibition of oculomotor neurons resulting from secondary vestibular volleys is blocked by picrotoxin (but not by strychnine). This suggests that GABA is the inhibitory transmitter released from secondary vestibular fibres on to oculomotor neurons.

Glycine also depresses oculomotor neurons, and this depression is blocked by strychnine, which does not affect the action of GABA. However, the action of glycine has not been related to a synaptic inhibition in the oculomotor nucleus.

Rubral Neurons

Glycine, GABA, 5-HT, NA, DA, and in most cases ACh, depress the firing of rubral neurons, but the possible significance of these actions has not been investigated. Glutamate, and on some cells ACh, have excitant actions. The synaptic excitation of rubral neurons evoked by impulses in brachium conjunctivum fibres is not affected by dihydro-β-erythroidine or atropine, indicating that the excitatory transmitter is not ACh. This synaptic excitation is blocked by LSD, suggesting that the transmitter is the same as that released from optic nerve terminals on to lateral geniculate neurons.

Chapter

3

CEREBELLUM

THE cerebellum is thought to function as a type of computer concerned particularly with the smooth and effective control of movement. It integrates and organizes the afferent information from the muscles, skin, fascia, and joints from all parts of the body, and by means of the processed information carried by the efferent pathways to the spinal cord, basal ganglia, and cerebral cortex, it modifies or participates in the control of movement. The most important anatomical and physiological findings on the cerebellum have been discussed by Eccles, Ito, and Szentágothai (1967). The following introduction is a summary of some points in their book, which should be consulted for a fuller understanding of the functions of cerebellar neurons. Many references have been omitted for the sake of conciseness, but may be obtained from Eccles and others (1967).

CEREBELLAR CORTEX

The cerebellar cortex consists of three well-defined layers—the molecular layer (outermost), the ganglionic or Purkinje cell body layer, and the granular layer (innermost). Five distinct types of neurons have been identified morphologically (and to some extent physiologically) in these layers. The sole output of the cerebellar cortex is the inhibition produced by Purkinje axons which terminate on neurons in Deiters' and the intracerebellar nuclei. Inputs to the cerebellum are varied. Mossy fibres originate in the spinal cord, external cuneate nucleus, lateral reticular nucleus, inferior olive and vestibular nuclei, and project to the granular layer where they synapse with granule-cell dendrites and Golgi-cell dendrites and terminals (cerebellar isles or glomeruli). Climbing fibres also originate in the brain-stem (for example, the inferior olive) and project to the Purkinje-cell layer, where they arborize on the dendritic ramifications of Purkinje cells. Characteristically, each mossy fibre branches considerably, forming synaptic contacts with many granule neurons, whereas a climbing fibre typically innervates the dendrites of a single Purkinje cell, branches to basket, Golgi, or other Purkinje cells being rare. *Fig.* 31 summarizes the main features of the complex, but orderly, neuronal architecture of the cerebellar cortex. Monoaminergic afferents to the cerebellum are not shown (*see* p. 94).

The axons of granule cells ascend to the molecular layer, where they bifurcate to give rise to the parallel fibres which lie strictly parallel to the longitudinal axis of the folium. Synaptic excitation from mossy fibres evokes excitatory impulses in granule cells and hence, parallel fibres, which in turn excite the dendrites of all the neurons in their path. These are the outer stellate, basket, Purkinje, and Golgi cells, all of which are inhibitory. Granule cells fire Golgi neurons not only by parallel fibres, but also by terminations on Golgi-cell dendrites in the glomeruli of the granule-cell layer.

The axons of Golgi cells terminate exclusively as inhibitory synapses on granule-cell dendrites in glomeruli. The other two inhibitory interneurons—basket and outer stellate cells—are located predominantly in the molecular layer and mediate inhibition on to Purkinje cells via axosomatic and axodendritic synapses respectively (*Fig.* 31). The dendritic tree of basket and stellate cells is confined to the transverse plane of the folium, whereas that of Golgi cells expands equally in all directions.

Purkinje Cells

Purkinje cells are highly differentiated neurons arranged in a single layer. The pear-shaped Purkinje cell body gives rise to a large dendritic tree which ascends in the molecular layer and arborizes strictly in the transverse plane of the folium. The axon runs radially towards the medullary lamina, giving off abundant recurrent collaterals. Although the termination sites of these recurrent collaterals have not been established with certainty, they are thought to terminate on basket, Golgi, and other Purkinje cells (*Fig.* 31).

The criteria which have been used to identify Purkinje cells electrophysiologically are: neurons recorded at depths of 150-400 μm. below the surface of the folium, which can be antidromically activated by stimulation of the fastigial nucleus; orthodromically excited by 'in-line' surface stimulation of parallel fibres of the same folium; inhibited by 'off-line' surface stimulation; excited by stimulation of the inferior olive; and which exhibit 'inactivation' responses during spontaneous firing at rates of about 20–100 spikes/sec. (Granit and Phillips, 1956, 1957; Andersen, Eccles, and Voorhoeve, 1964; Crawford, Curtis, Voorhoeve, and Wilson, 1966, a, b, c; Eccles, Llinás, and Sasaki, 1966b, d, e, f).

Depressant Amino-acids

There is good evidence that GABA is an inhibitory transmitter in the cerebellar cortex. First, since GABA is released from

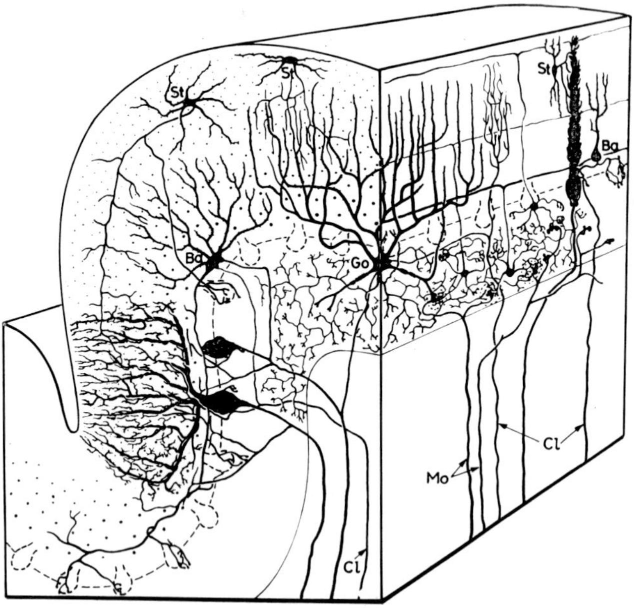

Fig. 31. Stereodiagram illustrating the five main types of neurons in the cerebellar cortex and the two types of afferent fibres (excluding monoaminergic projections)—the climbing (Cl) and mossy (Mo) fibres—with their main interconnexions. Cerebellar glomeruli and some Purkinje cell bodies are indicated by dotted lines. Interneurons are abbreviated as St (stellate cells), Ba (basket cells), and Go (Golgi cells). *See* text for description. (Eccles and others, 1967.)

Purkinje axons on to Deiters' neurons (Chapter 2), it is likely that GABA is also released on to Purkinje cells from recurrent collaterals of Purkinje axons (*see* Eccles and others, 1967, Ch. IX; Larramendi and Victor, 1967). However, pharmacological evidence for this is lacking. The main source of GABA-mediated inhibition of Purkinje cells is probably basket and outer stellate cells, a view supported by convincing pharmacological evidence.

GABA readily depressed the firing of Purkinje cells and other unidentified neurons in the mammalian cerebellar cortex (Krnjević and Phillis, 1963a; Hoffer, Siggins, and Bloom, 1969; Curtis, Duggan, Felix, and Johnston, 1970; Kawamura and Provini, 1970;

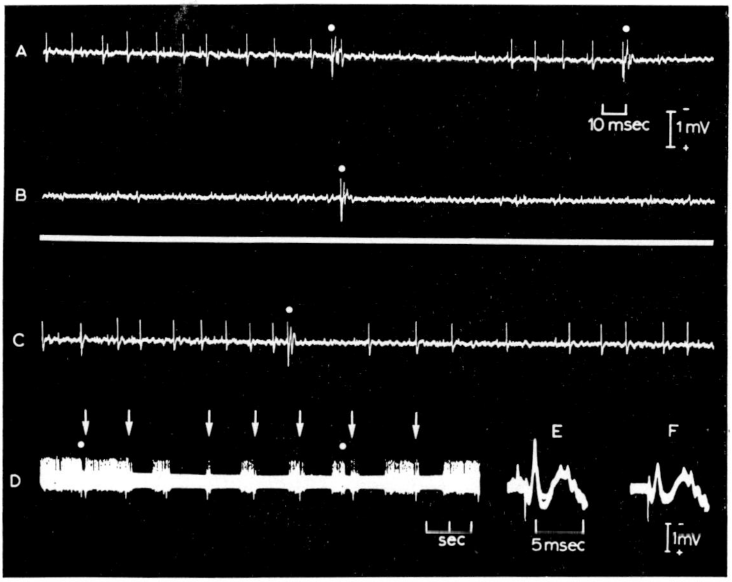

Fig. 32. Effects of GABA on Purkinje cell firing. A, B, C, Records of the spontaneous firing of a Purkinje cell, interrupted by climbing fibre responses (white dots) before, during, and after the administration of GABA (50 nA). D, Brief pulses of GABA (300 nA for 10 msec.) were ejected as indicated by the arrows, and illustrate the rapid action of GABA. E, Antidromic spike elicited by juxtafastigial stimulation (200/sec., near threshold strength; some stimuli were therefore ineffective). F, Antidromic invasion is blocked by GABA (50 nA). (Kawamura and Provini, 1970.)

Curtis, Duggan, Felix, Johnston, and McLennan, 1971). The depression of Purkinje cells by GABA resulted from a hyperpolarization and an increase in conductance of the membrane (Siggins, Oliver, and others, 1971). Several other ω-amino-acids (including glycine) also depressed the firing of Purkinje cells, but all were less effective than GABA except γ-amino-β-hydroxybutyrate, which was equally potent (Kawamura and Provini, 1970; Curtis, Duggan, Felix, Johnston, and McLennan, 1971). *Fig.* 32 illustrates the depressant action of GABA on the spontaneous (*Fig.* 32A–D) and antidromic (*Fig.* 32E, F) firing of a Purkinje cell. The rapid time course of action of GABA can be seen in *Fig.* 32D. Climbing fibre responses were almost always insensitive to GABA.

Stellate and basket cells induce long-duration (up to 200 msec.) IPSPs in Purkinje cells (Eccles and others, 1966a, f). Strychnine affected neither the inhibition of Purkinje cells by basket and stellate axons, nor the depressant action of GABA on Purkinje cells (Andersen, Eccles, Løyning, and Voorhoeve, 1963; Crawford, Curtis, Voorhoeve, and Wilson, 1963a; Kawamura and Provini, 1970; Curtis, Duggan, Felix, Johnston, and McLennan, 1971; Curtis and Felix, 1971b; Bisti, Iosif, Marchesi, and Strata, 1971). Although earlier studies indicated that relatively low doses (up to 1·1 mg./kg.) of picrotoxin had no effect on basket cell inhibition (Crawford and others, 1963a) or on the depression of Purkinje cells by GABA (Kawamura and Provini, 1970), more recent studies have indicated that both picrotoxin and bicuculline can block the depression by GABA and basket cell inhibition of Purkinje cells (Curtis, Duggan, Felix, and Johnston, 1970; Curtis, Duggan, Felix, Johnston, and McLennan, 1971; Curtis and Felix, 1971b; Bisti, Iosif, and Strata, 1971; Bisti, Iosif and others, 1971). Bicuculline blocked the action of GABA in concentrations which had little or no effect on the depression by glycine or NA (Curtis, Duggan, Felix, Johnston, and McLennan, 1971). *Fig.* 33 illustrates the blocking action of bicuculline on the inhibition of firing of a Purkinje cell produced by 'off-line' stimulation of the cerebellar cortex. Stimulation of the cerebellar surface inhibited the firing of the Purkinje cell for approximately 50 msec. (*Fig.* 33A), and this inhibition was completely blocked by bicuculline administered electrophoretically (*Fig.* 33B). Recovery occurred a few minutes after terminating the bicuculline current (*Fig.* 33C). However, strychnine, which blocked the depressant action of β-alanine on this neuron (not illustrated), either increased or had no action on the synaptic inhibition (*Fig.* 33D, E, F).

Similar observations have been made in the amphibian cerebellum (Woodward, Hoffer, Siggins, and Oliver, 1971). Experiments on two different species of frogs revealed that electrophoretically administered GABA mimicked the stellate cell inhibition of Purkinje cells, producing a depression of Purkinje cell firing, a hyperpolarization of the membrane, and an increase in its conductance. Both picrotoxin and bicuculline, but not strychnine, antagonized the action of GABA on Purkinje cells, and blocked the inhibitory effects of cortical stimulation on Purkinje cell firing.

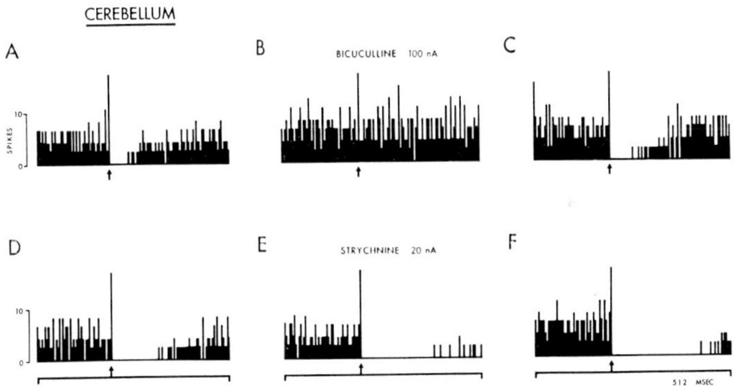

Fig. 33. Effects of bicuculline and strychnine on basket cell inhibition of a Purkinje cell. The 'off-line' cerebellar cortex was stimulated electrically (onset of stimulus indicated by an upward arrow; 0.2 msec. pulse, 6·2 V for A–C, 6·0 V for D–F). The number of spikes in 128 consecutive periods of 4 msec. were counted for a total of 20 sweeps at intervals of 1·8 sec. A, Control. B, During the electrophoretic administration of bicuculline (5 m*M* in 165 m*M* NaCl; 100 nA for 3 min.). C, 4 min. after terminating the bicuculline ejection. D, New control. E, During the administration of strychnine (10 m*M* in 165 m*M* NaCl; 20 nA for 6 min.). F, 7 min. after terminating the administration of strychnine. (Curtis and Felix, 1971b.)

The levels of GABA in the monkey cerebellar cortex are highest in the molecular layer, somewhat lower in the granular layer, and much lower in the white matter (Hirsch and Robins, 1962). GABA is taken up (though weakly) by slices of the rat cerebellar cortex (Hökfelt and others, 1970), and this GABA content is accumulated predominantly by nerve terminals (Iversen and Bloom, 1972). In various mammals the Purkinje cell layer is particularly rich in GABA and the enzymes which synthesize and destroy GABA (van Gelder, 1965a; Kuriyama, Haber, Sisken, and Roberts, 1966; Roberts and Kuriyama, 1968; Knyihár and Csillik, 1970).

Relatively high concentrations of GABA are localized in or on Purkinje cells compared with spinal ganglion cells, spinal motoneurons, and cerebral Betz cells (Otsuka and others, 1971). Electron microscopy combined with autoradiographic studies has demonstrated further the accumulation of radioactive GABA on stellate and basket cells, somewhat lower radioactivity on Purkinje cells, as well as some activity in nerve-endings of basket cells which terminate on Purkinje cells (Hökfelt and Ljungdahl, 1970, 1971b, 1972; Iversen and Bloom, 1971). Autoradiographs demonstrating some of these observations are shown in *Fig.* 34.

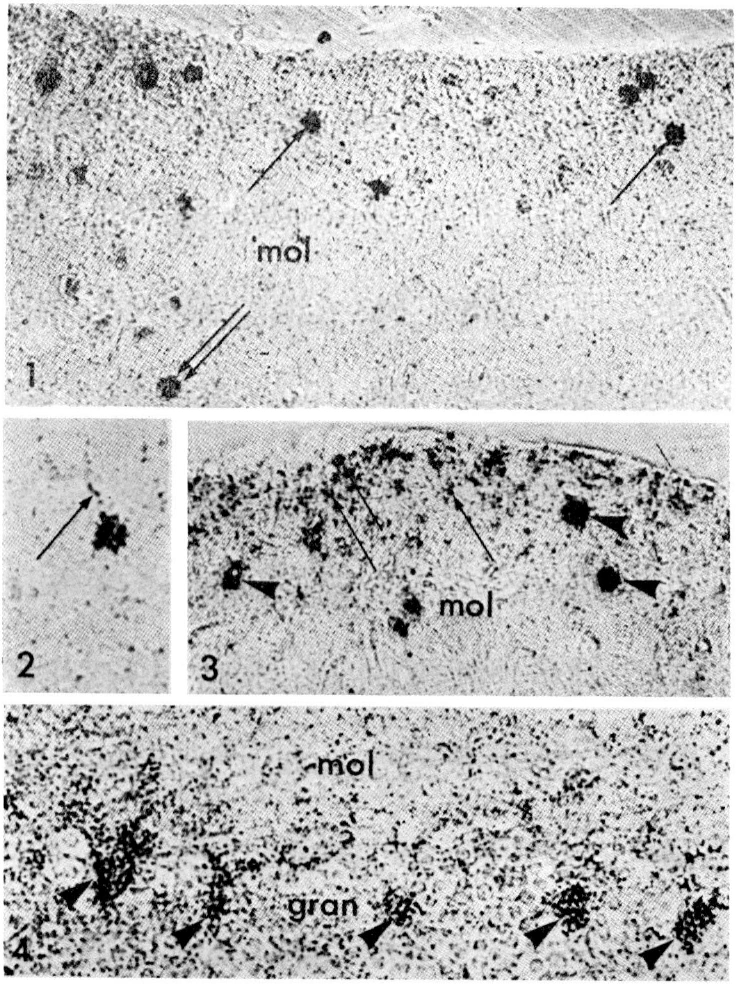

All autoradiographs were obtained from slices of rat cerebellar cortex incubated with [^{3}H]GABA. Strong accumulations of grains can be seen in the molecular layer, especially the superficial region (*Fig.* 34, 1), and these accumulations usually occur over cell bodies (*Fig.* 34, 2). In addition to large accumulations over cell bodies, smaller accumulations can be seen in the superficial region of the molecular layer (*Fig.* 34, 3). Large accumulations of grains at regular intervals can also be seen in the granular layer close to the Purkinje cell layer (*Fig.* 34, 4).

All these results provide convincing evidence that GABA is an inhibitory transmitter in the cerebellar cortex, most likely to be released from basket and stellate axons on to Purkinje cells.

EXCITANT AMINO-ACIDS

At least two types of excitatory pathways terminate on Purkinje cells—climbing fibres (which originate in the brain-stem) and parallel fibres (which are the axons of granular cells). The terminal arborization of the climbing fibre corresponds to the dendritic ramifications of a single Purkinje cell, forming the most powerful single excitatory input yet discovered in the C.N.S. (Eccles and others, 1966d). Parallel fibres form excitatory 'crossing-over' synapses on the dendritic spines of Purkinje cells, and the convergence from the parallel fibres on to each Purkinje cell is so dense (about 200,000 synapses) that this afferent excitation also has a high safety factor (Eccles and others, 1967). Glutamate, aspartate, and other acidic amino-acids have been shown to have powerful excitant actions of a rapid time course on cerebellar cortical neurons, including Purkinje cells (Krnjević and Phillis,

---

Fig. 34. Autoradiographs from slices of rat cerebellar cortex incubated with labelled GABA. 1, Strong accumulations of grains (single arrow) are seen in the molecular layer (mol), mainly in its superficial part. One accumulation (double arrows) is localized deeper. Under most of these accumulations cell bodies can be identified. Approximately ×300. 2, Accumulation of grains over a cell body and possibly over a cell process (single arrow) in the superficial zone of the molecular layer. Approximately ×560. 3, In addition to a few large accumulations (arrow heads) over cell bodies, smaller accumulations are seen in the superficial zone of the molecular layer (mol). Approximately ×330. 4, Large accumulations of grains (arrow heads) at rather regular intervals are seen in the granular layer (gran) close to the Purkinje cell layer. Approximately ×385. (Hökfelt and Ljungdahl, 1970.)

1963a; Crawford and others, 1966; McCance and Phillis, 1968; Kawamura and Provini, 1970; Curtis, Duggan, Felix, Johnston, and McLennan, 1971). Since ACh is unlikely to mediate either of these types of synaptic excitation (*see* p. 100), glutamate or a related amino-acid may be released by climbing or parallel fibres on to Purkinje cells. There is no direct evidence for this, however, other than the observation that dicarboxylic amino-acids excite Purkinje cells, and that the cerebellar cortex contains relatively high levels of glutamate (Johnson and Aprison, 1971).

Noradrenaline

The effects of NA apparently depend on the area of the cerebellum investigated. For example, NA depressed the firing of almost all cerebellar cortical neurons (cat) in the crus I and II of the neocerebellar hemisphere, but excited two-thirds and depressed the remainder of cells in the flocculus (Yamamoto, 1967). In Larsell's lobuli V and VI (and to some extent, the cerebellar hemispheres) NA depressed approximately three-fifths of the Purkinje cells tested, having little or no action on the remainder (Kawamura and Provini, 1970; cf. Curtis, Duggan, Felix, Johnston, and McLennan, 1971). In the rat cerebellum, however, NA depressed the firing of virtually all Purkinje cells tested, although it had little action on the interspike interval and climbing fibre bursts (Hoffer, Siggins, and Bloom, 1971). There is now extremely good evidence that NA is an inhibitory transmitter on cerebellar Purkinje cells (*see* Bloom, Hoffer, and Siggins, 1972). All the important evidence for this has been obtained from studies of rats.

Histochemical studies of the rat cerebellum have revealed the presence of catecholamine-containing nerve-fibres in the Purkinje cell and molecular layers both *in vitro* (Hökfelt and Fuxe, 1969) and *in vivo* (Bloom and others, 1971). Autoradiographic studies further showed the presence of labelled NA on fine unmyelinated axons and small axons making synaptic contact both with the surface of large Purkinje dendrites and with small dendritic spines (Bloom and others, 1971). Particularly intense labelling was found over some preterminal axons containing large granular vesicles, and after treatment with 6-hydroxydopamine (which led to a complete depletion of chemically detectable NA in the cerebellar cortex), degenerating nerve terminals were again found to contact larger dendritic trunks and spines. These studies indicated that endogenous NA is contained within axons which synapse with Purkinje cell dendrites and spines (Bloom and others, 1971).

The depressant action of NA on Purkinje cells was reversibly blocked by the adrenergic antagonist, MJ-1999, and potentiated by desmethylimipramine, which is known to block re-uptake of NA in the peripheral nervous system (Hoffer, Siggins, and Bloom, 1971). Studies involving intracellular recording during extracellular drug administration revealed that, contrary to the action of GABA, which *increased* membrane conductance during the hyperpolarization, NA hyperpolarized the membrane and either *decreased* or had no action on the conductance (Siggins, Oliver and others, 1971). Examples of this are illustrated in *Fig.* 35C.

The depressant action of NA on Purkinje cells has been postulated to be mediated by trans-synaptic activation of a metabolic system (Siggins, Hoffer, and Bloom, 1971; Siggins, Oliver and others, 1971). Electrophoretically administered cyclic AMP and its synthetic dibutyryl derivative mimicked the action of NA on Purkinje cells, except that the latter produced a depression of a slower time course (*Fig.* 35C). In contrast, two other adenine nucleotides, ATP and 5′-adenosine monophosphate, usually had no effect or excited Purkinje cells, probably owing to chelation of divalent cations. Methyl xanthines (theophylline and aminophylline), which inhibit the enzymatic breakdown of cyclic AMP by phosphodiesterase in the peripheral nervous system, markedly potentiated the depressant actions of both NA and cyclic AMP on Purkinje cells. These observations suggested that NA exerts its depressant effect by stimulating the activity of adenyl cyclase, the enzyme which converts ATP to cyclic AMP (Siggins, Hoffer, and Bloom, 1971; Siggins, Oliver and others, 1971). Nicotinate and the prostaglandins E_1 and E_2 (but not $F_{1\alpha}$, $F_{2\alpha}$, linoleic, or linolenic acids), which have been reported to reduce cyclic AMP levels in the peripheral nervous system [possibly by inhibiting adenyl cyclase (Horton, 1969)], blocked the action of NA, but not that of cyclic AMP or GABA on Purkinje cells. This supports the postulate that NA depression is mediated through activation of adenyl cyclase. The excitant actions of the prostaglandins on Purkinje cells were interpreted as being due to antagonism of a tonic noradrenergic inhibition, since after the cerebellum had been depleted of NA by 6-hydroxydopamine, prostaglandins E_1 and E_2 rarely increased Purkinje cell firing compared with untreated animals. Treatment with 6-hydroxydopamine also specifically increased the mean discharge rate of Purkinje cells, supporting the conclusion that the NA-containing afferents to the cerebellar cortex exert a tonic inhibitory bombardment (Hoffer, Siggins, Woodward, and Bloom, 1971).

It is relevant to note at this stage that in many peripheral sympathetically innervated tissues the effect of NA is thought to

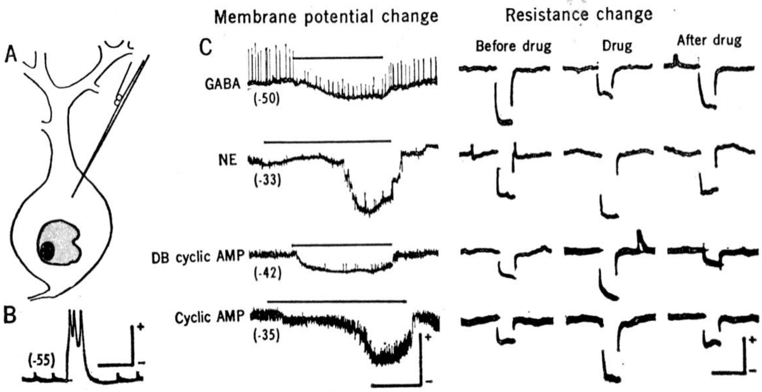

Fig. 35. Intracellular recordings from cerebellar Purkinje cells of the rat. A, Schematic diagram showing the recording arrangement. The intracellular electrode extends beyond the orifices of the two extracellular pipettes used for ejecting the drugs. B, Multispiked spontaneous climbing fibre discharge recorded intracellularly from a Purkinje cell. Numbers in parentheses in both B and C indicate the resting membrane potential (mV). *Calibrations*: 20 msec. and 25 mV. C, Changes in membrane potential and membrane resistance of four different Purkinje cells in response to GABA, noradrenaline (NE), dibutyryl cyclic AMP (DB cyclic AMP), and cyclic AMP, each ejected with currents of 100–150 nA. All records of a particular row refer to the same cell. *Calibrations*: 10 sec. and 20 mV for NE, DB cyclic AMP, and cyclic AMP, and 5 sec. and 10 mV for GABA. The effective input resistance was estimated by the amplitude of pulses resulting from the passage across the membrane of a brief constant current (1 nA) pulse, before, during, and after the administration of the respective substances (1 mV = 1 megohm). Discontinuities in the fast transients of the pulses resulted from the loss of high frequencies (>10 kHz) and from the chopped nature of the frequency-modulated magnetic tape recording used. All 'pulse' records were graphically normalized to the same baseline level. *Calibration*: 80 msec. and 15 mV for all pulse records. (Siggins, Oliver and others, 1971.)

be mediated metabolically by cyclic AMP, as NA increases the synthesis of this nucleotide by stimulating the activity of adenyl cyclase (Sutherland, Robison, and Butcher, 1968; Robison, Butcher, and Sutherland, 1971). Both adenyl cyclase and cyclic AMP are found in high concentrations in the grey matter of the brain, particularly in the cerebellum, and these levels can be even further increased *in vitro* by NA, histamine, and 5-HT (Rall and Gilman, 1970).

The origin of the NA-containing fibres is thought to be in the brain-stem, predominantly the locus coeruleus, a rich source of NA-containing neurons which project to many parts of the brain,

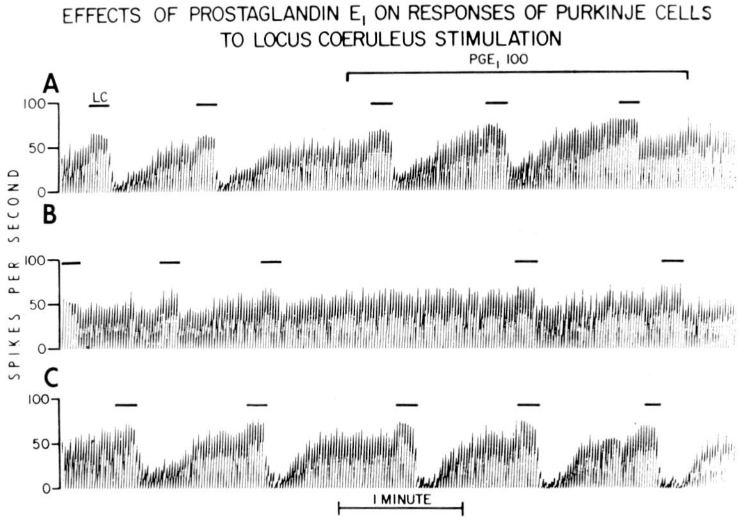

Fig. 36. Antagonism by prostaglandin E_1 (PGE_1) of the synaptic inhibition of a Purkinje cell evoked by stimulating locus coeruleus (LC). The chart recordings in A, B, C, represent continuous records of the firing frequency of the same cell. The short horizontal bars represent repetitive stimulation of LC (6 V, 0·1 msec. duration rectangular pulses, 100 pulses at 10/sec.). LC stimulation produced almost maximal inhibition of firing, and this action was blocked by electrophoretically administered PGE_1 (100 nA). Recovery occurred several minutes after the current ejecting prostaglandin was terminated. The apparent increase in firing during LC stimulation reflects the addition of the stimulus artefact to the Purkinje cell discharge and does not indicate synaptic excitation. (Bloom and others, 1972.)

including the cerebellum (Dahlström and Fuxe, 1965a, b; Loizou, 1969; Fuxe and others, 1970; Olson and Fuxe, 1971; Ungerstedt, 1971). Single or repetitive stimulation of locus coeruleus inhibited the firing of most Purkinje cells, the latency of inhibition varying from 50 to 290 msec. (Siggins, Hoffer, Oliver, and Bloom, 1971). The inhibition decayed with high rates of repetitive stimulation (greater than 50/sec.), and was observed less frequently when the stimulating electrode was positioned slightly outside the locus coeruleus. The membrane resistance of Purkinje cells increased or remained unchanged during the late hyperpolarization evoked by stimulation of locus coeruleus. Thus, the characteristics of this inhibition were similar to those of the depressant action of NA, and supported the evidence for the presence of a slowly conducting ('C-fibre' type) noradrenergic inhibitory pathway from the locus coeruleus to cerebellar Purkinje cells (Siggins, Hoffer and others, 1971).

Direct pharmacological evidence for this was obtained from the findings that the inhibitory effects of stimulation of locus coeruleus were prevented by electrophoretically administered prostaglandins E_1 and E_2, by chronic pretreatment with 6-hydroxydopamine, and by acute treatment with reserpine and alpha-methyltyrosine (Hoffer, Siggins, Oliver, and Bloom, 1971, 1972; Siggins, Hoffer and others, 1971; Bloom, Hoffer, and Siggins, 1972). *Fig.* 36 illustrates the reversible antagonism by electrophoretically administered prostaglandin E_1 of the synaptic inhibition of a Purkinje cell evoked by repetitive stimulation of locus coeruleus. An example of the blocking action of alpha-methyltyrosine on Purkinje cell inhibition and the recovery of this inhibition following administration of DOPA, are shown in *Fig.* 37. In addition, the phosphodiesterase inhibitor, papaverine, potentiated both the depressant actions of NA and cyclic AMP on Purkinje cells, as well as the synaptic inhibition evoked by stimulating locus coeruleus (Bloom, Hoffer, and Siggins, 1972; Hoffer and others, 1972).

Lastly, the recent development of an immunofluorescence histochemical method for localizing cyclic AMP intracellularly in various types of tissue (including brain) (Wedner, Hoffer, Steiner, Parker, and Bloom, 1972) has strengthened considerably the evidence for the type of noradrenergic transmission in the cerebellum described above (Bloom, Hoffer, Battenberg, Siggins, Steiner, Parker, and Wedner, 1972). Thus, there is a wealth of biochemical, histochemical, pharmacological, and physiological evidence that NA is an inhibitory transmitter on Purkinje cells of the rat cerebellum, and that the mechanism underlying this inhibition is entirely different from that underlying synaptic transmission mediated by amino-acids. It is now important to determine whether this type of noradrenergic inhibition is confined to the cerebellum or whether it occurs throughout the C.N.S.

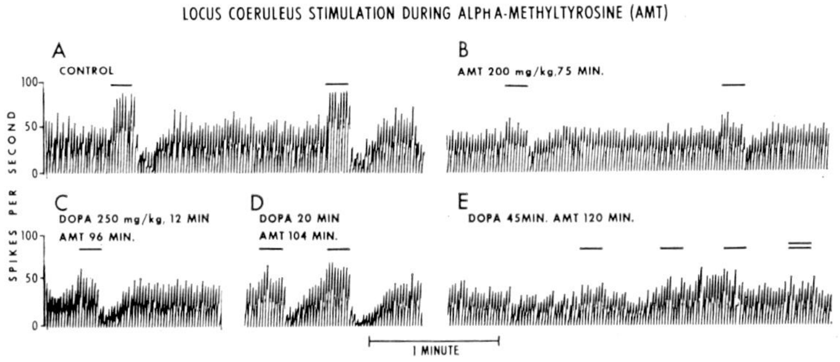

Fig. 37. Blockade by alpha-methyltyrosine (AMT) of Purkinje cell inhibition evoked by stimulation of locus coeruleus (LC), and recovery of this inhibition by DOPA injection. To deplete the NA content of adrenergic terminals, the rat was injected with reserpine (1 mg./kg. intravenously) 75 min. prior to the record shown in A. The records of firing frequency in A–E were obtained at successive times from the same Purkinje cell, the stimulating electrode remaining in the same position in LC. Single solid lines above all records indicate LC stimulation with 50 pulses of 10 V, 0·1 msec. duration, at 10/sec. (The apparent increase in firing rate during LC stimulation is the result of the summation of the stimulus artefact with the true action potentials.) A, Reduction of discharge rate evoked by LC stimulation. B, Blockade of inhibition from LC 75 min. after an injection of AMT, an inhibitor of NA synthesis. C, D, Complete recovery of the synaptic inhibition following the administration of DL-DOPA. E, Fading of LC-induced inhibition 45 min. after the DOPA injection, presumably due to the rapid metabolism of DOPA at a time when reserpine (3·5 hr.) and AMT (2 hr.) continued to deplete NA. At this time even stimuli of double strength (50 pulses of 20 V; double bar above record) had no effect on the Purkinje cell firing frequency (Bloom, Hoffer, and Siggins, 1972.)

Other Monoamines

5-HT may also be a transmitter in the cerebellar cortex, but there is no direct pharmacological evidence for this. The cerebellar cortex contains very fine, varicose 5-HT-containing nerve-fibres which innervate mainly the molecular layer, especially of the anterior lobe, and are distributed predominantly in the transverse plane of the folium parallel to the surface (Hökfelt and Fuxe, 1969). These fibres presumably originate in the raphe nuclei (Dahlström and Fuxe, 1965a). Electrophoretically administered 5-HT excited one-half and depressed the other half of Purkinje cells tested in an early study of the rat cerebellum (Hoffer and others, 1969), but inconsistently depressed only a small minority of Purkinje cells in the cat (Kawamura and Provini, 1970). However, results from subsequent studies suggest that the predominant action of 5-HT on Purkinje cells (rat) is depression (Bloom, Hoffer, Siggins and others, 1972).

DA often weakly depressed the firing of Purkinje cells (Kawamura and Provini, 1970). However, DA-containing nerve-fibres have not been observed in the cerebellum, and the role of this catecholamine, if any, in the cerebellum is unknown. Similarly, although histamine depressed the firing of Purkinje cells (Siggins, Hoffer, and Bloom, 1971) the significance of this is unknown, as the cerebellum contains little histamine (Taylor and others, 1972).

Acetylcholine

ACh excited a proportion of Purkinje cells (Crawford and others, 1963b, 1966; McCance and Phillis, 1964, 1968; Hoffer and others, 1969; Kawamura and Provini, 1970), but different opinions have been expressed about the possible significance of this.

Crawford and others (1966) observed that ACh excited the majority (75 per cent) of Purkinje cells tested in the superficial layer of the folium, excitation often being preceded by a depression of the spontaneous firing. The excitation by ACh, which was of relatively slow onset and long duration, was prolonged by neostigmine and reduced by both atropine and dihydro-β-erythroidine. On the basis of the actions of the cholinergic antagonists and the relative potencies of various cholinomimetics, the excitatory cholinoceptive receptors on Purkinje cells were classified as being mainly muscarinic. However, since atropine and dihydro-β-erythroidine (up to 1 mg./kg. intravenously) had no effect on cerebellar cortical fields evoked by impulses in mossy, climbing, and parallel fibres, it was concluded that ACh is unlikely to be an excitatory transmitter in the cerebellum (Crawford and others, 1966).

McCance and Phillis (1968) observed that ACh excited only about one-third of the Purkinje cells tested. In particular, neurons (including Purkinje cells) situated near the cerebellar surface were generally insensitive to ACh, the proportion of cells excited by ACh increasing with depth. Moreover, the excitation by ACh was

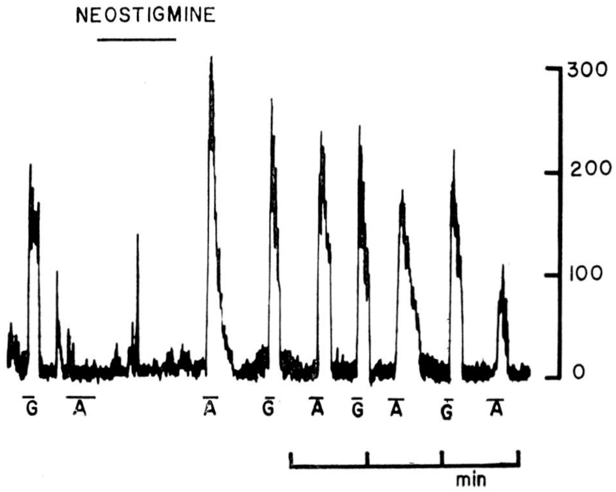

Fig. 38. Chart recording of the firing frequency of a Purkinje cell, showing the actions of L-glutamate (G, 40 nA) and ACh (A, 60 nA) before and after an ejection of neostigmine (40 nA). *Ordinate*: firing frequency (spikes/sec.). *Abscissa*: time (in minutes). (McCance and Phillis, 1968.)

sometimes not apparent until the cell had been treated with a cholinesterase inhibitor, as in *Fig.* 38. As was observed by Crawford and others (1966), atropine and dihydro-β-erythroidine (up to 1 mg./kg.) had no action on the surface field potentials evoked by climbing and parallel fibres, and it was concluded that Purkinje cells are probably insensitive to ACh, the observed excitation being indirectly mediated by neighbouring cholinoceptive granule cells (McCance and Phillis, 1968).

Biochemical and histochemical studies of the cerebellum have also provided variable results. The mammalian cerebellum contains relatively little ACh and choline acetyltransferase (MacIntosh, 1941; Feldberg and Vogt, 1948; Hebb and Silver, 1956; Silver, 1967; Goldberg and McCaman, 1967), but high concentrations of AChE, which vary considerably between species in both quantity and distribution (Burgen and Chipman, 1951; Koelle, 1954; Gerebtzoff, 1959; Snell, 1961; Friede and Fleming, 1964; Austin and Phillis, 1965; Silver, 1967). Nevertheless, although the significance of all these reports (considered collectively) is difficult to evaluate, a number of studies suggest that ACh may be a transmitter in the granular cell layer of the cerebellum.

Granule Cells

Granule cells may be recognized by their small-amplitude action potentials recorded at depths of 400–700 μm. below the surface of the folium (granular layer). They do not fire antidromically to stimulation of the juxtafastigial region as do Purkinje cells, nor synaptically to a parallel fibre volley as do most Golgi cells, but may fire antidromically to stimulation of parallel fibres (Eccles and others, 1966c). Despite these criteria, the spike potentials of a single granule cell are difficult to isolate clearly from the background of spike potentials generated by the densely packed adjacent small granule cells. This has increased the difficulty of conducting and interpreting pharmacological experiments on granule cells, using relatively coarse micropipettes (Crawford and others, 1966; McCance and Phillis, 1968).

ACETYLCHOLINE

Crawford and others (1966) observed no action of ACh or other cholinomimetics on granule cells, although ACh ejected from the same micropipette excited Purkinje cells, and DLH excited both granule and Purkinje cells. The lack of effect of atropine and dihydro-β-erythroidine on field potentials evoked by stimulating mossy afferents (which terminate exclusively in the granular layer) further supported the conclusion that granule cells have no cholinoceptive receptors.

McCance and Phillis (1968), however, observed that ACh excited many granule cells, the sensitivity to ACh increasing with depth. This excitation by ACh was considered to be mediated mainly by nicotinic receptors. Anticholinesterases frequently potentiated, whereas antagonists of ACh reduced both the excitant action of ACh on granule cells and the spontaneous firing of ACh-sensitive granule cells. The spontaneous activity in the granular layer was attributed to incoming signals in mossy fibres, as it was usually absent in chronically isolated cerebellar areas. In particular, dihydro-β-erythroidine blocked the excitation by both ACh and repetitive stimulation of areas in the mesencephalic reticular formation/nucleus reticularis oralis complex of some granular neurons (*Fig.* 39).

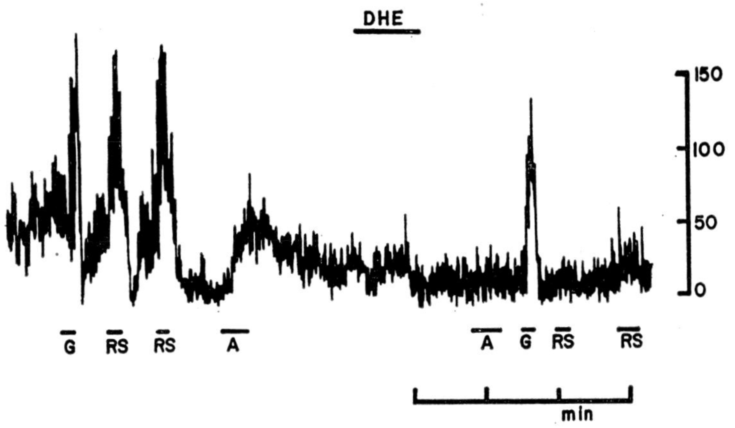

Fig. 39. Antagonism by dihydro-β-erythroidine (DHE) of the excitant actions of ACh and stimulation of the reticular formation on a granule cell. The chart recording shows the change in firing frequency (spikes/sec., *ordinate*) with time (in minutes, *abscissa*). The excitant action of ACh (A, 50 nA) and repetitive stimulation (10/sec.) of the mesencephalic reticular formation/nucleus reticularis oralis complex (RS), were blocked by DHE (80 nA), which had little action on the excitation by L-glutamate (G, 40 nA). (McCance and Phillis, 1968.)

Different experimental procedures probably account for the different observations of Crawford and others (1966) and McCance and Phillis (1968). Crawford and others (1966) studied mainly the superficial layers of the cerebellar cortex (up to 600 μm. deep), where ACh-sensitive cells were rarely observed by McCance and Phillis (1968). The latter made electrode penetrations up to 6 mm. deep, and encountered progressively higher proportions of ACh-sensitive cells with increasing depth. A subsequent analysis of the distribution of cholinoceptive cells in the cerebellar cortex, recorded in electrode penetrations through several folia, confirmed the suggestion that it is the granule cells which are sensitive to ACh (McCance, personal communication).

The increase in sensitivity to ACh of the deeper granule cells correlated well with other observations that the granular layer stains more intensively for AChE in the deeper layers of the cerebellar cortex (Phillis, 1968a). Furthermore, cerebellar afferent fibres from the pons pass through the middle peduncle (Jansen and Brodal, 1954), which also contains the highest proportion of AChE-containing afferent fibres (Phillis, 1968a). Relatively few afferent fibres in the inferior cerebellar peduncle (through which most afferents from the medullary reticular formation pass) stain for AChE (Phillis, 1968a), and interestingly, the facilitatory effects of stimulation of the medullary reticular formation were insensitive to antagonists of ACh (McCance and Phillis, 1968).

Other studies of the feline cerebellar cortex have also revealed that AChE is present mainly in the granular layer, partly associated with the terminals of mossy fibres (Snell, 1961; Friede and Fleming, 1964; Kása, Joó, and Csillik, 1965). Choline acetyltransferase is distributed mainly in the molecular and granular layers (Goldberg and McCaman, 1967). A method for the possible microscopic localization of ACh synthesis in the cerebellar cortex revealed activity mainly in the glomeruli of the granular layer, but also in parallel fibres, the supraganglionic plexus in the molecular layer and the Purkinje cell bodies (Knyihár and Csillik, 1970). Although the mammalian cerebellum contains relatively little ACh (MacIntosh, 1941; Hebb and Silver, 1956), most of it is located in nerve-endings, the proportion of ACh contained by endings from the molecular and granular layers varying between species (Israël and Whittaker, 1965). Electrical stimulation of the cerebellar surface has been shown to increase the release of ACh from the cerebellar cortex (Phillis and Chong, 1965). Thus, the bulk of the evidence suggests that ACh is a transmitter in the cerebellar cortex, possibly released from some mossy fibres which originate in the mesencephalon and pons (McCance and Phillis, 1968). Nevertheless, the pharmacological evidence for this should be extended to determine the precise origin of these cholinergic afferents, and whether single stimuli evoke excitatory responses from granule cells that can be *reversibly* blocked by dihydro-β-erythroidine. In addition, the effects of atropine on such synaptic excitation should be investigated.

AMINO-ACIDS

The insensitivity of some granule cells to ACh, as well as the lack of effect of cholinergic antagonists on synaptic responses evoked by impulses in some mossy fibres, indicate that not all mossy fibres are cholinergic (Crawford and others, 1966; McCance and Phillis, 1968). This is perhaps not surprising considering the diverse origins (spinal cord, cuneate nucleus, reticular formation, vestibular nuclei) of mossy fibres (Eccles and others, 1967). It is

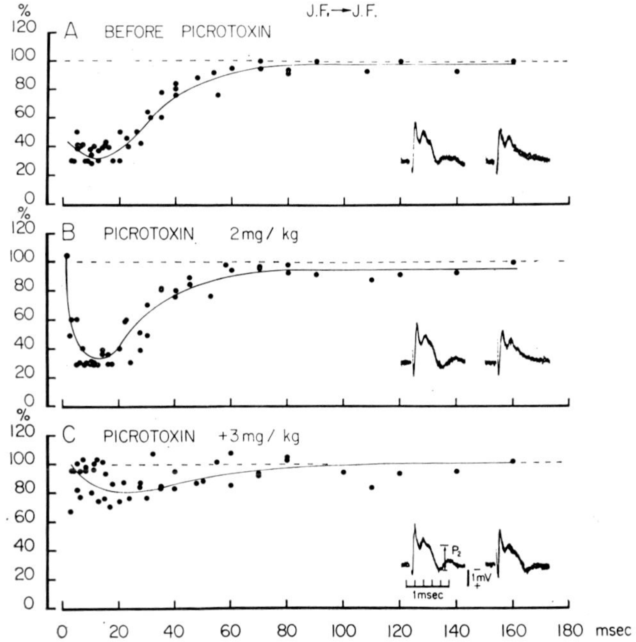

Fig. 40. Blocking action of picrotoxin on inhibition produced by Golgi cells. The amplitude of the positive (P_2) potential reflects the proportion of granule cell axons conducting nerve impulses. The potentials evoked by a juxtafastigial (J.F.) stimulus are inhibited by a preceding J.F. stimulus. The curves show the time course of inhibition of granule cells, using different intervals between the J.F. stimuli. The specimen records beneath each curve show the control P_2 potential and that inhibited at 10, 10, and 12 msec. A, Control curve. Little change was observed in the inhibition after an intravenous injection of picrotoxin (2 mg./kg.) (B). However, after an additional dose of picrotoxin (3 mg./kg.) 20 min. later (C) the inhibition was virtually blocked. (Bisti, Iosif and others, 1971.)

also interesting that deeper granule cells have longer and more dendrites than those located superficially (Eccles and others, 1967; p. 45). It is possible that the excitant actions of DLH and glutamate on granule cells (Crawford and others, 1966; McCance and Phillis, 1968) are related to a transmitter action of an acidic amino-acid on granule cells. The presence of relatively high concentrations of glutamate in the cerebellum (Johnson and Aprison, 1971) is consistent with this hypothesis.

Neutral amino-acids have not been tested on granule cells, but recent pharmacological experiments using convulsants suggest that GABA is an inhibitory transmitter released on to granule cells from Golgi axon terminals (Bisti, Iosif and others, 1971; Bisti, Iosif, and Strata, 1971). Golgi cells readily accumulate radioactive GABA (Hökfelt and Ljungdahl, 1972). The inhibition of granule cells following Golgi cell activation is resistant to strychnine (Andersen and others, 1963; Bisti, Iosif and others, 1971), but can be blocked by intravenous picrotoxin and bicuculline (Bisti, Iosif and others, 1971; Bisti, Iosif, and Strata, 1971). *Fig.* 40 illustrates the blocking action of picrotoxin on the synaptic inhibition produced by Golgi axons. These findings should now be extended by investigating the actions of various depressant amino-acids and convulsants on the firing and synaptic inhibition of single granule cells.

Monoamines

Electrophoretically administered 5-HT excited cerebellar neurons identified electrophysiologically as granule cells (Bloom, Hoffer and others, 1972). Both fluorescence histochemical and autoradiographic studies of the rat cerebellum have revealed the distribution of 5-HT mainly within the granular cell layer in structures resembling the 'rosette' configuration characteristic of mossy fibre terminals. The mossy fibre terminals are generally quite large and filled with spherical agranular vesicles, and a few large granular vesicles. Moreover, 2–4 days after lesions were made in the dorsal and median raphe nuclei, degeneration in the cerebellar cortex was most marked within the granule-cell layer. These findings suggest that a proportion of mossy fibre terminals release 5-HT as a transmitter (Bloom, Hoffer and others, 1972).

The relatively high density of NA-containing nerve-fibres in the granular layer of the rat cerebellum (Hökfelt and Fuxe, 1969) raises the possibility that NA is also a transmitter acting on granule cells, but further studies are required to investigate this possibility.

Basket, Outer Stellate, and Golgi Cells

Although these three different types of inhibitory interneurons may be readily identified morphologically, the only criterion for differentiating between them in electrophysiological experiments is provided by the depth below the surface of the cortex at which the sharply localized spike potentials can be recorded (Eccles and others, 1966a). Since all of these interneurons receive excitatory inputs from the parallel fibres, it is reasonable to suppose that they are all activated by the same excitatory transmitter. However, just as it is possible that granule cells receive different excitatory transmitters released from different mossy fibres, it is also possible that the axons of different populations of granule cells release different transmitters. It is noteworthy that the diameter of parallel fibres increases with depth below the surface of the folium (Fox and Barnard, 1957; Fox, Siegesmund, and Dutta, 1964), indicating a wide range of conduction velocities for these fibres.

Few definitive conclusions can be made about the possible transmitters acting on stellate, basket, and Golgi neurons, as these cells have not been identified in microelectrophoretic experiments. ACh is unlikely to be a transmitter released from parallel fibres in the more superficial folia, as cholinergic antagonists had no action on the field potentials evoked by surface stimulation of the cerebellar cortex (Crawford and others, 1966; McCance and Phillis, 1968). Furthermore, an indirect analysis of the chemical sensitivity of basket cells by studying basket-cell–Purkinje-cell interactions with two multi-barrel micropipettes suggested that basket cells in superficial folia were insensitive to cholinomimetics (Crawford and others, 1966). If ACh is released synaptically from mossy fibres on to a proportion of the deeper granule cells (McCance and Phillis, 1968), it may also be an excitatory transmitter on those Golgi neurons whose dendrites form part of the same glomeruli as the cholinoceptive granule cells.

GABA is likely to be an inhibitory transmitter on those basket and Golgi cells which receive collaterals from Purkinje axons (Dale's principle, 1934), but this has yet to be shown directly.

INTRACEREBELLAR NUCLEI

Purkinje cells terminate somatotopically on neurons in the intracerebellaı nuclei, the medial cortical zones (vermis) projecting to

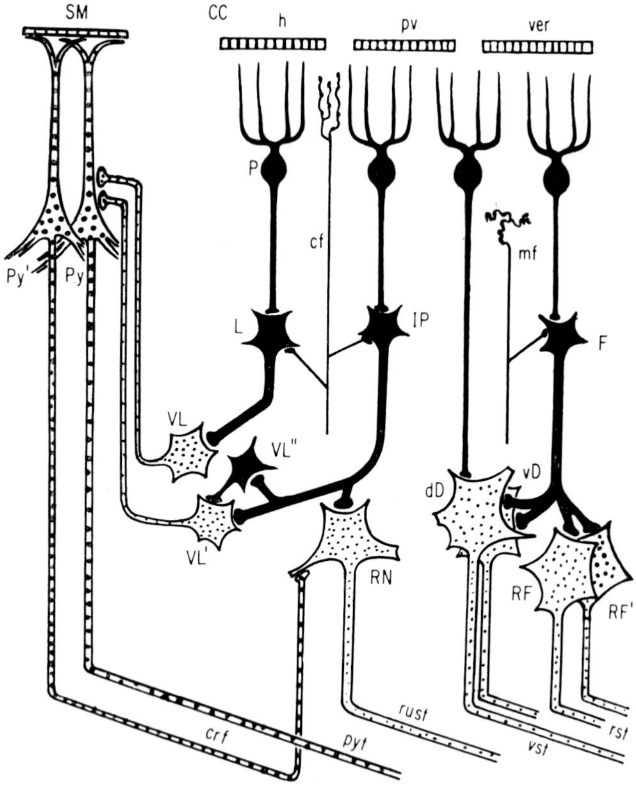

Fig. 41. Schematic diagram of the efferent pathways of the cerebellum and their main connexions. Abbreviations: Cerebellar cortex (CC); Cerebellar hemisphere (h); Paravermis (pv); Vermis (ver); Purkinje cells (P); Climbing fibre (cf); Mossy fibre (mf); Lateral nucleus cell (L); Interpositus nucleus cell (IP); Fastigial Nucleus cell (F); Cells of nucleus ventralis lateralis of the thalamus (VL, VL′, VL″); Red nucleus cell (RN); Dorsal Deiters' neuron (dD); Ventral Deiters' neuron (vD); Neurons of the pontomedullary reticular formation (RF, RF′); Sensorimotor cortex (SM); Fast pyramidal tract cell (Py); Pyramidal cell of origin of the corticorubral fibres (Py′); Corticorubral fibres (crf); Pyramidal tract (pyt); Rubrospinal tract (rust); Vestibulospinal tract (vst); Reticulospinal tract (rst). *See* text for description. (Eccles and others, 1967.)

the fastigial nuclei, the intermediate zones (paravermal cortices) to the interpositus nuclei and the lateral zones (hemispheres) to the lateral (dentate) nuclei (Jansen and Brodal, 1958). This is illustrated diagrammatically in *Fig.* 41. Purkinje axons mediate inhibition monosynaptically on to these deep nuclei neurons (Ito,

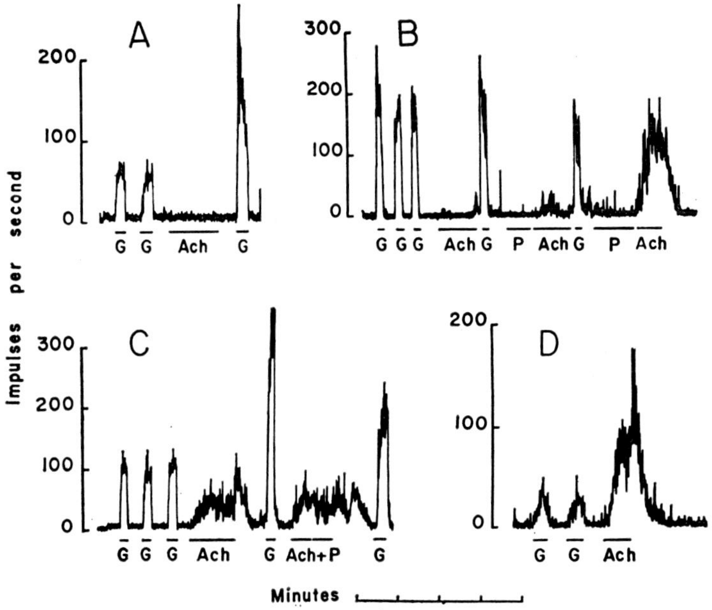

Fig. 42. Responses of four intracerebellar nuclei neurons to L-glutamate (G) and acetylcholine (Ach). Chart recordings, showing the change in firing frequency with time. A, Cell of n. interpositus; glutamate (100 nA) and ACh (100 nA). B, Cell of dentate nucleus; glutamate (100 nA), ACh (80 nA) and physostigmine (P), (80 nA followed by 120 nA). C, Cell of n. interpositus; glutamate (80 nA), ACh (80 nA), and physostigmine (20 nA). D, Cell of n. interpositus; glutamate (100 nA), ACh (100 nA). (Chapman and McCance, 1967.)

Yoshida, and Obata, 1964; Ito, Yoshida, Obata, Kawai, and Udo, 1970). The efferent fibres from the fastigial nuclei terminate mainly in the medullary reticular formation, whereas those from the interpositus and lateral nuclei project mainly to the red nucleus and ventrolateral thalamic nucleus (Jansen and Brodal, 1958). All these efferent monosynaptic pathways are thought to be exclusively excitatory (Toyama and others, 1968; Uno, Yoshida, and Hirota, 1970; Ito, Udo, Mano, and Kawai, 1970).

Most intracerebellar nuclei neurons are large multipolar cells which have a relatively simple, but irregular, stellate arborization of the dendrites. In addition, numerous smaller cells with short axons (Golgi type II) are present (Eccles and others, 1967). Recurrent collaterals of mossy and climbing fibres synapse with intracerebellar nuclei neurons (Eccles and others, 1967; Matsushita and Ikeda, 1970a, b). Thus, although the synaptic arrangements in the cerebellar nuclei are far less understood than those of the cerebellar cortex, it is clear that there are at least three types of synapses on the large projective neurons of the deep nuclei—the terminals of Purkinje axons, local axons, and afferent fibre collaterals.

Transmitters in the intracerebellar nuclei have not been identified. However, since Purkinje axons mediate monosynaptic inhibition on to both Deiters' and intracerebellar nuclei neurons (Ito and Yoshida, 1966; Ito and others, 1966, 1968; Ito, Yoshida and others, 1970), and since it is almost certain that these axons release GABA on to Deiters' neurons (Chapter 2), it is highly likely that GABA is the inhibitory transmitter released from Purkinje axon terminals on to intracerebellar nuclei neurons. Studies of the release of GABA into the fourth ventricle (Obata and Takeda, 1969) and of the GABA content of the intracerebellar nuclei neurons (Otsuka and others, 1971) are consistent with this hypothesis.

Similarly, since mossy and climbing fibre collaterals terminate on intracerebellar nuclei neurons, the excitatory transmitters released on to these neurons may be those released on to granule and Purkinje cells in the cerebellar cortex. This raises the possibility that ACh, and even glutamate, may be transmitters in this area. ACh and glutamate excited the majority of neurons tested in the intracerebellar nuclei (Chapman and McCance, 1967; McCance, 1972). Some examples of this are illustrated in *Fig.* 42.

In a study of intracerebellar nuclei (interpositus and lateral) neurons, which were identified by antidromic stimulation of the brachium conjunctivum, the excitant action of ACh was found to be mediated by both muscarinic and nicotinic receptors, muscarinic receptors predominating (McCance, 1972). Carbamylcholine, acetyl-β-methylcholine, and nicotine each had strong excitant actions, and the excitation by ACh could be antagonized by either atropine or dihydro-β-erythroidine. The synaptic discharges of deep nuclei neurons evoked by stimulation of the brachium conjunctivum, brachium pontis, lateral reticular nucleus, and hind-limb nerves were unaffected by electrophoretically administered atropine in concentrations which blocked the excitation by ACh and sometimes depressed that by glutamate, indicating that these excitations are unlikely to be mediated by cholinergic synapses involving muscarinic receptors.

Intracerebellar nuclei neurons have not been tested with monoamines. However, the presence of few NA- and 5-HT-containing nerve terminals in the intracerebellar nuclei (Hökfelt and Fuxe, 1969) suggests that monoaminergic pathways probably terminate on relatively few neurons in this area.

Another Transmitter?

One of the unusual findings about the mammalian cerebellum is that it has surprisingly low levels of ACh, 5-HT, NA, and DA compared with other structures of the C.N.S. Thus, unless the excitant and depressant amino-acids discussed in this chapter account for the transmitters at the very large majority of synapses in the cerebellum, there must be another substance which has hitherto received little attention as a likely transmitter. Such reasoning led to the discovery of a factor in cerebellar extracts of various mammalian species which increased the electrical activity of the cerebellum when injected into the carotid artery (Crossland and Mitchell, 1956). This 'cerebellar factor' was later reported to be identical with ergothioneine, and the relatively high concentrations of ergothioneine in the mammalian cerebellum raised the possibility that ergothioneine is an important excitatory transmitter in the cerebellum (Crossland, Woodruff, and Mitchell, 1964; Crossland, Mitchell, and Woodruff, 1966).

Results of subsequent studies, however, have cast doubt on this view. Electrophoretically administered ergothioneine usually had no action, although it weakly excited a small proportion of neurons in the cerebellar and cerebral cortices (Krnjević, Randić, and Straughan, 1965; Crawford and others, 1966). The possibility that the negative results may have been due to the difficulty of expelling ergothioneine ions electrophoretically was raised by Avanzino, Bradley, Comis, and Wolstencroft (1966), who observed that ergothioneine administered by means of the micro-tap had relatively strong excitant actions on about one-quarter of the brain-stem neurons tested, whereas electrophoretically administered ergothioneine weakly excited only 10 per cent of the neurons studied in the same area. However, despite the possibility that ergothioneine may, in fact, strongly excite certain cortical neurons, recent neurochemical findings do not provide any more positive evidence that ergothioneine has a transmitter role in the brain (Briggs, 1972). Of several mammalian species studied, only the cerebellum of the ox contained ergothioneine, and only in relatively small amounts (Briggs, 1972). Furthermore, although ergothioneine was found in different regions of the bovine brain, this regional distribution was considerably different from that of other mammals, reported earlier by Crossland and others (1964, 1966). Various observations suggested that the 'cerebellar factor' is not ergothioneine (Briggs, 1972).

Thus, in conclusion, the possible significance of the original observations is no clearer than it was several years ago. It appears that ergothioneine does not have a role particular to the cerebellum, but its presence in different regions of the C.N.S. raises the possibility that it is involved in central synaptic functions, though not necessarily as a transmitter. The active principle of cerebellar extracts may nevertheless prove to be an excitatory transmitter. These problems require further investigation.

SUMMARY

Purkinje Cells

GABA is a powerful depressant of Purkinje cells, the depression resulting from a membrane hyperpolarization and an increase in membrane conductance. Glycine is a weaker depressant of Purkinje cells than GABA. In particular, the inhibition of firing of Purkinje cells resulting from impulses in basket and stellate cell fibres is blocked by bicuculline and picrotoxin, which also block the action of GABA, but do not affect that of glycine. Conversely, strychnine does not affect this synaptic inhibition or the depression by GABA, but blocks the action of glycine. In addition, results from neurochemical and autoradiographic studies of the cerebellar cortex indicate that GABA and its synthesizing and inactivating enzymes are located in or in the vicinity of Purkinje, stellate, and basket cells. There is thus very good evidence that GABA is the inhibitory transmitter released on to Purkinje cells from basket and stellate axons. Since Purkinje axons are known to release GABA on to dorsal Deiters' neurons, and since some Purkinje axon collaterals appear to terminate on other Purkinje cells, it is reasonable to suppose that some Purkinje cells receive a GABA-mediated inhibition from Purkinje axon collaterals. Direct evidence for this, however, is lacking.

Glutamate and structurally related amino-acids excite Purkinje cells, but much more data is required before an acidic amino-acid can be considered to be a transmitter on these cells. The excitatory transmitters released from climbing and parallel fibres on to Purkinje cells have not been identified, but are substances other than ACh (excitatory amino-acids ?), as antagonists of ACh do not affect the superficial field potentials evoked by stimulating the inferior olive and surface of the folium. Although ACh has been reported to have both excitant and depressant actions on Purkinje cells, the significance of this is not clear, as both types of effect, as well as the proportion of Purkinje cells influenced by ACh vary for a number of reasons, including the different species studied and the depth of recording below the cerebellar surface.

A large body of different types of findings provides excellent evidence that NA is a slow-acting inhibitory transmitter released on to Purkinje cells from axons which originate in locus coeruleus of the brain-stem. Indeed, the evidence represents a major breakthrough in the study of central noradrenergic transmission. NA depresses the firing of Purkinje cells, the depression being associated with a hyperpolarization and a *decrease* in membrane conductance. Several observations suggest that the depression by NA is mediated metabolically by cyclic AMP. The depressant action of NA is mimicked by cyclic AMP and its dibutyryl derivative, except that the nucleotides have a more rapid time course of action. Phosphodiesterase inhibitors potentiate the depression by NA and cyclic AMP, whereas prostaglandins E_1 and E_2 and nicotinate (which reduce cyclic AMP levels in the peripheral nervous system) block the action of NA, but not that of cyclic AMP or GABA, supporting the postulate that depression by NA is mediated through activation of adenyl cyclase. Fine nerve-fibres containing NA terminate on Purkinje dendrites and spines, and treatment with 6-hydroxydopamine (which depletes the NA content of the cerebellum) leads to degeneration of a proportion of terminals which synapse with Purkinje dendrites and spines. Stimulation of locus coeruleus, which is a rich source of many NA-containing fibre systems (including a projection to the cerebellum), inhibits the firing of Purkinje cells, and the characteristics of this inhibition as revealed by intracellular recording resemble the depressant action of NA on Purkinje cells. In particular, this inhibition can be prevented by electrophoretically administered prostaglandins E_1 and E_2, by chronic pretreatment with 6-hydroxydopamine, and by acute treatment with reserpine and alpha-methyltyrosine. Moreover, the phosphodiesterase inhibitor, papaverine, potentiates both the actions of NA (and cyclic AMP) and the synaptic inhibition evoked by stimulating locus coeruleus, on Purkinje cells.

Little is known about a possible role of other monoamines in the cerebellum. 5-HT, DA, and histamine have been observed to depress the firing of some Purkinje cells. The presence of 5-HT-containing nerve-fibres in the cerebellar cortex raises the possibility that 5-HT functions as a transmitter in this area, but this remains to be shown.

Granule Cells

ACh apparently has no action on granule cells in the superficial folia, but excites many of those in the deeper layers. Dihydro-β-erythroidine blocks the excitant actions of both ACh and the synaptic excitation of some granule cells evoked by repetitively stimulating certain sites in the pons and mesencephalon. This suggests that a proportion of the mossy fibres is cholinergic, a view supported by results from studies of the distribution of AChE and choline acetyltransferase in the cerebellum. Nevertheless, further studies from different laboratories are required to substantiate this postulate, in view of the existing controversy.

Glutamate excites granule cells, and the evidence that at least a proportion of mossy fibres is non-cholinergic raises the possibility that some mossy fibres release an excitatory amino-acid as a transmitter. However, there is no direct pharmacological evidence for this.

Although the actions of neutral amino-acids on granule cells have not been studied, GABA is likely to be the inhibitory transmitter released from Golgi axons on to granule cells, as Golgi inhibition of granule cells is blocked by picrotoxin and bicuculline, but unaffected by strychnine.

Recent investigations suggest that 5-HT is an excitatory transmitter released from a proportion of mossy fibres on to granule cells. The actions of other monoamines have not been tested on granule cells.

Basket, Outer Stellate, and Golgi Cells

Basket cells in superficial folia are insensitive to cholinomimetics. ACh is unlikely to be an excitatory transmitter released from parallel fibres on to these inhibitory interneurons, as cholinergic antagonists have no effect on field potentials evoked by stimulation of the cerebellar cortical surface. GABA is probably an inhibitory transmitter released from Purkinje axon collaterals on to some basket and Golgi cells, but this has yet to be demonstrated.

Intracerebellar Nuclei Neurons

Glutamate and ACh excite the majority of the different types of intracerebellar nuclei neurons, but these actions have not been related to synaptic excitation in the intracerebellar nuclei. The synaptic discharges of deep nuclei neurons to stimulation of the brachium conjunctivum, brachium pontis, lateral reticular nucleus, and hind-limb nerves are resistant to atropine, suggesting that the final terminations of these pathways on cerebellar neurons are not mediated by cholinergic synapses involving muscarinic receptors. Intracerebellar nuclei neurons have not been tested with other substances suspected to be transmitters. However, the evidence that Purkinje axons release GABA on to dorsal Deiters' neurons strongly suggests that Purkinje axons also release GABA on to deep nuclei neurons.

Chapter

4

DIENCEPHALON

THE diencephalon may be broadly subdivided into the thalamus and the hypothalamus. Until recently pharmacological studies of electrophysiologically identified diencephalic neurons had been confined to a few cell types in the thalamus. In the last two or three years, however, different areas of the hypothalamus have also been explored, and it is now possible to discuss with more certainty the possible transmitters that operate in particular nuclei of the hypothalamus.

THALAMUS

The thalamus, a large structure situated at the rostral border of the mesencephalon, consists of a number of major nuclei which serve (amongst other functions) as relay stations for sensory and other pathways that terminate in the cerebral cortex. The gross divisions of the thalamus include the anterior, medial, lateral, ventral, and pulvinar nuclei, as well as the metathalamic (lateral and medial geniculate) nuclei. Each nucleus itself may be subdivided into different morphologically distinct areas, consisting of long-axon projective cells and interneurons arranged in complex, but orderly, networks. Although it is convenient to describe thalamic nuclei as 'relay stations' as a first approximation, this is a vastly over-simplified concept; it is now recognized that some integration of the afferent signals to the cortex occurs at the thalamic level. As might be expected, many anatomical, physiological, pharmacological, biochemical, and histochemical similarities exist between the different thalamic nuclei. However, there are also interesting differences, and for this reason studies of the different nuclei will be described separately in this chapter. Several other findings related to the discussion here may be found in articles by Purpura and Yahr (1966), Tömböl (1966), Jones and Powell (1969a, b), and Phillis (1971).

Ventrobasal Complex and Nucleus Ventralis Lateralis

Although the effects of suspected transmitter substances have been studied in a number of different thalamic nuclei (*see* Phillis, 1971), investigations of the ventrobasal complex (mainly nuclei ventralis posterolateralis and posteromedialis) will be described in particular detail here, as this is the main area where thalamo-cortical relay neurons are located. Thalamo-cortical relay neurons may be identified electrophysiologically by their antidromic responses to stimulation of the somatosensory (pericruciate) cortex and on the basis of short-latency responses to impulses in one contralateral peripheral nerve of the fore limb (median, external radial, or ulnar, 6–8 msec. latency) or hind limb (tibial or common peroneal, 8–12 msec. latency) (Andersen, Eccles, and Sears, 1964).

Thalamo-cortical Relay Neurons

Acetylcholine

Electrophoretically administered ACh excited different proportions of neurons in various nuclei of the thalamus, the highest proportions of ACh-excited cells being in the ventrobasal complex and nucleus ventralis lateralis (Andersen and Curtis, 1964a, b; Davis, 1966; McCance, Phillis, Tebēcis, and Westerman, 1968; McCance, Phillis, and Westerman, 1968a; McLennan, Huffman, and Marshall, 1968), and possibly also nucleus lateralis dorsalis (McLennan, 1970a; p. 97). In particular, ACh excited the majority of thalamo-cortical relay neurons tested (Andersen and Curtis, 1964a, b; McCance, Phillis, and Westerman, 1968a). Two examples of this are illustrated in *Fig.* 43. Some of these studies were made of cats anaesthetized with pentobarbitone sodium, and since subanaesthetic doses of this barbiturate selectively reduce the sensitivity of thalamic neurons to ACh (Phillis and Tebēcis, 1967b; Tebēcis, 1972c; Tebēcis and Di Maria, 1972b; but cf. Andersen and Curtis, 1964a), it is possible that in the absence of anaesthesia, all thalamo-cortical relay neurons would be excited by ACh (*see* Tebēcis, 1972c).

Cholinesterase inhibitors potentiated the excitation by ACh and sometimes increased the rate of background firing of thalamo-cortical relay and other thalamic neurons. Several muscarinic and nicotinic cholinergic antagonists (including atropine and dihydro-β-erythroidine) blocked the actions of ACh, and a number of different cholinomimetics (including acetyl-β-methylcholine and nicotine) excited thalamic neurons, suggesting that these cells have both muscarinic and nicotinic properties (Andersen and Curtis, 1964a, b; Davis, 1966; McCance, Phillis and others, 1968; McCance, Phillis, and Westerman, 1968a). It should be pointed out, however, that from these early experiments, it is not really

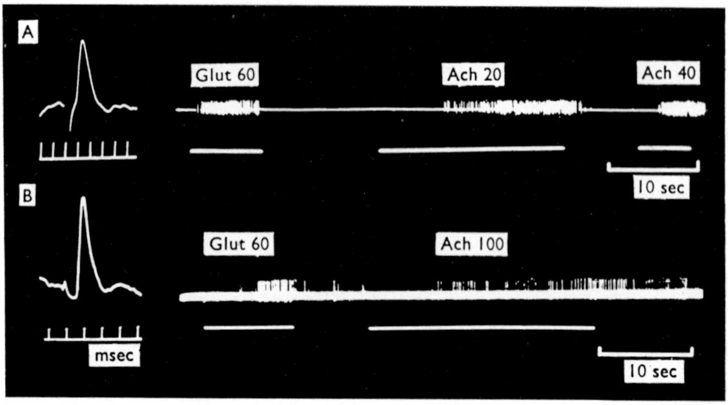

Fig. 43. Excitant actions of L-glutamate (Glut) and acetylcholine (ACh) on two different cells (A and B) identified as thalamo-cortical relay neurons. Stimulation of the pericruciate cortex evoked antidromic action potentials from these neurons, as shown on the left (msec. time calibration). The oscilloscope records on the right (10 sec. time calibration) show the firing evoked by electrophoretically administered glutamate and ACh, using the ejecting currents indicated (nA). (McCance, Phillis, and Westerman, 1968a, b.)

possible to conclude whether thalamic neurons possess different proportions of purely muscarinic and purely nicotinic receptors, or whether the cholinoceptive receptors are of a single type (intermediate between the two), as different antagonists of ACh were not administered to the same neuron. Recent studies of the medial geniculate nucleus, where the excitant action of ACh was completely blocked by both atropine and dihydro-β-erythroidine administered to the same neuron, have indicated that cholinoceptive receptors on medial geniculo-cortical relay neurons are of intermediate muscarinic-nicotinic type (Tebēcis, 1972c). This is probably also true of ventrobasal thalamic neurons, considering other similarities between these two thalamic areas in the actions of ACh and related compounds.

Attempts to relate the excitatory action of ACh to synaptic excitation in the thalamus have involved studies of 'sensory', cortico-thalamic, cerebello-thalamic, and reticulo-thalamic pathways.

a. Medial Lemniscal Pathway

The synaptic responses of thalamic neurons to afferent volleys in cutaneous and limb nerves were unaffected by atropine or dihydro-β-erythroidine in concentrations which readily reduced the excitant

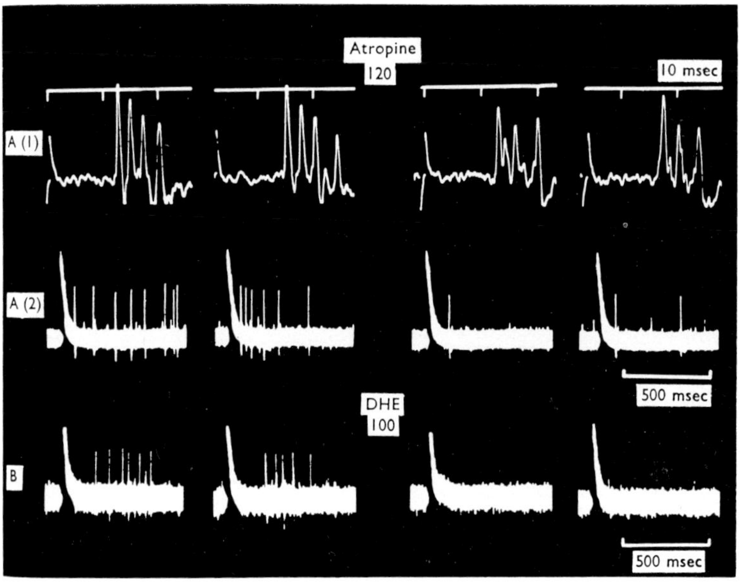

Fig. 44. Effects of antagonists of ACh on different types of synaptic excitation in the ventrobasal thalamus. A, Short-latency orthodromic responses to stimulation of the triceps nerve (1) and long-latency, long-duration responses of the same neuron to stimulation of the mesencephalic reticular formation (anterior + 3 mm., lateral + 3 mm., vertical — 1 mm.) (2) before and after the electrophoretic administration of atropine (120 nA for 90 sec.). Atropine had no obvious effect on the responses to stimulation of the triceps nerve, but almost abolished those evoked by reticular stimulation. This neuron could be excited by ACh before the administration of atropine. B, Responses of another ACh-sensitive neuron in the ventrobasal complex to stimulation of the reticular formation. Dihydro-β-erythroidine (DHE, 100 nA for 60 sec.) blocked this excitation. (McCance, Phillis, and Westerman, 1968a,b.)

action of ACh, indicating that the medial lemniscal pathway to the thalamus is non-cholinergic (Andersen and Curtis, 1964a, b; McCance, Phillis, and Westerman, 1968a). Such evidence obtained for one neuron is illustrated in *Fig.* 44. This conclusion is further supported by the recent finding that the transmitter released from medial lemniscal fibres is likely to be an excitatory amino-acid (Haldeman and others, 1972).

b. Cortico-thalamic Fibres

The excitatory transmitter released from corticofugal fibres on to thalamic neurons has not been identified. Electrophoretically administered antagonists of ACh did not affect the synaptic responses evoked by stimulating the somatosensory cortex (McCance, Phillis, and Westerman, 1968a). Although it is not possible to conclude whether these synaptic responses were evoked from impulses in corticofugal fibres or in thalamo-cortical axons which sent recurrent collaterals to the neurons recorded, the relatively dense projection of fibres from the cortex to the thalamus (Jones and Powell, 1968) suggests that cortico-thalamic fibres were stimulated (McCance, Phillis, and Westerman, 1968a) in some cases, at least. Thus, it is unlikely that the excitatory transmitter released from cortico-thalamic axon terminals is ACh.

c. Cerebello-thalamic Fibres

A small proportion of fibres from the cerebellum (lateral and interpositus nuclei) to the thalamus (nucleus ventralis lateralis; *see Fig.* 41) is thought to be cholinergic. The superior cerebellar peduncle, or brachium conjunctivum (through which the cerebello-thalamic fibres pass) contains both choline acetyltransferase and AChE (Feldberg and Vogt, 1948; Hebb and Silver, 1956; Phillis, 1968a). In the first pharmacological investigation of this pathway Frigyesi and Purpura (1966) observed that intracarotid injections of ACh facilitated and/or depressed, whereas atropine depressed the field potentials evoked in the ventrolateral nucleus following stimulation of the brachium conjunctivum. Subsequently, however, although electrophoretically administered ACh excited the majority of thalamic neurons, which were monosynaptically excited by stimulating the brachium conjunctivum (Davis, 1966; McCance, Phillis, and Westerman, 1968a; McLennan and others,

1968), only a small proportion of these orthodromic responses could be reduced by atropine and dihydro-β-erythroidine (McCance, Phillis, and Westerman, 1968a, b). The current view, therefore, is that few fibres projecting from the cerebellum to the thalamus can be cholinergic (Phillis, 1971).

d. Reticulo-thalamic Fibres

The most important cholinergic innervation of the mammalian thalamus is thought to originate in the brain-stem, and is possibly the cholinesterase-containing dorsal tegmental pathway described in rats by Shute and Lewis (1967). Antagonists of ACh often reduced or blocked the synaptic responses of thalamic neurons evoked by stimulating the mesencephalic reticular formation (McCance, Phillis, and Westerman, 1968a). *Fig.* 44 shows the responses of an ACh-excited thalamic cell to stimulation of the triceps nerve (*Fig.* 44A(1)) and mesencephalic reticular formation (*Fig.* 44A(2)). Electrophoretically applied atropine had no effect on the short-latency response to peripheral nerve stimulation, but almost blocked the long-latency, long-duration response to brain-stem stimulation. Similarly, electrophoretic dihydro-β-erythroidine blocked the synaptic discharge of another thalamic neuron following stimulation of the reticular formation (*Fig.* 44B).

Despite these studies, however, the precise origin and course of the cholinergic afferents to the thalamus have not been established. Interpretation of the responses evoked by stimulating (especially repetitively) the reticular formation is fraught with difficulties, particularly as the responses are usually of long latency and duration, and may therefore indicate the involvement of either a single long-chain polysynaptic pathway, or more than one pathway. The label 'reticular formation' conveniently refers to a complex mass of adjacent nerve nuclei and fibre tracts, and the spread of stimulus to neighbouring regions is difficult to exclude, even though the position of the tip of the stimulating electrode can be verified histologically (a procedure which is rarely done routinely in any case). In addition, recovery from the effect of atropine has not always been demonstrated. Nevertheless, the results of the pharmacological investigations provide evidence that ACh is an excitatory transmitter acting on some thalamic neurons, the cholinergic fibres probably originating somewhere in the brain-stem. Depressant actions of ACh on thalamic neurons have also been observed (Andersen and Curtis, 1964a; McCance, Phillis, and Westerman, 1968a), but these have not been related to synaptic inhibition in the thalamus.

The results of biochemical, histochemical, and ACh-release studies of the mammalian thalamus are consistent with the view that ACh is a transmitter in this structure. ACh, choline acetyltransferase, and AChE are present in the thalamus of various mammalian species (MacIntosh, 1941; Feldberg, 1945; Feldberg and Vogt, 1948; Burgen and Chipman, 1951; Koelle, 1954; Zetler and Schlosser, 1955; Hebb and Silver, 1956; Shute and Lewis, 1967), and all three substances are unevenly distributed in different regions of the thalamus in cats and monkeys (Phillis, Tebēcis, and York, 1967b; Fahn and Côté, 1968b; Myers and Beleslin, 1970). ACh is released spontaneously at different rates from some thalamic nuclei, the rate of release being highest in the ventral regions (Myers and Beleslin, 1970). The efflux of ACh from the dorsal and ventrobasal complex of the thalamus was increased by stimulating various nervous pathways, including the eyes with flashes of light, limb nerves, and the mesencephalic reticular formation (Phillis and others, 1968b). Thus, the cholinergic afferents to the thalamus may represent part of the 'reticular arousal system' (*see* Shute and Lewis, 1967; McCance, Phillis, and Westerman, 1968a; Phillis and others, 1968b; Phillis, 1971).

Excitatory Amino-acids

Relatively high amounts of glutamate are present in the thalamus of the cat (Johnson and Aprison, 1971). L-Glutamate and structurally related amino-acids have been shown to excite thalamocortical relay cells (for example, *Fig.* 43) and other thalamic neurons (Andersen and Curtis, 1964a, b; Steiner and Ruf, 1966; McCance, Phillis, and Westerman, 1968a; McLennan and others, 1968; Curtis, Duggan, Felix, Johnston, and McLennan, 1971; Curtis and others, 1972). Although most reports on the actions of amino-acids on central neurons have implied that glutamate is a universal excitant, all neurons being fairly similar in sensitivity to the amino-acid, McLennan and others (1968) first showed that thalamic neurons are differentially sensitive to glutamate. Neurons in the superficial areas of the thalamus were relatively insensitive to glutamate when compared with DLH and DL-*N*-methyl-aspartate (which have not been found to occur naturally in the C.N.S.), whereas cells in the deeper areas were strongly excited by glutamate. Still deeper neurons were again less sensitive to the amino-acid. Moreover, the neurons which were particularly sensitive to glutamate were also particularly sensitive to ACh, and were identified as belonging to nucleus ventralis lateralis, both histologically and by their responses to stimulation of the brachium conjunctivum. These observations have been recently confirmed (Haldeman and others, 1972). McLennan and others (1968) suggested that thalamic neurons have both unspecific receptors, which are affected by all excitant amino-acids, and specific receptors (which may be identical with the excitatory synaptic receptors) for glutamate. This, of course, may also apply to other structures of the C.N.S., but systematic studies of the relative sensitivity of different identified neurons to excitant amino-acids have not been made in most central regions (cf. cuneate neurons; Chapter 2).

More direct pharmacological evidence that glutamate is an excitatory transmitter in the thalamus has recently been obtained using glutamate analogues that can reduce the excitation evoked by excitatory amino-acids (glutamate, aspartate, and to a lesser extent, DLH) without affecting that by ACh (McLennan and others, 1971; Haldeman and others, 1972; cf. Curtis and others, 1972). In particular, the diethyl ester of glutamate sometimes blocked both the synaptic excitation evoked by stimulating hind limb nerves and the action of glutamate on neurons in the ventrobasal complex, in concentrations which had little or no effect on those of ACh and DLH (McLennan and others, 1971; Haldeman and others, 1972). *Fig.* 45 illustrates such an action of L-glutamic acid diethyl ester.

These findings are obviously of considerable importance as evidence that an excitatory amino-acid is likely to be the transmitter released from medial lemniscal fibres on to neurons in the ventrobasal complex of the thalamus. However, it is not possible to conclude that this transmitter is glutamate, as neither glutamate diethyl ester nor a number of other substances tested could distinguish between glutamate and aspartate as excitants (Curtis and others, 1972). A specific antagonist of glutamate has yet to be discovered. In conclusion, it is important to remember that different populations of medial lemniscal fibres may release different transmitters.

Depressant Amino-acids

The mammalian thalamus contains relatively high concentrations of GABA and the enzymes for its metabolism (Albers and

Brady, 1959; Salvador and Albers, 1959; Kržalić, Mandić, and Mihailović, 1962; Lovell, Elliott, and Elliott, 1963; van Gelder, 1965b; Shank and Aprison, 1970; Okada and others, 1971). In the monkey thalamus, GABA is distributed unevenly (Fahn and Côté, 1968a). Of the depressant amino-acids studied, pharmacological evidence has been obtained that GABA is an inhibitory

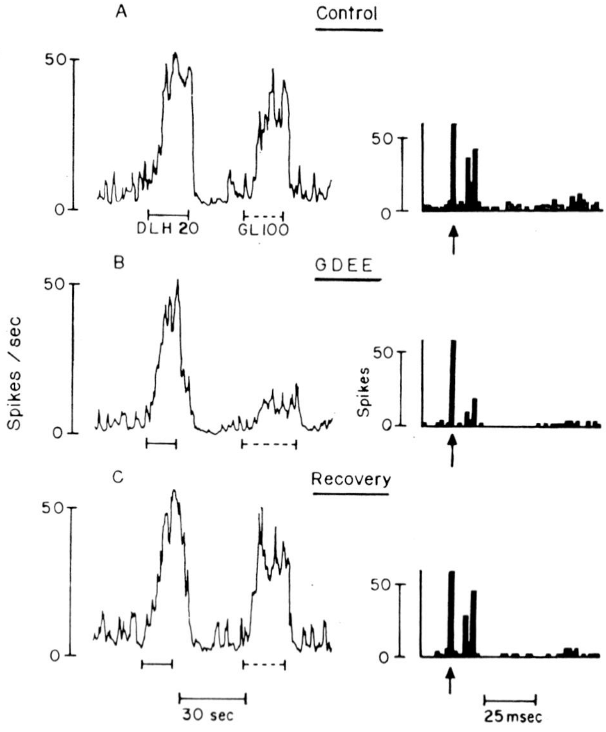

Fig. 45. Antagonism by L-glutamic acid diethyl ester (GDEE) of the action of L-glutamate and the synaptic excitation evoked by stimulating a hind limb nerve on a neuron in the ventrobasal complex of the thalamus. The records of the left show the excitant actions of DLH (20 nA) and glutamate (GL; 100 nA) on the firing frequency of the neuron. Those on the right are computed histograms showing the summed evoked responses (50 sweeps) of the neuron to stimulation of the contralateral tibial nerve (0·1 msec., 3 V, once every 4 sec., stimulus artefact indicated by arrows). The responses usually consisted of two spikes. A, Before; B, During the application of GDEE (160 nA for 6 min.); C, 6 min. after cessation of the GDEE current. (Haldeman and others, 1972.)

transmitter in the ventrobasal complex of the thalamus (Curtis, Duggan, Felix, and others, 1971; Duggan and McLennan, 1971; Curtis and Tebēcis, 1972). GABA was almost invariably a more effective depressant of thalamo-cortical relay neurons than glycine, and the action of GABA was blocked by bicuculline, which did not affect that of glycine. Initially, it was reported that although intravenous bicuculline reduced the inhibition of the synaptic discharge evoked from a peripheral nerve by a preceding volley to the same

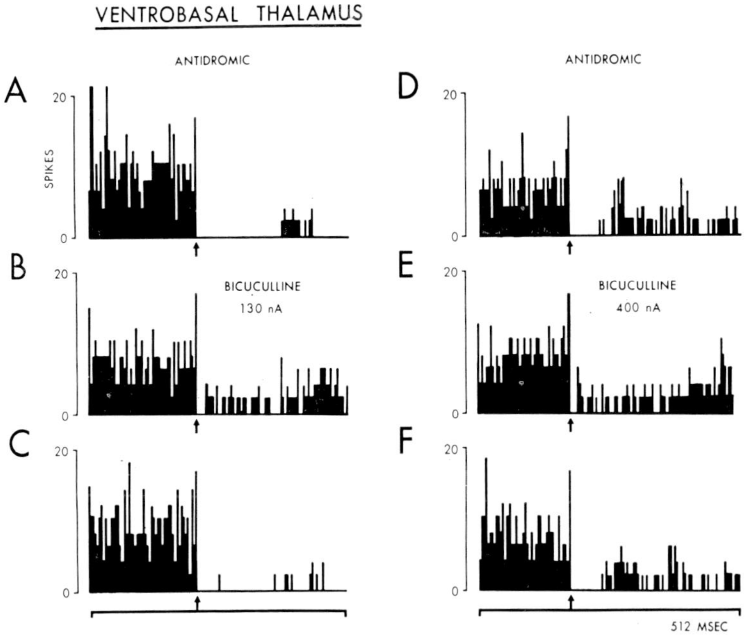

Fig. 46. Effects of bicuculline on the inhibition of two thalamo-cortical relay neurons (A–C and D–F) evoked by stimulating the primary somatosensory cortex (onset of stimulus indicated by an upward arrow). Firing was maintained by DLH ejected continuously. Each histogram illustrates the number of action potentials in 128 bins (each of 4 msec. duration), counted for a total of 40 sweeps, once every 2 sec. A, Control. B, 5 min. after a current of 130 nA began to eject bicuculline (5 m*M* in 165 m*M* NaCl), which was administered for a total of 9 min. C, 8 min. after the bicuculline current had been terminated. D, Control. E, 13 min. after a current of 400 nA began to eject bicuculline for 15 min. F, 17 min. after termination of this current. *Ordinates*: number of spikes. *Abscissae*: time, 512 msec. (Curtis and Tebēcis, 1972.)

nerve, electrophoretically administered bicuculline *prolonged* the inhibition of thalamo-cortical relay neurons produced by a single afferent stimulus; this latter observation was explained in terms of a reduction of 'post-anodal exaltation' or a facilitation of inhibitory interneurons in the vicinity of the thalamo-cortical relay neuron recorded (Duggan and McLennan, 1971). Subsequently, however, electrophoretically administered bicuculline was observed to reduce the inhibition of DLH-evoked firing of thalamo-cortical relay neurons following antidromic activation (Curtis and Tebēcis, 1972). Such results obtained from two different neurons are illustrated in *Fig.* 46. In both cases bicuculline blocked the action of GABA in concentrations which reduced the synaptic inhibition, and recovery occurred when the bicuculline ejections had been terminated (not illustrated).

Strychnine had no effect on postsynaptic inhibition in the ventrobasal thalamus (Andersen and others, 1963; Duggan and McLennan, 1971) and blocked the action of glycine without affecting that of GABA on thalamo-cortical relay neurons (Curtis and others, 1971b; Duggan and McLennan, 1971). Thus, the long-lasting IPSP which follows antidromic excitation of thalamo-cortical relay neurons (Andersen, Eccles, and Sears, 1964) is likely to be mediated by GABA (Curtis, Duggan, Felix, and others, 1971; Duggan and McLennan, 1971; Curtis and Tebēcis, 1972). This prolonged inhibition, frequently cyclic, is thought to result from impulses in recurrent collaterals of thalamo-cortical axons, which terminate on inhibitory interneurons in the ventrobasal thalamus (Andersen, Eccles, and Sears, 1964; Tömböl, 1966; Jones and Powell, 1969a, b).

Monoamines

The monoamines, 5-HT, NA, and DA, as well as the enzyme systems for their metabolism, are present in the thalamus of various mammals, including man (Vogt, 1954; Bogdanski, Weissbach, and Udenfriend, 1957; Axelrod, Albers, and Clemente, 1959; Bertler and Rosengren, 1959; Sano, Gamo, Kakimoto, Taniguchi, Takesada, and Nishinuma, 1959; Shimizu, Morikawa, and Okada, 1959; Bertler, 1961; Kuntzman, Shore, Bogdanski, and Brodie, 1961; McGeer, McGeer, and Wada, 1963; Manocha, Shantha, and Bourne, 1967), and at least a proportion of this 5-HT and NA content (in rats) is contained within nerve-fibres which originate in the pons and medulla (Fuxe, 1965;

Andén and others, 1966; Fuxe and others, 1970; Ungerstedt, 1971). 5-HT is released spontaneously at different rates from various (but not all) areas of the monkey thalamus (Myers and Beleslin, 1970). These findings provide a sound basis for considering that at least 5-HT and NA are transmitters in the mammalian thalamus.

NA had excitant or depressant, or both actions on various unidentified thalamic neurons (Phillis and Tebēcis, 1967c). Of particular interest, however, was the observation that this catecholamine depressed the firing of most thalamic neurons tested in the superficial layers, but facilitated as often as it depressed the firing of neurons in the deep layers (ventrobasal complex). Accordingly, the firing of thalamo-cortical relay neurons was either facilitated or depressed by NA. Excitant and depressant actions of NA on two different thalamo-cortical relay neurons are illustrated in *Figs.* 47 and 48 respectively.

5-HT depressed the firing of most thalamic neurons, including thalamo-cortical relay cells, although it also excited a small proportion (Andersen and Curtis, 1964b; Phillis and Tebēcis, 1967c). An interesting point worth mentioning now is that 5-HT did not appear to have two different (dose-dependent) types of depressant action on ventrobasal thalamic neurons, as it did not selectively affect the short-latency synaptic responses of thalamic neurons to cutaneous volleys, in low concentrations (Andersen and Curtis, 1964b). It will be clear from the next section that this

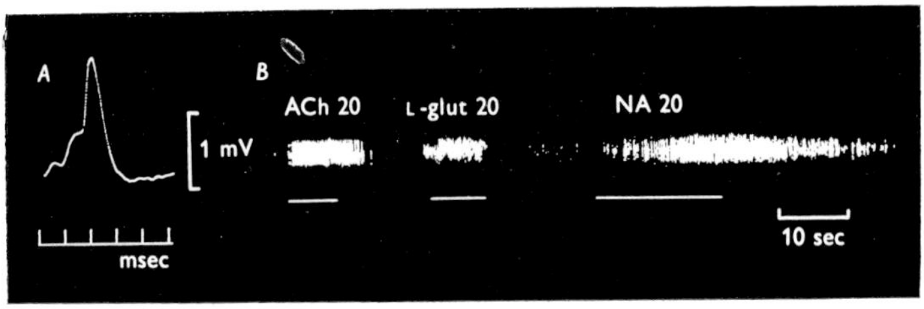

Fig. 47. Excitant actions of ACh (20 nA), glutamate (L-glut, 20 nA), and NA (20 nA) on a thalamo-cortical relay neuron in the ventrobasal thalamus. A, Antidromic spike evoked by stimulating the sensorimotor area of the ipsilateral cortex. B, Oscilloscope records showing the firing of this neuron elicited by ACh, glutamate, and NA. Excitation by NA usually had a longer time course of action than that by the other two substances. Many spikes were too faint to be seen clearly on the film. (Phillis and Tebēcis, 1967c.)

differs from the lateral geniculate nucleus, where 5-HT in low concentrations has an unusual selective depressant action on orthodromic firing, whereas in higher concentrations it also has a

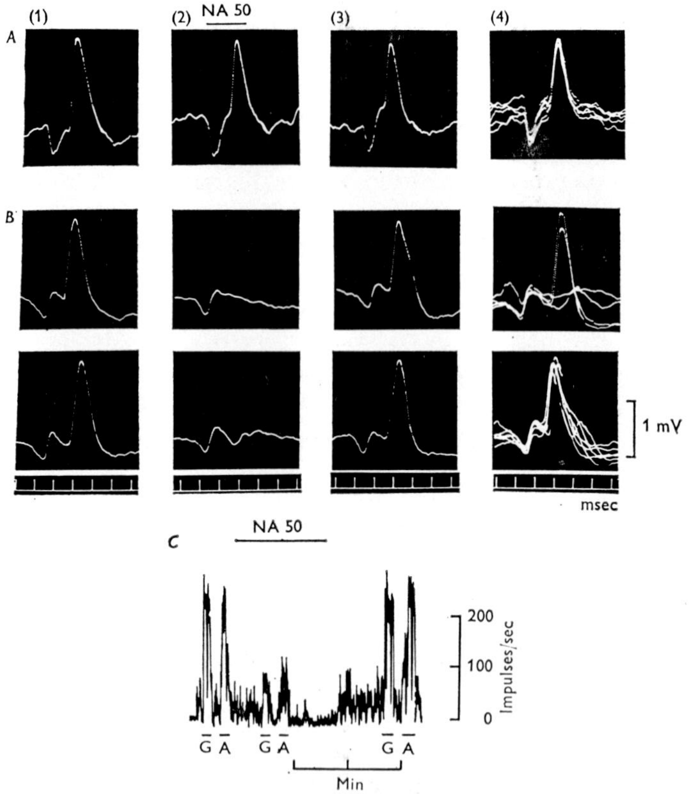

Fig. 48. Depressant actions of NA (50 nA) on different types of firing of a thalamo-cortical relay neuron of the ventrobasal thalamus. This unit was activated both antidromically by stimulating the ipsilateral sensorimotor cortex (A) and orthodromically from the brachium conjunctivum (B). Each pair of records in B represents consecutive responses. A (1), B (1), Control responses. NA (50 nA) depressed the orthodromic response [B (2)], but not the antidromic response [A (2)]. The records in A (3) and B (3) were obtained several minutes after termination of the NA application. A (4), Repetitive stimulation at 100/sec. B (4), Repetitive stimulation at 30/sec. (first record) and 10/sec. (second record). The chart recording in C shows the action of NA (50 nA) on the excitation by L-glutamate (G, 50 nA) and ACh (A, 50 nA) of the same cell. (Phillis and Tebēcis, 1967c.)

postsynaptic depressant action comparable to that observed elsewhere in the C.N.S.

DA depressed the firing of the large majority of thalamic neurons tested at all depths of the thalamus, including thalamo-cortical relay neurons (Phillis and Tebēcis, 1967c). The depression by DA was usually of a more rapid time course than that by NA. In contrast to the action of NA, DA had no excitant action on thalamic neurons.

The presence of 5-HT and NA in nerve-fibres in the thalamus, together with the biochemical and pharmacological findings, suggest that both monoamines are transmitters in the thalamus. 5-HT is likely to be predominantly an inhibitory transmitter, whereas NA may be both an inhibitory and an excitatory transmitter at different thalamic sites. However, the actions of these monoamines have yet to be related to synaptic activity in the thalamus. Although electrophoretic strychnine and intravenous picrotoxin reduced the depressant action of NA and the inhibitory effects of repetitive stimulation of the mesencephalic reticular formation on some thalamic neurons (Phillis and Tebēcis, 1967c), the specificity of these actions was not determined and no firm conclusions may be made about the transmitter liberated synaptically. The differences between the actions of NA and DA indicate that these catecholamines probably act on different receptors, but since there is no histochemical evidence for the presence of DA within thalamic nerve-fibres, the role of this catecholamine in the thalamus is not clear.

Lateral Geniculate Nucleus

The dorsal lateral geniculate nucleus has a relatively simple cellular architecture, consisting mainly of 'principal cells' whose axons project to the visual cortex, and short-axon cells or interneurons (O'Leary, 1940). These may be further subdivided into various cell types on the basis of the morphology and orientation of the dendrites and terminals (Guillery, 1966, 1969; Szentágothai, Hámori, and Tömböl, 1966). Several cellular laminae may be distinguished, and the ordered geometrical arrangement of the optic tract, the dendrites and cell bodies of geniculate neurons, as well as of the optic radiation, have facilitated the interpretation of evoked potentials recorded in the lateral geniculate nucleus (Bishop, 1953; Bishop, Jeremy, and McLeod, 1953; Bishop and McLeod, 1954; Bishop and Davis, 1960; and many subsequently).

Geniculo-cortical Relay Neurons

Lateral geniculo-cortical relay neurons, or principal cells of the lateral geniculate nucleus, can be identified electrophysiologically by their antidromic responses to stimulation of the optic radiation, and orthodromic responses to stimulation of the optic tract, optic nerve, or eyes with light.

5-Hydroxytryptamine

5-Hydroxytryptamine has a unique action on lateral geniculate neurons, in addition to its 'conventional' actions. This will be described in some detail, as it may also prove to be important elsewhere in the C.N.S. Interest in the possible functional importance of tryptamine derivatives in the lateral geniculate nucleus originated from the experiments of Evarts, Landau, Freygang, and Marshall (1955) and Bishop, Field, Hennessy, and Smith (1958), who showed that intracarotid injections of LSD and bufotenine blocked the synaptic responses evoked in the lateral geniculate nucleus by stimulating the optic nerve. It was concluded that LSD blocked the interaction between the excitatory transmitter released from optic nerve terminals and subsynaptic receptors on lateral geniculate neurons (Bishop, Burke, and Hayhow, 1959). In the first microelectrophoretic study of lateral geniculate neurons, Curtis and Davis (1962) observed that LSD had actions similar to those of 5-HT and a number of other tryptamine and diethylamide derivatives. All these substances depressed the orthodromic firing of geniculate neurons evoked by impulses in optic nerve fibres without affecting the excitant action of glutamate or antidromic invasion. It seemed, therefore, that 5-HT and LSD had no postsynaptic depressant action on lateral geniculate neurons, but either blocked the release of the excitatory transmitter from optic nerve terminals, or blocked its access to receptors on geniculate neurons (Curtis and Davis, 1962).

In subsequent studies, however, 5-HT also depressed the spontaneous firing, the excitant actions of glutamate and ACh, as well as antidromic invasion of lateral geniculate neurons, suggesting a postsynaptic depressant action of 5-HT (Phillis and others, 1967a; Satinsky, 1967). The apparent discrepancies were recently explained in terms of two different types of depressant action of 5-HT, which are dose-dependent (Tebēcis and Di Maria, 1972b). In low concentrations, 5-HT readily depressed the orthodromic firing of lateral geniculate neurons without affecting their sensitivity to ACh and excitant amino-acids, or antidromic invasion; in higher concentrations, 5-HT also depressed chemical and antidromic excitation. Results from one lateral geniculo-cortical relay neuron are illustrated in *Fig*. 49.

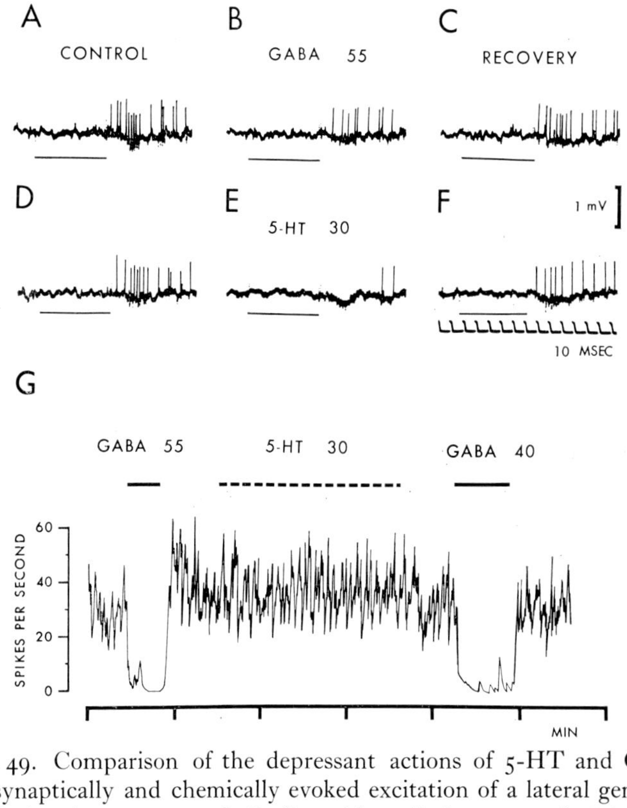

Fig. 49. Comparison of the depressant actions of 5-HT and GABA on synaptically and chemically evoked excitation of a lateral geniculo-cortical relay neuron. A–F, Repetitive discharges evoked by light flashes (once every 3 sec., duration indicated by horizontal bars) to the ipsilateral eye. A, D, Control responses. B, 1 min. after the beginning of an ejection of GABA (55 nA), which was continued for a further 30 sec. C, 40 sec. after the current ejecting GABA had been terminated. E, 2 min. after the beginning of an ejection of 5-HT (30 nA). F, Approximately 2·5 min. after the current ejecting 5-HT had been terminated. G, Recording of the firing frequency of the same neuron illustrating the actions of GABA (55 and 40 nA) and 5-HT (30 nA) on firing evoked by L-glutamate (—90 nA). (Tebēcis and Di Maria. 1972b.)

Stimulation of the ipsilateral eye with a flash of light evoked a synaptic discharge, which was slightly reduced by GABA (55 nA) (*Fig.* 49A, B, C), but almost blocked by 5-HT (30 nA) (*Fig.* 49D, E, F). The firing evoked by glutamate, however, was unaffected by 5-HT in the same concentration which almost blocked the orthodromic firing, although GABA (55 nA and 40 nA) had a marked depressant action (*Fig.* 49G). Assuming that glutamate increases, whereas GABA decreases, neuronal postsynaptic excitability, these results indicate that 5-HT in a concentration attained by 30 nA had no postsynaptic depressant action, but depressed the synaptic responses by impulses in optic nerve fibres. In higher concentrations (50 nA) 5-HT depressed the firing by glutamate (not illustrated), indicating a postsynaptic depressant action.

Two types of depressant action were also observed with 4-hydroxytryptamine and LSD on lateral geniculate neurons (Tebēcis and Di Maria, 1972b). The selective depression of synaptic responses by these substances in low concentrations is best explained as either a block of release of the transmitter from optic nerve terminals, or as a block of its postsynaptic action on lateral geniculate neurons (Curtis and Davis, 1962; Tebēcis and Di Maria, 1972b; cf. Bishop and others, 1958, 1959). Whether this action of 5-HT is of functional importance, however, is unknown.

The postsynaptic depression of lateral geniculate neurons produced by higher concentrations of 5-HT may well be related to an inhibitory transmitter role for this tryptamine derivative (Phillis and others, 1967a; Tebēcis and Di Maria, 1972b), in view of the histochemical evidence for 5-HT-containing nerve-fibres which originate in the brain-stem and project to the lateral geniculate nucleus (Fuxe, 1965; Andén, Dahlström and others, 1966; Ungerstedt, 1971). Since 5-HT depressed the firing of all lateral geniculo-cortical relay neurons tested (Curtis and Davis, 1962; Phillis and others, 1967a; Satinsky, 1967; Tebēcis and Di Maria, 1972b), it cannot be the excitatory transmitter released from optic nerve terminals on to lateral geniculate neurons, a conclusion reached earlier by the paucity of 5-HT in the optic nerve (Amin, Crawford, and Gaddum, 1954; Bogdanski and others, 1957).

In unanaesthetized cats 5-HT excited a small proportion of lateral geniculate neurons which could not be activated antidromically by stimulating the optic radiation, and were therefore considered to be interneurons (Satinsky, 1967; Tebēcis and Di Maria, 1972b). The significance of this observation is not clear.

Catecholamines

NA depressed the firing of the majority of geniculo-cortical relay neurons tested (Curtis and Davis, 1962; Phillis and others, 1967a; Tebēcis and Di Maria, 1972b). Although Satinsky (1967) reported that NA consistently excited these neurons, the number of neurons tested (4) is too small to make generalizations. However, these observations at least indicate that two types of geniculo-cortical relay neurons are sensitive to NA, as is true of thalamo-cortical relay neurons (Phillis and Tebēcis, 1967c; *see Figs.* 47, 48). The presence of NA-containing nerve-fibres in the lateral geniculate nucleus, which originate in the brain-stem (Fuxe, 1965; Olson and Fuxe, 1971; Ungerstedt, 1971) suggest that NA is released synaptically on to geniculate neurons from these fibres (Phillis and others, 1967a; Tebēcis and Di Maria, 1972b).

DA depressed the firing of geniculo-cortical relay neurons (Curtis and Davis, 1962; Phillis and others, 1967a; Tebēcis and Di Maria, 1972b). Both DA and NA appeared to have only a post-synaptic depressant action on lateral geniculate neurons (Tebēcis and Di Maria, 1972b; cf. Curtis and Davis, 1962). The predominantly depressant actions of these catecholamines on lateral geniculate neurons, and their low content in the optic nerve (Vogt, 1954; Cobbin, Leeder, and Pollard, 1965) indicate that neither substance can be the *excitatory* transmitter released from optic nerve terminals. However, the evidence that a small proportion of ganglion cells in the retina of a number of different mammals contains DA, but not NA (Chapter 8) raises the intriguing possibility that a small proportion of the inhibition from the retina may be mediated by DA released synaptically on to lateral geniculate neurons. This possibility has received little attention, but certainly warrants further investigation. Otherwise, there is no evidence to suggest that DA may be a transmitter in the lateral geniculate nucleus. There is no evidence for DA-containing nerve-fibres which arise from the brain-stem, as is true for NA.

Acetylcholine

ACh excited geniculo-cortical relay and most unidentified lateral geniculate neurons (Curtis and Davis, 1963; Phillis and others, 1967b; Satinsky, 1967; Steiner, 1968; Tebēcis and Di

Maria, 1972b). The weak excitant action of ACh originally observed on lateral geniculate neurons (Curtis and Davis, 1963) was probably due to the use of barbiturate anaesthesia, as powerful excitation by ACh was observed using volatile anaesthetics (Phillis and others, 1967b) or unanaesthetized animals (Satinsky,

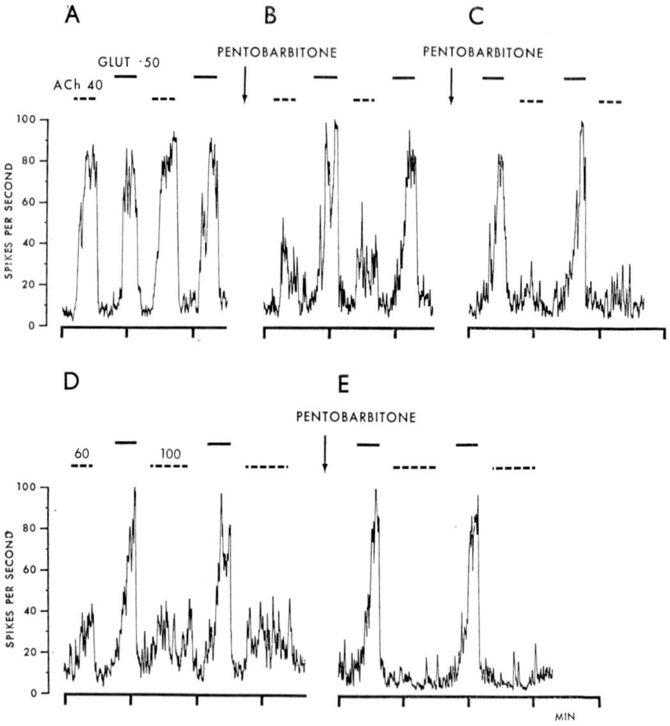

Fig. 50. Effect of intravenously administered pentobarbitone sodium on the excitation of a lateral geniculo-cortical relay neuron by ACh and L-glutamate, in an unanaesthetized cat. ACh (40 nA, broken horizontal line) and glutamate (—50 nA, solid horizontal line) were administered at regular intervals and the firing frequency of the neuron was plotted continuously. A, Control. B, Approximately 3 min. after pentobarbitone sodium (2·1 mg./kg., infused over a period of about 70 sec.). C, Approximately 3 min. after another dose of pentobarbitone sodium (2·1 mg./kg.). D, Approximately 8 min. after C, ACh was ejected for longer periods with currents of 60 and 100 nA. E, Approximately 1 min. after a final dose of pentobarbitone sodium (2·1 mg./kg.). (Tebēcis and Di Maria, 1972b.)

1967; Tebēcis and Di Maria, 1972b). In particular, low intravenous doses of pentobarbitone sodium into unanaesthetized animals selectively reduced the sensitivity of lateral geniculate neurons to ACh (Tebēcis and Di Maria, 1972b), as observed elsewhere in the thalamus (Phillis and Tebēcis, 1967b; Tebēcis, 1972c). *Fig.* 50 illustrates such results from a lateral geniculocortical relay neuron.

One difference between lateral geniculate neurons and those of the ventrobasal complex and medial geniculate nucleus is that lateral geniculate neurons are usually excited by ACh with a time course comparable to that of the excitatory amino-acids (*Fig.* 50, for example). In the ventrobasal complex and medial geniculate nucleus, the excitation by ACh is generally of a slower time course (several seconds to 1 min. for maximal excitation and for recovery; *see* appropriate sections in this chapter). The greater concentration of AChE in the lateral geniculate nucleus compared with the other thalamic areas (Phillis and others, 1967b) could account for the rapid termination of the excitant action of ACh on lateral geniculate neurons, but does not explain the rapid time for maximal excitation. Otherwise, the effects of other cholinomimetics, cholinesterase inhibitors, and antagonists of ACh on lateral geniculate neurons (Curtis and Davis, 1963; Phillis and others, 1967b) resembled those on neurons in the ventrobasal complex and medial geniculate nucleus.

Direct pharmacological evidence that ACh is an excitatory transmitter in the lateral geniculate nucleus was obtained by Phillis and others (1967b), who showed that repetitive stimulation of the mesencephalic reticular formation frequently facilitated the firing of lateral geniculate neurons, and that this increase in excitability as well as the excitant action of ACh, were blocked by the curariform agent, benzoquinonium. Results illustrating this are shown in *Fig.* 51. Stimulation of the reticular formation facilitated the excitant action of ACh and glutamate on a lateral geniculate neuron (*Fig.* 51A), but after benzoquinonium had been administered electrophoretically, the excitation by reticular stimulation and ACh, but not that by glutamate, were virtually blocked (*Fig.* 51B).

However, as has been discussed for the cholinergic afferents to the ventrobasal thalamus (p. 121), the precise origin and course of the cholinergic pathways to the lateral geniculate nucleus are not known. These pathways may be the cholinesterase-containing fibres described in rats by Shute and Lewis (1967). Further support for the view that the main cholinergic innervation of the lateral geniculate nucleus arises in the brain-stem was obtained by Deffenu, Bertaccini, and Pepeu (1967), who found a relationship between the ACh levels in the lateral geniculate nucleus and the degree of 'arousal'. The concentrations of ACh in the lateral geniculate nucleus were higher in cats which exhibited a greater synchronization of the EEG, suggesting that the ACh content is largely dependent on the activity of cholinergic fibres from the reticular formation.

Although the mammalian lateral geniculate nucleus contains appreciable quantities of ACh, choline acetyltransferase, and AChE (MacIntosh, 1941; Feldberg and Vogt, 1948; Hebb and Silver, 1956; Hebb, 1957; Cobbin and others, 1965; Phillis and others, 1967b; Deffenu and others, 1967), there is extremely little of any of these substances in the optic nerve (Loewi and Hellauer, 1938; MacIntosh, 1941; Feldberg and Vogt, 1948; de Roetth,

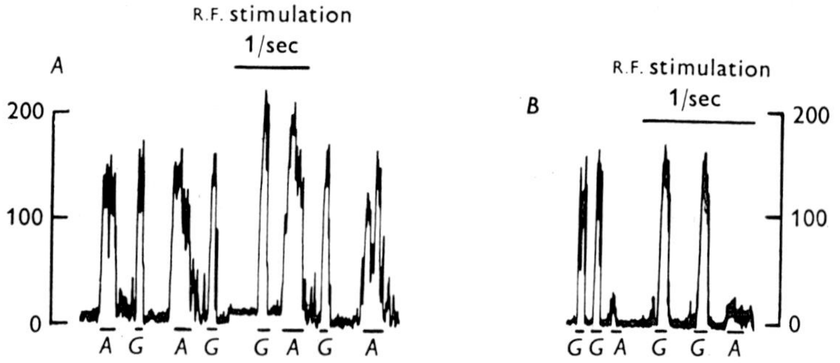

Fig. 51. Antagonism by benzoquinonium of the excitant action of ACh and repetitive stimulation of the mesencephalic reticular formation (R.F. stimulation) on a lateral geniculate neuron. A, R.F. (anterior +3 mm., lateral +3 mm., vertical —1 mm.) stimulation (1/sec.) enhanced the actions of ACh (A, 40 nA) and L-glutamate (G, 40 nA). B, After benzoquinonium (100 nA) had been administered for 1·5 min., the action of ACh (but not that of glutamate) and the facilitatory effect of R.F. stimulation were largely blocked. The actions of ACh and R.F. stimulation recovered at approximately the same time after the current ejecting benzoquinonium had been terminated (not illustrated). *Ordinate*: firing frequency (spikes/sec.). The duration of R.F. stimulation in B represents 2 min. (Phillis and others, 1967b.)

1951; Hebb and Silver, 1956), indicating that ACh is unlikely to be the major excitatory transmitter released from optic nerve terminals on to lateral geniculate neurons. This is supported by the pharmacological observations that antagonists of ACh had no effect on the synaptic responses of geniculate neurons evoked by light and optic nerve stimulation (Curtis and Davis, 1963; Phillis and others, 1967b; Tebēcis, unpublished observations). Although removal of the eyes and/or visual cortex resulted in relatively large decreases of ACh in the lateral geniculate nucleus of the rabbit three weeks later (Miller, Heller, and Moore, 1969), this need not necessarily mean that the ACh had been contained in the optic nerves, as trans-synaptic degeneration could have occurred in this time. The experiments do, however, indicate that at least a proportion of the ACh content of the lateral geniculate nucleus depends on optic nerve fibres being intact. Since a substantial proportion of ACh remained following enucleation and ablation of the visual cortex, this ACh must represent a cholinergic innervation from areas other than the retina and visual cortex, and is likely to be the reticular formation in view of the results discussed above. Similarly, although intracarotid injections of ACh, physostigmine, atropine, and dihydro-β-erythroidine have been shown to modify postsynaptic responses evoked in the lateral geniculate nucleus following stimulation of the optic nerve (David, Murayama, Machne, and Unna, 1963), these results can be explained in terms of effects on a cholinergic projection system from the brain-stem which controls the level of excitability of lateral geniculate neurons. In conclusion, the evidence indicates that the main excitatory transmitter released from optic nerve terminals is not ACh, although the presence of a small proportion of cholinergic fibres in the optic nerve cannot be excluded.

A small proportion of lateral geniculate neurons was depressed by ACh (Phillis and others, 1967b; Satinsky, 1967), but this action has not been related to synaptic inhibition in the lateral geniculate nucleus. Depressant actions of ACh were recorded predominantly in the most superficial layers of the lateral geniculate nucleus, and some of these neurons could be activated synaptically (1–3 msec. latency) by stimulating the visual cortex (Tebēcis, unpublished observations). However, the chemical identity of the excitatory transmitter released by corticofugal fibres (Widén and Ajmone-Marsan, 1960; Bishop, Burke, and Davis, 1962) on to lateral geniculate neurons is unknown.

Excitatory Amino-acids

Glutamate, aspartate, and other acidic amino-acids excited geniculo-cortical relay and unidentified lateral geniculate neurons (Curtis and Davis, 1962, 1963; Phillis and others, 1967a, b; Satinsky, 1967; Tebēcis and Di Maria, 1972b; Curtis and others, 1972; Morgan, Vrobová, and Wolstencroft, 1972), but it is not clear whether any one of these amino-acids functions as a transmitter in the lateral geniculate nucleus. The excitatory transmitter released from optic nerve terminals on to lateral geniculate neurons is unlikely to be glutamate, as little of this amino-acid is present in the optic nerve (Johnson and Aprison, 1971). Moreover, although the differential sensitivity of lateral geniculate neurons to glutamate and aspartate correlated with the distance of the receptive field from the centre (Morgan and others, 1972), neither L-glutamate nor L-aspartate can be the optic nerve transmitter if 5-HT and LSD act by blocking the postsynaptic action of this transmitter on lateral geniculate neurons (Bishop and others, 1959; Curtis and Davis, 1962; Tebēcis and Di Maria, 1972b), since the tryptamine derivatives and LSD blocked the orthodromic firing evoked by optic nerve impulses in concentrations which had no effect on the excitation by these amino-acids (Curtis and Davis, 1962; Tebēcis and Di Maria, 1972b). Similar conclusions may be made for other excitatory amino-acids, including D-glutamic acid, D-aspartic acid, L-cysteic acid, D-cysteic acid, *N*-acetyl-L-aspartic acid and L-cysteine sulphinic acid (Tebēcis, unpublished observations).

In addition, possible antagonists of excitant amino-acids in concentrations which reduced the excitation by glutamate, aspartate, and DLH, did not affect the spontaneous firing of lateral geniculate neurons (much of which probably results from impulses in optic nerve fibres), but the full significance of this is not clear as technical reasons may account for this lack of effect (Curtis and others, 1972). Thus, the bulk of the evidence suggests that the major optic nerve transmitter is probably not glutamate or aspartate. However, the possibility that one of these amino-acids *is* the transmitter cannot be excluded, particularly if 5-HT and LSD block the *release* of the transmitter from optic nerve terminals (Tebēcis and Di Maria, 1972b). Of course, one or more of these excitatory amino-acids may well function as transmitters released from other pathways on to lateral geniculate neurons.

Depressant Amino-acids

The lateral geniculate nucleus (of the cat) appears to be unique amongst the thalamic nuclei in that it contains high levels of taurine (Guidotti, Badiani, and Pepeu, 1972), in addition to its relatively high GABA content (Okada and others, 1971). Not only does the lateral geniculate nucleus contain 2–3 times as much taurine as the medial geniculate nucleus or the ventrobasal complex of the thalamus, but these taurine levels are the highest of any region in the cat brain, excluding the pineal and pituitary glands (Guidotti and others, 1972). The possibility that GABA and taurine are therefore involved in synaptic functions in the lateral geniculate nucleus has recently been investigated (Curtis and Tebēcis, 1972).

GABA was almost invariably a more effective depressant of lateral geniculo-cortical relay neurons than glycine, and the action of GABA was blocked by bicuculline in concentrations which had no effect on that of glycine (Curtis and Tebēcis, 1972). In particular, bicuculline reduced or blocked the inhibition of geniculo-cortical relay neurons produced by stimulating both the visual cortex and optic nerve. Results from one geniculo-cortical relay neuron are illustrated in *Fig.* 52. The neuron was excited antidromically by stimulating the visual cortex and orthodromically by stimulating the optic nerve. This is one of the rare cases in which glycine was a more effective depressant that GABA (*Fig.* 52A), but the results are illustrated to demonstrate the selectivity of antagonism by bicuculline. Electrophoretically administered bicuculline had no effect on the depressant action of glycine, but almost blocked that of GABA, and induced periods of cessation of firing (*Fig.* 52B). These effects were fully reversible (*Fig.* 52C). Similarly, the inhibition of firing produced by stimulating both the visual cortex and optic nerve were reversibly reduced by bicuculline in concentrations which suppressed the action of GABA without affecting that of glycine (*Fig.* 52D, E, F).

Conversely, electrophoretically administered strychnine blocked the action of glycine, but had little or no effect on that of GABA or on the synaptic inhibition of lateral geniculo-cortical relay neurons evoked by 'antidromic' and 'orthodromic' stimulation (Curtis and Tebēcis, 1972).

Taurine was a more effective depressant of lateral geniculo-cortical relay neurons than glycine, but was usually either less effective or comparable in potency to GABA (*Fig.* 53). The depression by taurine was blocked by both strychnine and bicuculline. Thus, since the synaptic inhibitions of geniculo-cortical relay neurons studied were resistant to strychnine, they could not have been mediated by taurine (Curtis and Tebēcis, 1972). The situation in the lateral geniculate nucleus therefore appears to be similar to

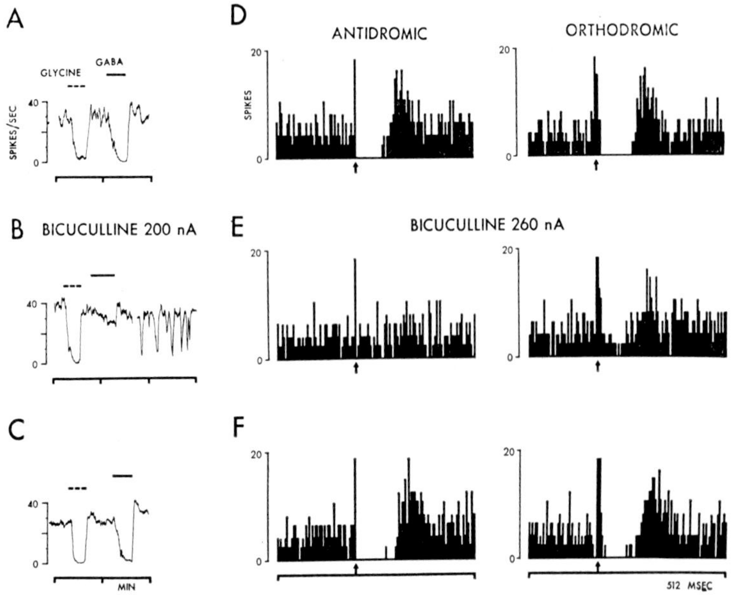

Fig. 52. Effects of bicuculline on a lateral geniculo-cortical relay neuron, fired continuously by DLH. A–C, Firing frequency of the neuron showing the depressant actions of electrophoretically administered glycine (10 nA) and GABA (22 nA), the periods of ejection being indicated by horizontal broken and solid lines. A, Control. B, 5 min. after a current of 200 nA began to eject bicuculline (5 m*M* in 165 m*M* NaCl) into the vicinity of the neuron for a total period of 13·5 min. C, 20 min. after this current was terminated. D–F, Synaptic inhibition evoked by electrical stimulation of the ipsilateral visual cortex (ANTIDROMIC, 0·1 msec., 25 V, indicated by the arrow) and the contralateral optic nerve (ORTHODROMIC, 0·2 msec., 5 V, indicated by the arrow). The histograms represent the number of action potentials in 128 consecutive periods of 4 msec., counted for a total of 40 sweeps (once every 2 sec.). D, Controls. E, 13 min. (antidromic) and 6 min. (orthodromic) after a current of 260 nA began to eject bicuculline near the neuron for a total period of 14·5 min. F, 17 and 22 min. after terminating this current. (Curtis and Tebēcis, 1972.)

that in the ventrobasal complex of the thalamus. The inhibition of lateral geniculo-cortical relay neurons resulting from impulses in recurrent collaterals (of geniculo-cortical axons) which terminate on inhibitory interneurons (Bishop and Davis, 1960; Fuster, Creutzfeldt, and Straschill, 1965; Burke and Sefton, 1966a, b; cf. Vastola, 1959) is probably mediated by GABA.

The role of taurine in the lateral geniculate nucleus remains a mystery. Its high concentration in this nucleus (Guidotti and others, 1972), together with the pharmacological findings (Curtis and Tebēcis, 1972) raise the possibility that taurine may be a transmitter mediating an inhibition sensitive to both strychnine and bicuculline in the lateral geniculate nucleus. Alternatively, the presence of taurine in various types of tissue, including non-nervous structures (Tallan, 1962), may mean that the taurine levels in the lateral geniculate nucleus are not primarily involved in synaptic inhibition, but are more important in the metabolism

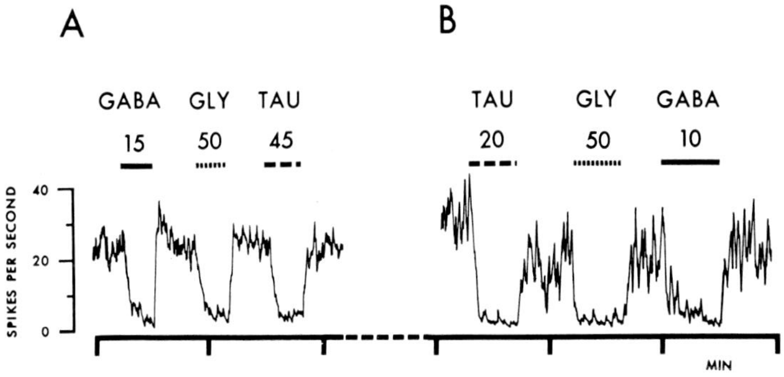

Fig. 53. Comparison of the depressant actions of GABA, glycine (GLY), and taurine (TAU) on the firing of two different lateral geniculate neurons (A and B) identified as geniculo-cortical relay cells. All three amino-acids were administered electrophoretically with cationic currents for the periods indicated by the horizontal bars (figures refer to ejecting currents, in nA). The chart recordings show the firing frequency with time. A, The spontaneous firing was increased and maintained at 20–30 spikes/sec. by DLH (—5 nA) ejected continuously. The antidromic spike of this neuron followed repetitive stimulation of the optic radiation at rates of up to 625/sec. B, Firing was maintained by DLH (—6·5 nA). The antidromic spike followed up to 500/sec. Both neurons were recorded in animals anaesthetized with pentobarbitone sodium. (Curtis and Tebēcis, unpublished observations.)

of sulphur-containing amino-acids. This problem has yet to be resolved.

Optic Nerve Transmitter

From the above discussion, it is clear that the main excitatory transmitter released from optic nerve fibres on to lateral geniculate neurons is not ACh or any of the better-known monoamines (5-HT NA, DA), and is unlikely to be the excitatory amino-acids (glutamate, aspartate) thought to be transmitters elsewhere in the C.N.S. This suggests that the optic nerve transmitter is either an unusual substance hitherto not considered to be a likely transmitter, or a substance which has not yet been characterized chemically. Studies with LSD suggest that the excitatory transmitter released from brachium conjunctivum fibres on red nucleus neurons (p. 79) is the same as the optic nerve transmitter. It is interesting in this respect that the optic nerve of the rabbit has been reported to contain the highest levels of ergothioneine of various C.N.S. regions studied (Crossland and others, 1964, 1966), and ergothioneine has been shown to excite central neurons (Avanzino, Bradley, and others, 1966). Unfortunately, a study of the distribution of ergothioneine in the ox brain (Briggs, 1972) has not revealed any pattern peculiar to the optic nerve or any other region in the brain. Moreover, electrophoretically administered ergothioneine did not excite any of the few lateral geniculo-cortical relay neurons tested in cats anaesthetized with pentobarbitone sodium (Tebēcis, unpublished observations). Thus, the identity of the optic nerve transmitter still remains an open question.

Medial Geniculate Nucleus

The medial geniculate nucleus, the thalamic nucleus of the auditory pathway, consists of two main types of neurons—principal cells whose axons project to the auditory cortex, and short-axon or Golgi type II cells—although further subdivisions may be made as for the lateral geniculate nucleus. The medial geniculate nucleus may be divided into dorsal, ventral, and medial areas, and within each division several discrete, relatively homogeneous populations of neurons may be distinguished (Morest, 1964). The major afferent projection to the medial geniculate nucleus (medial and ventral divisions) originates in the inferior colliculus (Woollard and Harpman, 1940; Moore and Goldberg, 1963; Jones and Rockel, 1971), although some afferents also arise in many other structures, including other areas of the brain-stem, cerebellum, and cerebral cortex (Ades, 1941; Mickle and Ades, 1954; Walther and Rasmussen, 1960; Altman and Carpenter, 1961; Morest, 1965; Goldberg and Moore, 1967; Majorossy and Réthelyi, 1968; Jones and Powell, 1969b). The principal efferent outflow projects to the auditory cortex (Mettler, 1932; Ades, 1941; Rose and Woolsey, 1949; Locke, 1961). The synaptic organization of the medial geniculate nucleus is similar

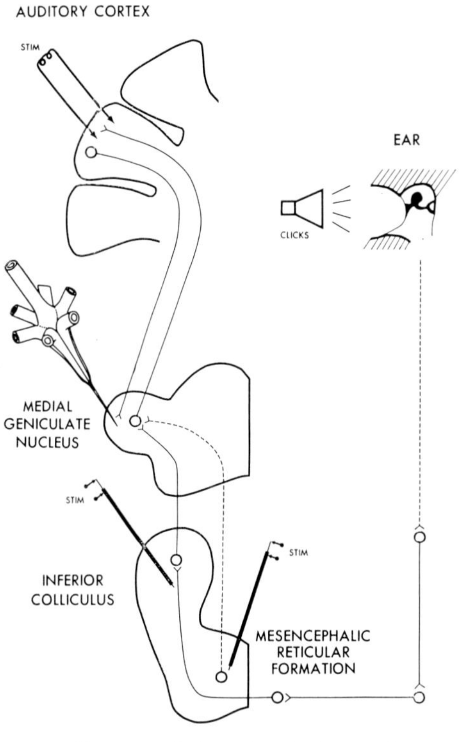

Fig. 54. Schematic diagram of the stimulating and recording arrangement that has been used in pharmacological studies of single medial geniculate neurons. The pathways in which the number of synapses involved is unknown are represented by broken lines. A grid of several stainless steel needles, insulated except for the tips, is used to stimulate the auditory cortex, and concentric bipolar stimulating electrodes are inserted into the inferior colliculus and mesencephalic reticular formation. In addition, sound stimuli can be delivered to the ears by earphones attached to hollow ear-bars. (Tebēcis, 1972c.)

to that of the lateral geniculate nucleus, the most important synaptic sites being on the dendrites, arranged in synaptic clusters (Majorossy and Réthelyi, 1968; Jones and Powell, 1969a, b; Jones and Rockel, 1971).

Geniculo-cortical Relay Neurons

Medial geniculo-cortical relay neurons (principal cells) may be identified electrophysiologically by their antidromic responses to stimulation of the acoustic radiation, and orthodromic responses to stimulation of the inferior colliculus or ears with clicks and tones. The stimulating and recording arrangement used in pharmacological studies of single medial geniculate neurons is shown diagrammatically in *Fig.* 54.

Acetylcholine

Virtually all medial geniculo-cortical relay neurons were excited by ACh, the excitation sometimes being either preceded or followed by a depression (Tebēcis, 1970a, 1972c). An example of this is given in *Fig.* 55, which illustrates the relatively slow time course of excitation by ACh. Studies with muscarinic and nicotinic cholinomimetics and antagonists of ACh indicated that geniculo-cortical relay neurons have both muscarinic and nicotinic properties, the muscarinic properties usually predominating. Since both atropine and dihydro-β-erythroidine could block completely the excitation of the same relay cell by ACh, it was possible to conclude that the cholinoceptive receptors are not purely muscarinic and purely nicotinic in type (distributed differently on the one neuron) but of mixed muscarinic-nicotinic type (Tebēcis, 1972c; cf. ventrobasal thalamic and lateral geniculate neurons). Cholinesterase inhibitors potentiated the excitant action of ACh, and in higher concentrations sometimes blocked this excitation, similar to the effects of atropine and dihydro-β-erythroidine (Tebēcis, 1970a, 1972c). The possibility that ACh is an excitatory transmitter in the medial geniculate nucleus was suggested by the finding that ACh excited most relay cells, whereas it had various or no actions on many unidentified medial geniculate neurons. Various forms of synaptic excitation in the medial geniculate nucleus were therefore tested with antagonists of ACh and anticholinesterases.

a. Reticulo-geniculate Fibres

As observed in the ventrobasal complex (McCance, Phillis, and Westerman, 1968a) and lateral geniculate nucleus (Phillis and others, 1967b), stimulation of the mesencephalic reticular formation excited a proportion of medial geniculate neurons, and this synaptic excitation and the excitant action of ACh were reversibly blocked by antagonists of ACh, and potentiated by cholinesterase inhibitors (Tebēcis, 1970b, 1972c). Results from one medial geniculo-cortical relay neuron are illustrated in *Fig.* 56. Stimulation of the auditory cortex evoked an antidromic action potential which was sometimes followed by a synaptic spike (*Fig.* 56A), whereas a longer-latency synaptic response was evoked from the same neuron following single stimuli to the mesencephalic reticular formation (*Fig.* 56B). ACh excited the neuron, the excitation being of slow time course (*Fig.* 56C). Electrophoretically administered atropine, in concentrations which did not affect the spontaneous firing or the ortho- and antidromic responses evoked from the cortex, blocked the excitation by ACh and stimulation of the mesencephalic reticular formation, and these remained blocked for several minutes after the atropine ejection had been terminated (*Fig.* 56D, E, F). Recovery from the effect of atropine lasted many minutes (*Fig.* 56G, H, I).

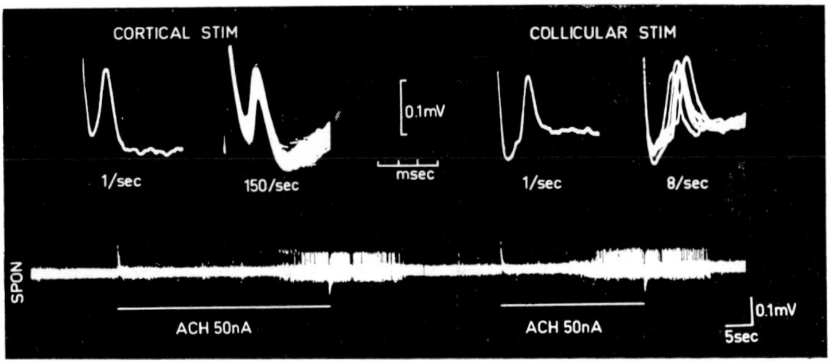

Fig. 55. Excitant action of acetylcholine (ACH) on a medial geniculate neuron identified as a geniculo-cortical relay cell. The records on the top left show that stimulation of the auditory cortex evoked an action potential of short and constant latency, and which followed repetitive stimuli at 150/sec., thus being characterized as an antidromic spike. The records on the right show that stimulation of the inferior colliculus evoked a similar spike from the neuron, but which varied in latency to repetitive stimuli. The spike failed to follow collicular stimulation at frequencies higher than 10/sec. The 'moving' film record illustrates the firing produced by electrophoretically administered ACh (50 nA) on the background activity of this neuron. (Tebēcis, 1970a.)

The most successful region from which cholinergic responses could be evoked from medial geniculate neurons was the mesencephalic reticular formation, although the precise localization of the appropriate afferent fibres was not determined (Tebēcis, 1972c). However, the position of the tip of the stimulating electrode was in the vicinity of the cuneiform nucleus, and it is

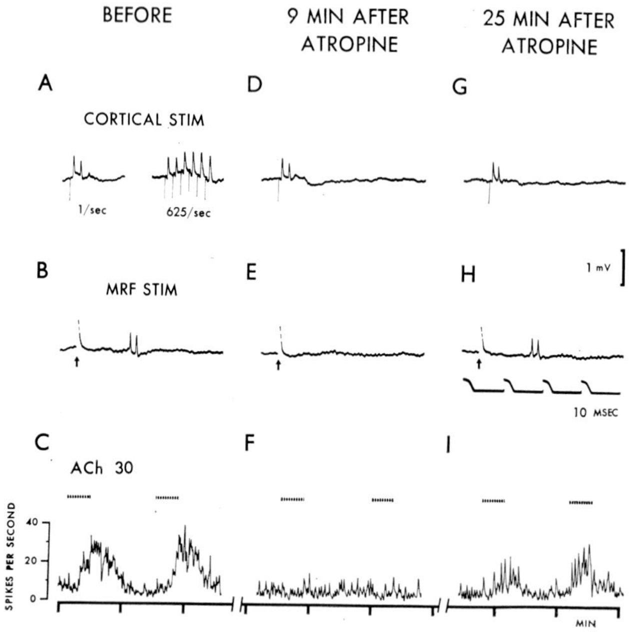

Fig. 56. Antagonism by atropine of the excitant action of ACh and synaptic excitation from the mesencephalic reticular formation (MRF; anterior −2 mm., lateral +2 mm., vertical −3 mm.) of a medial geniculo-cortical relay neuron in an unanaesthetized cat. A, Stimulation (8 V, 0·1 msec. duration) of the auditory cortex (area A1) evoked an antidromic spike which was sometimes followed by an orthodromic spike. B, Stimulation of the MRF (7·5 V, 0·1 msec. duration, 1/sec., onset of stimulus indicated by an upward arrow) evoked a synaptic response of 13–16 msec. latency. C, Chart recording showing the increase in firing rate by electrophoretic ejections of ACh (30 nA). A, B, C, Shortly before atropine was administered electrophoretically. D, E, F, Approximately 9 min. after atropine (10 m*M* in 165 m*M* NaCl; 50–150 nA) had been administered for a total duration of 10 min. G, H, I, 25 min. after the current ejecting atropine had been terminated. (Tebēcis, 1972c).

noteworthy that histochemical studies of the rat brain have revealed a cholinesterase-containing projection from the cuneiform nucleus to the geniculate nuclei (Shute and Lewis, 1967). The results indicated that ACh is an excitatory transmitter in the feline medial geniculate nucleus, most probably released from fibres which originate in or traverse the mesencephalon (Tebēcis, 1972c). Evidence for the presence of ACh, choline acetyltransferase, and AChE in the mammalian medial geniculate nucleus (MacIntosh, 1941; Feldberg, 1945; Feldberg and Vogt, 1948; Hebb and Silver, 1956; Shute and Lewis, 1967) is compatible with this view.

b. Auditory Afferents

Although it was originally thought that a proportion of fibres from the inferior colliculus to the medial geniculate nucleus is cholinergic (Tebēcis, 1970b), these responses were probably evoked by the spread of stimulus to the neighbouring reticular formation (possibly the cuneiform nucleus, which lies immediately below the colliculi), as responses of monosynaptic latency evoked by stimulating the inferior colliculus, and responses evoked by clicks, were not affected by antagonists of ACh (Tebēcis, 1972c). Thus, the auditory afferent fibres which terminate on medial geniculate neurons are probably largely or wholly non-cholinergic.

c. Cortico-geniculate Fibres

Orthodromic responses of monosynaptic latency evoked from medial geniculate neurons following stimulation of the auditory cortex were resistant to antagonists of ACh (*Fig.* 56, for example), or ACh depressed the firing of these cells, indicating that the excitatory transmitter released from corticofugal fibres on to medial geniculate neurons is unlikely to be ACh (Tebēcis, 1972c). As mentioned earlier for the ventrobasal complex, from these experiments it is impossible to conclude whether the orthodromic responses of monosynaptic latency were evoked by stimulation of cortico-geniculate fibres, or of relay cell axons which sent recurrent collaterals to the geniculate neurons from which recordings were made. However, since the corticofugal projection to the thalamus is relatively dense (Jones and Powell, 1968), it is probable that some of the responses, at least, were mediated by cortico-geniculate fibres (cf. McCance, Phillis, and Westerman, 1968a). The observation that some polysynaptic responses of medial geniculate neurons following stimulation of the auditory cortex could be blocked by atropine (Tebēcis, 1970b; 1972c) is most readily explained by assuming an indirect activation of geniculate neurons via cholinergic fibres from the brain-stem.

d. Spontaneous Firing

Atropine and dihydro-β-erythroidine had no apparent effect on the 'spontaneous' firing of medial geniculate neurons (*Fig.* 56, for example), suggesting that the tonic excitatory bombardment of these neurons is largely or wholly mediated by a transmitter other than ACh (Tebēcis, 1972c). The origin of these fibres is unknown, but since the spontaneous activity of medial geniculate neurons was similar in both intact and brain-stem transected cats, it is reasonable to conclude that the excitatory neurons are located rostral to the pons.

e. Cholinergic Inhibition ?

The depressant action of ACh, which sometimes preceded or followed the excitation of medial geniculo-cortical relay cells, was never reduced by atropine and dihydro-β-erythroidine in concentrations which blocked the excitant action of ACh (Tebēcis, 1972c). However, ACh depressed the firing of a number of unidentified medial geniculate neurons in the absence of any excitation, and this depression was blocked by both atropine and dihydro-β-erythroidine, and enhanced by the cholinesterase inhibitors (Tebēcis, 1970a, 1972c). It is possible that these latter cholinoceptive receptors are related to synaptic inhibition in the medial geniculate nucleus, but this possibility requires investigation (*see* Phillis, 1971).

MONOAMINES

The monoamines, 5-HT, NA, and DA, depressed the firing of the majority of unidentified medial geniculate neurons, although NA, and to a lesser extent 5-HT, also excited small proportions (Tebēcis, 1967, 1970c). Subsequently, it was found that 5-HT routinely depressed the firing of medial geniculo-cortical relay cells in cats anaesthetized with halothane and nitrous oxide (Tebēcis, unpublished observations). In particular, 5-HT did not selectively depress the orthodromic firing of medial geniculate neurons following click stimulation, but depressed all forms of excitation as expected of a substance that has only postsynaptic depressant effects. This is similar to the ventrobasal complex of the thalamus (Andersen and Curtis, 1964b; Phillis and Tebēcis, 1967c), but in contrast to the lateral geniculate nucleus (Tebēcis and Di Maria, 1972b), further supporting the suggestion that the transmitter released from optic nerve terminals is an 'unusual' substance.

The presence of 5-HT- and NA-containing nerve-fibres in the medial geniculate nucleus that arise in the brain-stem (Fuxe, 1965; Fuxe and others, 1970; Olson and Fuxe, 1971; Ungerstedt, 1971) raises the possibility that these two monoamines are transmitters in this nucleus (Tebēcis, 1967, 1970c). Repetitive stimulation of the mesencephalic reticular formation often inhibited the firing of

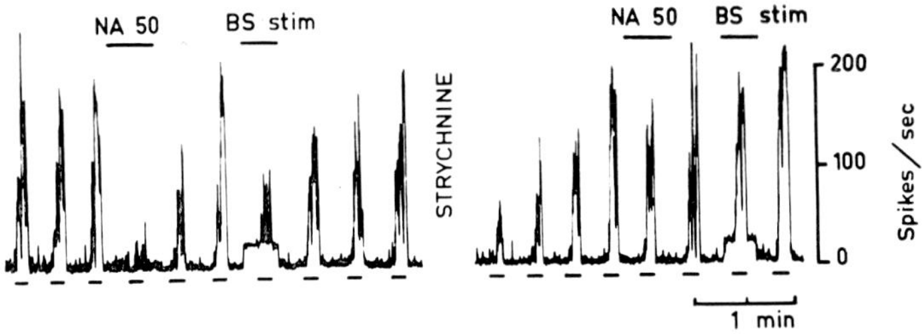

Fig. 57. Effect of strychnine on the depressant action of NA and the inhibitory effect of brain-stem (BS) stimulation on the firing of a medial geniculate neuron. The chart recording shows the firing frequency of the neuron with time. The horizontal bars below the trace represent electrophoretic ejections of L-glutamate (50 nA). The gap in the records represents an application of strychnine (60 nA for 1 min.). Both the action of NA (50 nA) and that of BS stimulation (anterior +3 mm., lateral +3 mm., vertical −1 mm.; 9 V at 30/sec.) were reduced after strychnine. (Tebēcis, 1967.)

medial geniculate neurons, and this synaptic inhibition, as well as the depressant actions of 5-HT and NA, were blocked by strychnine (Tebēcis, 1967, 1970c). Results from one such neuron are illustrated in *Fig.* 57. The significance of this observation is not clear in view of the abundant evidence that strychnine is a relatively specific antagonist of glycine on central neurons (*see* other chapters). In the medial geniculate nucleus, also, strychnine blocked the action of glycine in lower concentrations than required to reduce the depression by NA and 5-HT (Tebēcis, 1970c). It is therefore possible that the high concentrations of strychnine used to reduce the synaptic inhibition evoked from the brain-stem (Tebēcis, 1967) resulted in non-specific blockade. On the other hand, since strychnine has been shown to reduce the depressant actions of NA and 5-HT in concentrations which do not affect that of GABA (Tebēcis, 1970c), it is nevertheless possible that strychnine interfered with monoaminergic pathways (Tebēcis, 1967). This problem requires reinvestigation, preferably employing single stimuli to the brain-stem to evoke inhibition in the medial geniculate nucleus.

Amino-acids

Glutamate, aspartate, and DLH excited, whereas GABA and glycine depressed the firing of medial geniculate neurons, including relay cells (Tebēcis, 1967, 1970a, b, c, 1972c). GABA was more effective than glycine as a depressant, and the depression by glycine was blocked by strychnine, which had no effect on that of GABA (Tebēcis, 1970c). However, the actions of these amino-acids have not been related to synaptic pathways in the medial geniculate nucleus, although it is likely that at least some of the amino-acid transmitters operating in this nucleus are identical to those operating at comparable pathways in the ventrobasal complex.

Hypothalamus

The hypothalamus is an extremely complex 'structure', consisting of a number of diverse groups of cells situated below the thalamus, in the walls and floor of the third ventricle, extending from the mammillary bodies to a region slightly rostral to the optic chiasm. Amongst the hypothalamic regions which can be recognized morphologically are the preoptic, supraoptic, paraventricular, suprachiasmatic, tuberal, lateral, posterior hypothalamic, and mammillary nuclei. However, it should be borne in mind that the boundaries of hypothalamic nuclei are ill defined, and both the number and terminology of such nuclei vary considerably between authors (cf. *Fig.* 61). This is not surprising considering the large number and diversity of complex functions (including regulation of body temperature, water balance, milk secretion, adenohypophyseal function, food intake, gastric acid secretion, emotion) in which the hypothalamus is involved. For this reason, studies of the actions of possible transmitters on unidentified hypothalamic neurons have provided little more than an indication of what some transmitters might be 'somewhere in the hypothalamus'. Pharmacological investigations of physiologically, or at least, anatomically identified hypothalamic neurons have been far more valuable in contributing to the understanding of the functions of different regions of the hypothalamus. Before dealing with identified hypothalamic neurons, it is useful first to mention some biochemical and histochemical studies involving ACh and the monoamines in the hypothalamus, as various findings suggest that these substances are involved in synaptic functions in this structure.

Acetylcholine and Monoamines in the Hypothalamus

The results of various biochemical and histochemical studies of different mammals have indicated that ACh and monoamines (5-HT, NA, DA, histamine), as well as the enzyme systems for their metabolism, are present in the hypothalamus (MacIntosh, 1941; Burgen and Chipman, 1951; Vogt, 1954; Hebb and Silver, 1956; Bogdanski and others, 1957; Shimizu and others, 1959; Kuntzman and others, 1961; Carlsson, Falck, and Hillarp, 1962; McGeer and others, 1963; Manocha and others, 1967; Martini and Meites, 1970). At least a proportion of the 5-HT and NA content of the hypothalamus is associated with nerve-endings (Kuhar, Shaskan, and Snyder, 1971). Although the role of histamine in the C.N.S. is not clear, it is interesting that the hypothalamus contains the highest levels of histamine and histamine methyltransferase of several brain structures analysed, and a proportion of this histamine content is associated with synaptosomes similar to those that store NA and 5-HT (Kuhar, Taylor, and Snyder, 1971; Taylor and others, 1972). Ascending 5-HT- and NA-containing nerve-fibres, which originate in the mesencephalon of the rat, project to the hypothalamus (Andén and others, 1966; Fuxe and others, 1970; *see also* Hökfelt and Ljungdahl, 1971b). In addition, the external layer of the median eminence and the proximal part of the infundibular stem are rich in DA-containing nerve terminals, which are thought to arise from the DA-containing cell bodies of the arcuate nuclei and the ventral portion of the anterior periventricular nuclei (Fuxe, 1964; Fuxe and others, 1970; Ungerstedt, 1971; Jonsson, Fuxe, and Hökfelt, 1972). Stimulation of certain hypothalamic sites, which produced sham rage, increased the concentration of NA and metabolites in ventricular perfusates in the cat (Phillipu, Heyd, and Burger, 1970; Sweet and Reis, 1971). Lesions which interrupted the ascending fibres arising in the midbrain led to a decline in the content of 5-HT, but not that of NA, in the hypothalamus of the cat (Parent,

Saint-Jacques, and Poirier, 1969). The spontaneous release of 5-HT in the hypothalamus of the rhesus monkey occurred predominantly in the rostral and caudal portions, whereas ACh was released principally in the caudal area (Myers and Beleslin, 1970). It has been suggested that the regulation of body temperature is due to a fine balance in the release of 5-HT and NA in the rostral region, whereas ACh may be involved in both regulation of body temperature and ingestive behaviour (Myers and Beleslin, 1970).

These are only some of the biochemical and histochemical studies suggesting functions for ACh and the monoamines in the hypothalamus. There is little point in reviewing all such findings in a monograph of this type, as insufficient data are available to correlate the results of the various studies, especially as there are likely to be differences between species. Nevertheless, these findings provide a strong basis for considering that ACh and certain monoamines function as transmitters in particular areas of the mammalian hypothalamus.

The first pharmacological study demonstrating direct actions on hypothalamic neurons was made by Bloom, Oliver, and Salmoiraghi (1963), who observed that ACh, NA, and 5-HT excited or depressed the firing of many unidentified hypothalamic neurons. ACh excited and depressed approximately equal proportions of neurons, whereas NA and 5-HT had mainly depressant actions. Similar studies were made subsequently of neurons in the lateral and ventromedial hypothalamus (Oomura, Ooyama, Yamamoto, Ono, and Kobayashi, 1969). ACh increased more often than it decreased the mean discharge frequency of neurons in the lateral hypothalamus, whereas the reverse was true for neurons in the ventromedial hypothalamus. NA excited most cholinoceptive neurons in the lateral hypothalamus, but excited as often as it depressed the firing of ventromedial hypothalamic neurons. 5-HT usually depressed the firing of neurons in the lateral hypothalamus. In addition, this was the first study revealing excitant actions of glutamate and depressant actions of GABA on hypothalamic neurons (Oomura and others, 1969). Only recently have the effects of suspected transmitters been related to functionally distinct types of hypothalamic neurons.

Thermally Sensitive Hypothalamic Neurons

The activity of thermally sensitive neurons in the preoptic and anterior regions of the hypothalamus correlated well with the effects of ACh, NA, and 5-HT (Beckman and Eisenman, 1970). Thermodetectors (defined as hypothalamic neurons which increased their firing exponentially with increasing temperature) were relatively insensitive to ACh, NA, and 5-HT, but were excited by cationic currents (including Na^+), suggesting a lack of aminergic synaptic input to these cells. Thermosensitive interneurons were relatively insensitive to current flow alone, but often responded to the amines. In general, ACh excited, NA depressed, and 5-HT had no action on warm-sensitive interneurons. Cool-sensitive interneurons were unaffected by ACh, excited by NA, and (in 1 case) depressed by 5-HT. It would be interesting to know whether these generalizations would hold for a large sample of hypothalamic neurons of these areas, as this may provide a clue to the mechanisms involved in temperature regulation.

Supraoptic and Paraventricular Neurons

A unique feature of the hypothalamic neuroendocrine system of vertebrates, originally suggested by Scharrer (1928), is that certain cells behave both as neurons and hormone-secreting cells. Neurosecretory cells of the supraoptic and paraventricular nuclei give rise to unmyelinated nerve-fibres, which project to the neurohypophysis and apparently release both transmitter substances (unknown) and hormones (antidiuretic hormone and oxytocin) (Bargmann and Scharrer, 1970; *see also* Cross and Green, 1959; Kandel, 1964; Brooks, Ishikawa, Koizumi, and Lu, 1966; Yagi, Azuma, and Matsuda, 1966; Yamashita, Koizumi, and Brooks, 1970; Barker, Crayton, and Nicoll, 1971a; Dreifuss, Kalnins, Kelly, and Ruf, 1971; Dyball, 1971; Dreifuss and Kelly, 1972a; Koizumi and Yamashita, 1972). The axons of these cells terminate close to blood capillaries and contain both neurosecretory granules (the main intracellular store for neurohypophyseal hormones) and synaptic vesicles (Palay, 1957; Gerschenfeld, Tramezzani, and de Robertis, 1960; Rufener and Dreifuss, 1970; Christ and Bak, 1970).

Stimulation of vagal and carotid sinus nerve afferents results in both excitation and inhibition of the firing of supraoptic neurons, and this activation is thought to control the release of hormones from neurosecretory cells (Brooks, Ushiyama, and Lange, 1962; Cross and Silver, 1966; Dyball and Koizumi, 1969; Barker and others, 1971a; Dyball, 1971). Although various studies have suggested that adrenergic inhibition and cholinergic excitation control the release of antidiuretic hormone, only recently have the direct effects of ACh and NA on single supraoptic and paraventricular neurons been investigated (Dreifuss and Kelly, 1970, 1972b; Barker and others, 1971b, c; Cross, Moss, and Urban, 1971; Moss, Dyball, and Cross, 1971).

Acetylcholine

There is disagreement as to the predominant action of ACh on supraoptic and paraventricular neurons. ACh excited the majority of neurosecretory cells identified by antidromic stimulation of the pituitary stalk in rats and rabbits, and this excitation was mediated mainly by nicotinic receptors, as it could be readily blocked by dihydro-β-erythroidine, but antagonized only with difficulty by atropine (Dreifuss and Kelly, 1970, 1972b; Cross and others, 1971; Moss and others, 1971). In a study of a larger sample of supraoptic neurons in both rats and cats, however, ACh depressed the firing of most neurons, and excited only a small proportion (Barker and others, 1971b, c). Studies with other cholinomimetics (carbamylcholine, acetyl-β-methylcholine, nicotine) and antagonists

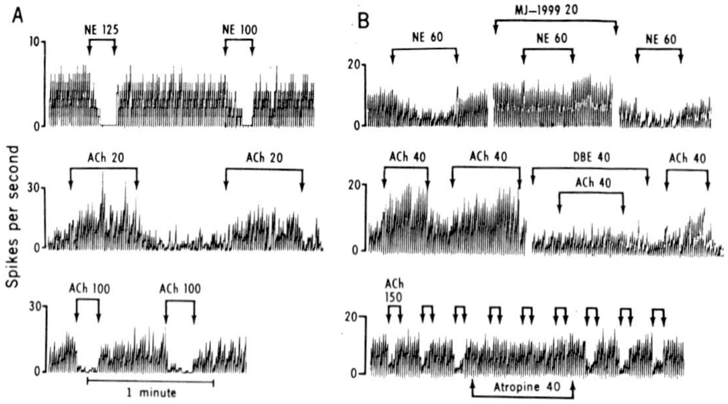

Fig. 58. Responses of neurons in the supraoptic nucleus to noradrenaline (NE) and ACh. The records illustrate the change in firing frequency (spikes/sec.) with time, using 1-sec. integrations. The values of the currents (nA) used to eject the substances electrophoretically are indicated by the figures. A, Actions of NA and ACh on three different, antidromically identified neurosecretory cells. B, Records obtained from three different supraoptic neurons which could not be activated antidromically. Abbreviation: DBE (dihydro-β-erythroidine). (Barker and others, 1971b.)

(atropine, dihydro-β-erythroidine) indicated that the depressant action of ACh on neurosecretory cells was mediated by muscarinic receptors, whereas the excitant action was mediated by nicotinic receptors. Results from several supraoptic neurons are illustrated in *Fig.* 58 (Barker and others, 1971b).

The reasons for the discrepancies between laboratories are not clear, but there are several possibilities, including variability between species, differences between supraoptic and paraventricular neurons, the criteria used for excitation and depression, and possibly differences between experimental techniques. Nevertheless, all agree that the cholinoceptive neurosecretory cells probably indicate the existence of cholinergic mechanisms involved in the release of hormones (Barker and others, 1971b, c; Dreifuss and Kelly, 1970, 1972b; Cross and others, 1971; Moss and others, 1971).

Noradrenaline

Biochemical, fluorescence histochemical, and electron microscopy combined with autoradiographic studies have revealed that the supraoptic and paraventricular nuclei are particularly rich in NA (which is associated with nerve terminals), but relatively poor in 5-HT (Bertler and Rosengren, 1959; Falck, 1964; Aghajanian and Bloom, 1967; Aghajanian and others, 1969). NA depressed the firing of almost all the neurosecretory cells tested, although excitation by the catecholamine has also been observed occasionally (Barker and others, 1971b, c; Cross and others, 1971; Moss and others, 1971). The depression by NA was reversibly blocked by MJ-1999, a β-adrenergic antagonist, and potentiated by desmethylimipramine, a blocker of NA uptake (Barker and others, 1971b, c). Some of these observations are illustrated in *Fig.* 58.

An evaluation of these data with previous findings suggests that NA-containing nerve terminals on neurosecretory cells represent noradrenergic inhibitory synapses (Barker and others, 1971c). 5-HT and DA also depressed the firing of neurosecretory cells (Barker and others, 1971c), but the significance of this is not known.

Recurrent Inhibition of Supraoptic Neurons

The firing of neurosecretory cells can be inhibited for several tens of msec. following stimulation of their axons, and this inhibition is thought to be the result of impulses in recurrent axon collaterals (Kandel, 1964; Yamashita and others, 1970; Barker and others, 1971a; Dreifuss and Kelly, 1972a; Koizumi and Yamashita, 1972). The identity of this transmitter, however, has escaped detection. Although ACh and NA depress the firing of many supraoptic neurons, neither substance can be the transmitter responsible for recurrent inhibition of these cells, as the recurrent

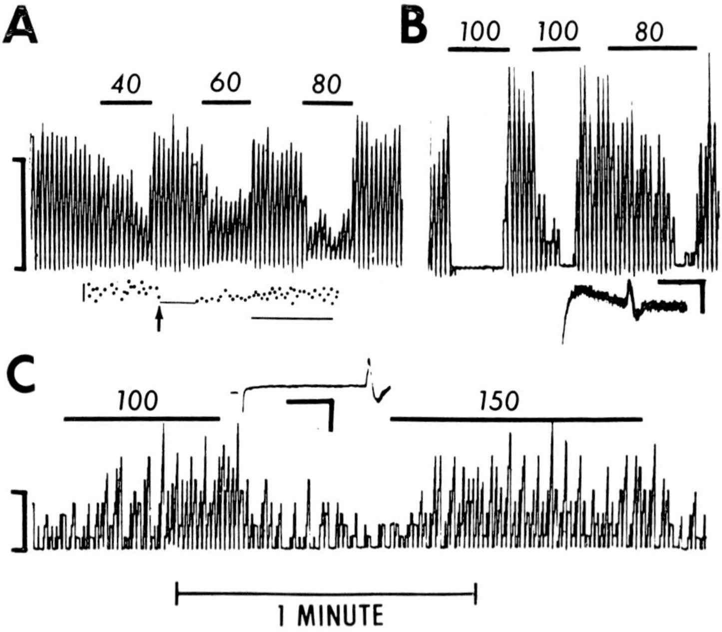

Fig. 59. Responses of three different supraoptic neurosecretory cells (A–C) to microelectrophoretically administered antidiuretic hormone (duration of administration indicated by horizontal bars, figures refer to ejecting currents, in nA). A, Dose-response depression of firing. This cell was inhibited recurrently (histogram below record; onset of stimulus indicated by an upward arrow; *calibration*: 5 counts/address; 100 msec.). Calibrations for the chart recordings in A and B are 0–20/sec. (ordinate) and 1 min. (abscissa). B, Depression of firing by antidiuretic hormone (100 nA and 80 nA) of a cell which was excited antidromically (antidromic spike below trace; *calibration*: 100 μV and 5 msec.). C, Excitation of a neurosecretory cell by antidiuretic hormone (100 and 150 nA). *Ordinate*: firing frequency (0–5/sec.). The antidromic spike is shown above the chart recording. *Calibration*: 4 mV and 2 msec. (Nicoll and Barker, 1971.)

inhibition was unaffected by atropine or pretreatment with 6-hydroxydopamine (Nicoll and Barker, 1971). Glycine and GABA readily depressed the firing of supraoptic neurons, GABA usually being more effective than glycine (Nicoll and Barker, 1971). Strychnine blocked the action of glycine, but did not reduce that of GABA or recurrent inhibition, ruling out the possibility that the inhibition is mediated by glycine. Picrotoxin and bicuculline blocked the action of GABA without affecting that of glycine, but apparently did not have a marked effect on recurrent inhibition; the convulsants only reduced this inhibition in concentrations which produced generalized seizures and bursting of neurosecretory cells. Although Nicoll and Barker (1971) could not rule out the possibility that GABA is involved in the recurrent inhibition of supraoptic neurons, they preferred the possibility that antidiuretic hormone might mediate this inhibition.

Electrophoretically administered antidiuretic hormone depressed the firing of the majority (80 per cent) of supraoptic neurons which were invaded antidromically or inhibited by stimulating the neurohypophysis. This was in contrast to cortical neurons, most (approximately 90 per cent) of which were *excited* by antidiuretic hormone (Nicoll and Barker, 1971). *Fig.* 59 illustrates the depressant actions of antidiuretic hormone on the firing of two different supraoptic neurons (A and B) which were inhibited (A) and antidromically activated (B) by stimulation of the neurohypophysis. The rare observation that antidiuretic hormone can excite some neurosecretory cells is shown in *Fig.* 59C.

The hypothesis that supraoptic nerve terminals release antidiuretic hormone synaptically on to other supraoptic neurons via axon collaterals is compatible with the fact that this hormone is released from supraoptic axons in the neurohypophysis (Nicoll and Barker, 1971). However, this view can only hold if recurrent inhibition is mediated directly, not via interneurons. Recently Koizumi and Yamashita (1972) recorded repetitive firing in the supraoptic nucleus following pituitary stimulation, suggesting the presence of interneurons which are excited by collaterals of supraoptic axons. Moreover, the presence of small cells (resembling Renshaw cells) has been observed in this area (*see* Discussion by Koizumi and Yamashita, 1972). Furthermore, if antidiuretic hormone is released as a transmitter from supraoptic nerve terminals, why do these terminals contain both neurosecretory granules and synaptic vesicles (Palay, 1957; Gerschenfeld and others, 1960; Christ and Bak, 1970; Rufener and Dreifuss, 1970)? Clearly, more experiments are required to test the hypothesis that antidiuretic hormone mediates synaptic inhibition in the supraoptic nucleus, which, nevertheless, is a provocative suggestion.

The observations that GABA is a strong depressant of supraoptic neurons, and that picrotoxin and bicuculline reduce recurrent inhibition of these neurons (Nicoll and Barker, 1971), strongly suggest that GABA is the transmitter mediating this inhibition, despite the finding that convulsive doses of the antagonists were required to reduce the inhibition. It is not uncommon to find GABA-mediated inhibitions elsewhere in the C.N.S. that require convulsant doses of picrotoxin and bicuculline (when administered intravenously) merely to reduce the synaptic inhibitions (*see* references in other chapters). The mammalian hypothalamus contains amongst the highest levels of GABA in the C.N.S. (Singh and Malhotra, 1962; Kržalić and others, 1962; Fahn and Côté, 1968a; de Feudis, Delgado, and Roth, 1970; Okada and others, 1971; *see also* Hökfelt and Ljungdahl, 1971b), a proportion of which is contained in nerve-endings (Hökfelt and others, 1970). In addition, GABA is taken up extremely readily by slices of the rat hypothalamus (Hökfelt and others, 1970), and a relatively large proportion of nerve terminals in the hypothalamus accumulate radioactive GABA (Iversen and Bloom, 1972). These findings support the suggestion that GABA is an inhibitory transmitter in this structure.

Neurons of the Suprachiasmatic Nucleus

The suprachiasmatic nucleus of the hypothalamus is especially interesting because it contains the highest density of 5-HT-containing nerve terminals of all the areas studied in the mammalian C.N.S. (Fuxe, 1965; Fuxe and others, 1970). In addition, studies involving electron microscopy and autoradiography have shown that nerve terminals in the suprachiasmatic nucleus accumulate labelled 5-HT (Aghajanian and Bloom, 1967). Moreover, lesions in the raphe nuclei have resulted in terminal degeneration in the suprachiasmatic nucleus (Aghajanian and others, 1969). Electrophoretically administered 5-HT depressed the spontaneous and glutamate-evoked firing of suprachiasmatic neurons, which could also be inhibited for long periods (50–150 msec.) by single stimuli to the raphe nuclei (Bloom, Hoffer,

Siggins and others, 1972). *Fig.* 60 illustrates some of these observations. Thus, although antagonism of synaptic inhibition in the suprachiasmatic nucleus by an antagonist of 5-HT has not yet been demonstrated, all the available findings strongly suggest that 5-HT is an inhibitory transmitter in this nucleus.

Steroid-sensitive Hypothalamic Neurons

Dexamethasone phosphate, a corticosteroid which suppresses the secretion of ACTH, had effects on a small proportion of unidentified neurons in the hypothalamus and midbrain (Ruf and Steiner, 1967; Steiner, Pieri, and Kaufmann, 1968; Steiner, Ruf, and Akert, 1969; Steiner, 1970). The firing of the large majority of these steroid-sensitive neurons was depressed by dexamethasone phosphate, although some cells were excited. ACh excited the majority of steroid-sensitive neurons, whereas NA and DA usually depressed their firing.

The significance of these observations would normally be difficult to evaluate, as the method of microelectrophoresis is notorious for enabling 'meaningful statistical' data to be derived from any number of unrelated observations. However, the studies

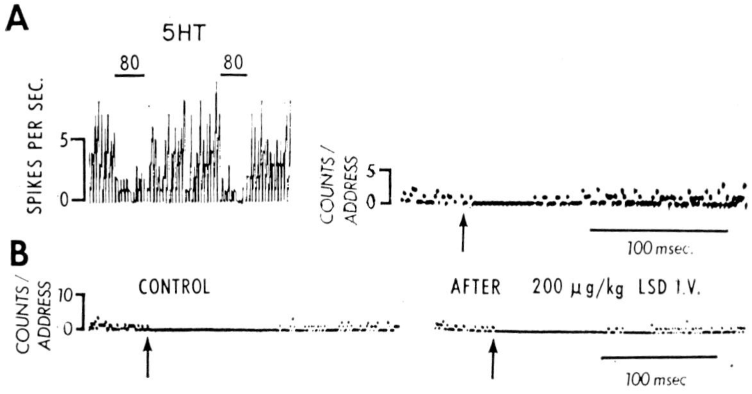

Fig. 60. Responses of a neuron in the cat suprachiasmatic nucleus to electrophoretic administration of 5-HT (A, *left*) and electrical stimulation of the brain-stem at the level of the nucleus centralis superior (A, *right*). Both 5-HT (80 nA) and stimulation of the raphe nuclei depressed the firing of the neuron. The poststimulus time histograms in B (from another suprachiasmatic neuron) show that there was no change in the synaptic inhibition from the raphe nuclei after a large intravenous dose of LSD. (Bloom, Hoffer, Siggins and others, 1972.)

with dexamethasone phosphate are particularly interesting because no steroid-sensitive neurons were found in the cerebral cortex, dorsal hippocampus, or thalamus. The locations of all dexamethasone-sensitive neurons were determined relatively accurately

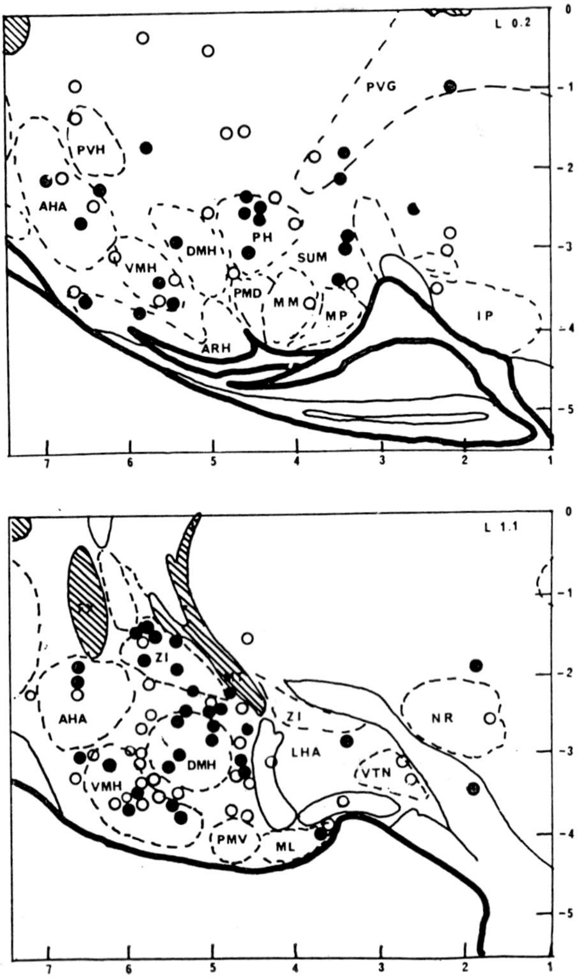

Fig. 61. Approximate locations of neurons depressed (●) and unaffected (○) by dexamethasone phosphate in parasagittal sections through the hypothalamus of the rat brain. Abbreviations are according to the atlas by De Groot (1959). Neurons in L 0–0·5 are projected on L 0·2; neurons in L 0·6–1·4 are projected on L 1·1 (Steiner and others, 1969.)

by marking the cells with dye deposited from the electrode, followed by histological studies of the brain. The approximate positions of a number of dexamethasone-sensitive and insensitive neurons studied in the rat hypothalamus are shown in *Fig.* 61 (Steiner and others, 1969). Dexamethasone-sensitive cells were widely scattered in the midbrain and hypothalamus, but appeared to occur predominantly within the central grey matter lining the third ventricle and aqueduct.

Synthetic ACTH excited the majority of dexamethasone-sensitive neurons. It was suggested that such cells might represent steroid receptors involved in the control of ACTH secretion, these neurons being sensitive to both hormones and transmitters (Ruf and Steiner, 1967; Steiner and others, 1968, 1969; Steiner, 1970; *see also* Schadé and van Wilgenburg, 1971).

SUMMARY

Thalamus

The pharmacology of thalamo-cortical relay neurons of the ventrobasal complex and the lateral and medial geniculate nuclei is similar in so many respects that the main findings for these three areas will be summarized together. *Fig.* 62 illustrates in diagrammatic form some of the connexions and the possible transmitters involved.

Acetylcholine

ACh excites the large majority of relay cells in the ventrobasal complex, and the lateral and medial geniculate nuclei, the cholinoceptive receptors being of intermediate muscarinic-nicotinic type. Pharmacological and other evidence indicates that ACh is an excitatory transmitter released on to thalamo-cortical relay neurons from certain fibres which originate in or traverse the ventral mesencephalon. The precise origin of this pathway is not known, but it may be the cholinesterase-containing dorsal tegmental fibres which arise in the cuneiform nucleus. Stimulation of the mesencephalic reticular formation excites some thalamic neurons, and this synaptic excitation, as well as the excitant action of ACh, can be reversibly blocked by muscarinic and nicotinic antagonists of ACh, and potentiated by cholinesterase inhibitors. This cholinergic projection from the brain-stem is thought to be part of the 'reticular arousal system', especially as stimulation of various central areas (including the reticular formation) can increase the release of ACh from the thalamus. Evidence for the differential distribution of ACh, choline acetyltransferase, and AChE, as well as for the differential release of ACh in various thalamic regions is consistent with the concept of cholinergic transmission at certain sites in the mammalian thalamus.

At least four other types of excitatory fibres terminating on thalamic neurons are non-cholinergic. Although the majority of neurons of the ventrolateral thalamic nucleus (monosynaptically excited by stimulating the brachium conjunctivum) are excited by ACh, few, if any, fibres from the cerebellum that terminate on these neurons can be cholinergic, as antagonists of ACh have little action on the synaptic responses evoked by impulses in fibres of the brachium conjunctivum. The excitatory transmitter released

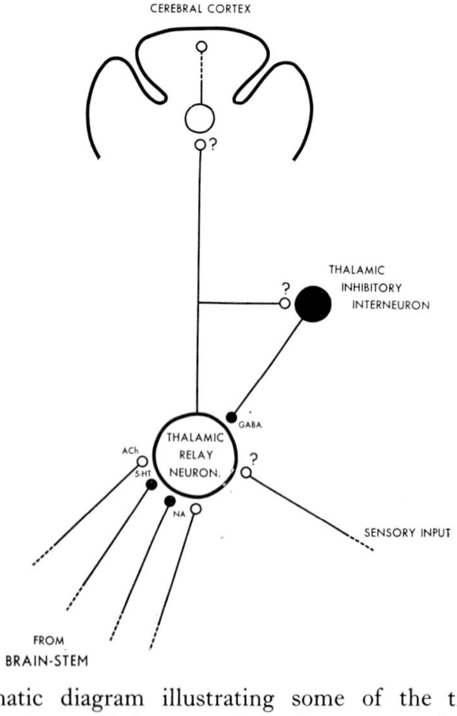

Fig. 62. Schematic diagram illustrating some of the transmitters likely to be operating on thalamo-cortical relay neurons of the ventrobasal complex, lateral geniculate nucleus, and medial geniculate nucleus. Excitatory terminals are represented by **O**, inhibitory by ●. *See* text for details.

on to ventrobasal and geniculate neurons from fibres which originate in the cerebral cortex is not ACh, as the synaptic excitation from impulses in corticofugal fibres is resistant to antagonists of ACh, or ACh depresses the firing of thalamic neurons excited by cortical stimulation. Similarly, pharmacological evidence supported by biochemical findings indicates that the excitatory transmitter released on to thalamic neurons from 'sensory' afferent fibres is non-cholinergic. The transmitter released from medial lemniscal fibres on to ventrobasal relay neurons is possibly the same as that released from auditory afferents on to medial geniculo-cortical relay cells, but appears to be quite different from the optic nerve transmitter released on to lateral geniculate neurons. The spontaneous firing of thalamic neurons is likely to be mediated largely or wholly by a transmitter other than ACh.

ACh depresses the firing of a number of thalamic neurons other than relay cells, and this depression can be blocked by both muscarinic and nicotinic antagonists, and potentiated by anti-cholinesterases. The possibility that this type of depression is related to synaptic inhibition in the thalamus remains to be investigated.

Amino-acids

Glutamate and other acidic amino-acids excite relay and other thalamic neurons, and some types of thalamic neurons (for example, those of the ventrolateral nucleus) appear to be more sensitive to glutamate than others. Certain analogues of glutamate have been shown to antagonize both the actions of excitatory amino-acids and the synaptic excitation evoked by stimulating peripheral nerves, providing evidence that glutamate or a related amino-acid is the transmitter released from medial lemniscal fibres on to thalamic neurons. It is possible that the non-cholinergic excitatory transmitter released from auditory afferent fibres projecting to the medial geniculate nucleus is also glutamate or a related amino-acid, but no experiments have been done to investigate this possibility. However, the transmitter released from optic nerve terminals on to lateral geniculate neurons is unlikely to be glutamate or a closely related amino-acid, if the mechanism of action of tryptamine derivatives and LSD is to block the post-synaptic action of the synaptically released transmitter, as these substances block the orthodromic firing of lateral geniculate neurons in concentrations which do not reduce the excitation by excitatory amino-acids. The optic nerve transmitter is possibly an unusual substance which has hitherto received little or no attention as a possible transmitter. The fibres mediating corticofugal excitation or spontaneous firing of thalamic neurons may well release an excitatory amino-acid as a transmitter, but this remains to be investigated.

GABA is more effective than glycine as a depressant of thalamo-cortical relay neurons of the ventrobasal complex and the lateral and medial geniculate nuclei. In the ventrobasal complex and lateral geniculate nucleus, the action of GABA is blocked by bicuculline in concentrations which have little or no action on that of glycine, and bicuculline also blocks the inhibition of relay cells following antidromic stimulation of relay cell axons. This synaptic inhibition is resistant to strychnine, which blocks the action of glycine in concentrations that have little effect on that of GABA. This type of inhibition has not been tested with convulsants in the medial geniculate nucleus, but the actions of the depressant amino-acids and strychnine on medial geniculate neurons resemble those on ventrobasal and lateral geniculate neurons, suggesting that similar inhibitory mechanisms operate in all three regions. Thus, there is good pharmacological evidence that GABA is the inhibitory transmitter released from inhibitory interneurons on to thalamic relay cells, as shown in *Fig.* 62. In addition, bicuculline blocks the inhibition of ventrobasal and lateral geniculate relay cells following *orthodromic* activation, but it is not certain whether the same or different inhibitory pathways utilizing GABA as the transmitter are involved.

Although the lateral geniculate nucleus contains relatively high levels of taurine, and taurine is a strong depressant of lateral geniculo-cortical relay cells, this amino-acid does not mediate the 'antidromic' inhibition shown in *Fig.* 62, as the action of taurine is blocked by both strychnine and bicuculline. Taurine may nevertheless be an inhibitory transmitter in the lateral geniculate nucleus, released from inhibitory fibres which are as yet undiscovered.

Monoamines

5-HT depresses the postsynaptic excitability of thalamo-cortical relay neurons of the ventrobasal complex and lateral and medial geniculate nuclei. In the lateral geniculate nucleus (but not in the medial geniculate or the ventrobasal complex) 5-HT also has another type of depressant action, which is apparent in concentrations which do not affect postsynaptic excitability. In low concentrations 5-HT blocks the orthodromic firing of lateral geniculate neurons resulting from impulses in optic nerve fibres without affecting the excitant actions of ACh and excitatory amino-acids, or antidromic invasion. This type of depression by 5-HT is best interpreted as either antagonism of the postsynaptic action of the excitatory transmitter released from optic nerve terminals, or as a block of release of this transmitter. However, it is not clear whether this type of depression by 5-HT is of functional importance in the lateral geniculate nucleus. The postsynaptic depressant action of 5-HT on relay cells of ventrobasal and geniculate nuclei is likely to be related to an inhibitory transmitter function for 5-HT at these sites, as 5-HT-containing fibres which originate in the brain-stem terminate on thalamic neurons (*Fig.* 62). However, although biochemical and histochemical studies are consistent with the concept of 5-hydroxytryptaminergic transmission in the thalamus, direct pharmacological evidence for such pathways has yet to be obtained.

NA excites and depresses different proportions of relay cells in the three thalamic areas studied, these actions probably being postsynaptic. Both histochemical and biochemical studies suggest that NA is a transmitter (both excitatory and inhibitory) released on to ventrobasal thalamic and geniculate neurons (*Fig.* 62). Pharmacological evidence for such pathways, however, is still to be obtained.

DA depresses the large majority of thalamic neurons, including relay cells, but since there is no histochemical evidence for DA-containing nerve-fibres in the thalamus, it is not known whether DA serves a synaptic function in this structure. The presence of DA in a very small proportion of retinal ganglion cells raises the interesting possibility that some dopaminergic fibres terminate on lateral geniculate neurons, but much more evidence is required before the significance of this observation can be evaluated.

Optic Nerve Transmitter

Various reasons were given why the major excitatory transmitter released from optic nerve terminals on to lateral geniculate neurons is unlikely to be ACh, 5-HT, NA, DA, glutamate, or aspartate. The chemical identity of this transmitter remains unknown.

Hypothalamus

ACh and the monoamines have excitant and depressant actions on various proportions of neurons in different hypothalamic nuclei. These observations, together with biochemical, histochemical, and release studies suggest that these substances are transmitters at certain sites in the hypothalamus, but pharmacological evidence for the pathways involved has not been shown. However, some interesting correlations between the actions of amines and certain types of hypothalamic neurons are worth summarizing.

Thermally Sensitive Hypothalamic Neurons

The actions of ACh, NA, and 5-HT have been correlated with the activity of thermally sensitive neurons in the preoptic and anterior regions of the hypothalamus. 'Thermodetectors' were apparently insensitive to these substances, but 'thermosensitive interneurons' responded in characteristic ways.

Supraoptic and Paraventricular Neurons

There is disagreement between laboratories whether ACh excites or depresses the majority of neurosecretory cells, although it is accepted that the excitation by ACh is mediated by nicotinic receptors, and the depression by muscarinic receptors. In either case, cholinoceptive neurosecretory cells are thought to indicate the existence of cholinergic mechanisms which control the release of hormones.

NA depresses the firing of most neurosecretory cells, and this depression can be blocked by MJ-1999 and potentiated by desmethylimipramine. It is thought that the depression by NA is related to noradrenergic inhibition in the supraoptic and paraventricular nuclei, but this has not yet been demonstrated.

The firing of neurosecretory cells is inhibited by stimulating their axons, and this inhibition is not mediated by ACh, NA, or glycine, as it is not affected by atropine, 6-hydroxydopamine, or strychnine. It has been suggested that antidiuretic hormone may be the inhibitory transmitter involved, as this hormone depresses the large majority of supraoptic neurons, whereas it excites most cortical neurons. However, the observations that GABA is a strong depressant of neurosecretory cells, and that bicuculline and picrotoxin block the actions of GABA and reduce the recurrent inhibition, suggest that GABA is more likely to be the inhibitory transmitter.

Suprachiasmatic Hypothalamic Neurons

5-HT depresses the firing of suprachiasmatic neurons inhibited by stimulating the raphe nuclei. The suprachiasmatic nucleus is particularly rich in 5-HT-containing nerve terminals which originate in the raphe nuclei. Thus, it is likely that 5-HT is an inhibitory transmitter released on to suprachiasmatic neurons from fibres arising in the raphe nuclei, but pharmacological evidence for this pathway is not complete.

Steroid-sensitive Hypothalamic Neurons

Dexamethasone phosphate, a corticosteroid which suppresses the secretion of ACTH, has been shown to affect (mostly depress, but sometimes excite) a small proportion of neurons in the hypothalamus which are usually excited by ACh and depressed by the catecholamines, NA and DA. Since no steroid-sensitive neurons were found in the cerebral cortex, hippocampus, or thalamus, the sensitivity of hypothalamic neurons to dexamethasone has been interpreted as representing steroid receptors involved in the control of ACTH secretion.

Chapter

5

BASAL GANGLIA

THE term 'basal ganglia' includes all the subcortical motor nuclei of the forebrain, but refers chiefly to the caudate nucleus, putamen, and globus pallidus. Although it is well known that the basal ganglia are involved in the control of movement and posture, the precise functions of the different regions are not clear. Few pharmacological studies of the basal ganglia have involved administering substances to identified neurons. However, it is useful to review the studies of unidentified neurons, as they have significantly contributed towards the identification of transmitter substances in the basal ganglia. The neostriatum (caudate nucleus and putamen) is of particular interest, as it is the only area in the C.N.S. (other than the retina—*see* Chapter 8) where DA has been shown to be a likely transmitter. The ideas on the possible role of DA in the basal ganglia prior to 1966 have been reviewed by Hornykiewicz (1966). The following discussion deals mainly with more recent studies pertinent to a transmitter function for DA.

CAUDATE NEURONS

The caudate nucleus is a large, relatively well-defined brain structure with a dense neuropile and small, closely packed neurons, thus increasing the difficulty in recording from single units for long periods. The caudate nucleus makes connexions with different areas of the cerebral cortex, thalamus, putamen, globus pallidus, substantia nigra, reticular formation, red nucleus, and other areas, some of which merge into each other without distinct boundaries. This increases the difficulty of interpreting caudate responses evoked by electrical stimuli which might have spread to different neighbouring areas. Nevertheless, neuropharmacological experiments, in conjunction with biochemical, histochemical, and other data, strongly suggest that DA is an inhibitory transmitter in the mammalian caudate nucleus.

Dopamine

The mammalian caudate nucleus contains the highest concentration of DA of any structure in the C.N.S. (Bertler and Rosengren, 1959; Bertler, 1961; Hornykiewicz, 1966), and at least a proportion of the striatal catecholamine content is contained within synaptosomes (Kuhar, Green, Snyder, and Gfeller, 1970). Both slices and homogenates of the rat corpus striatum have been shown to take up catecholamines far more rapidly than do many other brain structures, and DA had a greater affinity for striatal neurons than NA (Coyle and Snyder, 1969; Snyder, Kuhar, Green, Coyle, and Shaskan, 1970). Both DA-containing and NA-containing nerve-endings in the brain can take up either catecholamine, but whereas the affinity constants for noradrenergic endings seem to be the same in a large number of brain regions, they differ from the DA-containing endings of the striatum (Snyder and others, 1970).

Results from both biochemical and histochemical studies indicate that the DA-containing nerve terminals in the neostriatum are terminations of the axons of DA-containing cell bodies located in the substantia nigra, particularly as lesions in the substantia nigra and neighbouring areas have been shown to decrease the neostriatal DA content, and the DA levels in the substantia nigra increased after the neostriatum was removed (Andén, Carlsson, Dahlström, Fuxe, Hillarp, and Larsson, 1964; Bertler, Falck, Gottfries, Ljunggren, and Rosengren, 1964; Poirier and Sourkes, 1965; Andén and others, 1966; Hornykiewicz, 1966; Poirier, Sourkes, Bouvier, Boucher, and Carabin, 1966; Poirier, Singh, Boucher, Bouvier, Olivier, and Larochelle, 1967; Hökfelt and Ungerstedt, 1969; Moore, Bhatnagar, and Heller, 1971). *Fig.* 63 is a schematic diagram of a sagittal section through the rat brain, showing the locations of DA-containing neurons in the mesencephalon, and their projections. It can be seen that DA-containing nerve terminals are restricted to few brain regions, mainly the corpus striatum. This is in contrast to NA-containing nerve-fibres which have a much greater distribution (*Fig.* 101).

DA is released spontaneously from the caudate nucleus, and this efflux can be increased by chemical and electrical stimulation of various regions, including the rostral substantia nigra, nigro-striatal fibres, and the caudate nucleus itself (McLennan, 1964; McKenzie and Szerb, 1968; Portig and Vogt, 1969; Riddell and

Szerb, 1971; von Voigtlander and Moore, 1971 a, b; Besson, Cheramy, Feltz, and Glowinski, 1971).

In the first neuropharmacological studies of caudate neurons, electrophoretically administered DA depressed the firing of about one-half to two-thirds of unidentified neurons tested, and excited about one-tenth (Bloom, Costa, and Salmoiraghi, 1965, Herz and Zieglgänsberger, 1966, 1968; McLennan and York, 1967; York, 1967). The firing of caudate neurons was inhibited by

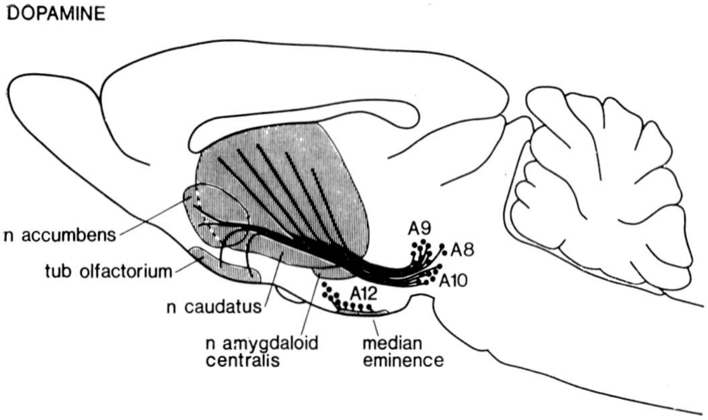

Fig. 63. Sagittal view of the rat brain, showing the DA-containing pathways diagrammatically. Sites of the DA-containing cell bodies are represented by the black dots; the nerve terminals, by the shaded areas. (Ungerstedt, 1971.)

stimulation of nucleus centromedianus of the thalamus, and both this inhibition and the depressant action of DA on caudate neurons were blocked by phenoxybenzamine (McLennan and York, 1967; York, 1967). Thus, in accordance with an earlier observation that stimulation of nucleus centromedianus increased the liberation of DA from the caudate nucleus (McLennan, 1964), it was postulated that there is an inhibitory dopaminergic pathway from the thalamus to the caudate nucleus (McLennan and York, 1967; York, 1967).

However, the extremely long latency (many seconds) of the phenoxybenzamine-sensitive inhibition evoked by thalamic stimulation (York, 1967) indicates that the results can be explained equally well in terms of activation of a dopaminergic pathway which originates elsewhere. A direct dopaminergic pathway from the thalamus to the caudate nucleus is unlikely because there is relatively little DA and no DA-containing cell bodies in the thalamus. The concentrations of DA in the caudate nucleus and substantia nigra are much higher than in the thalamus, and the nigro-neostriatal DA-containing neurons can account for most or all of the DA present in the neostriatum (Bertler and Rosengren, 1959; Andén and others, 1964, 1966). Moreover, some caudate neurons which were inhibited by thalamic stimulation were excited by DA (York, 1967).

Subsequently, pharmacological evidence has been obtained for the view that dopaminergic fibres to the caudate nucleus originate in the substantia nigra (Connor, 1968, 1970). Stimulation of the substantia nigra inhibited the firing of many caudate neurons, and those neurons inhibited by nigral stimulation were consistently depressed by DA. In particular, alpha-methyldopamine antagonized both the synaptic inhibition and the depression by DA, as shown in *Fig.* 64. The post-stimulus time histogram (*Fig.* 64A) shows that stimulation of the substantia nigra inhibited the firing of the caudate neuron for more than 50 msec. This inhibition was blocked by alpha-methyldopamine 6 min. after the beginning of the ejection (*Fig.* 64B), and recovery occurred after the ejection was terminated (*Fig.* 64C). Alpha-methyldopamine also blocked the depressant action of DA on the same neuron (*Fig.* 64D).

Since the DA-containing nigro-neostriatal fibres are of small diameter and unmyelinated (Andén and others, 1964, 1966) they are likely to be slowly-conducting, and the mean latency of the inhibition described by Connor (1968, 1970) would be compatible with a monosynaptic dopaminergic pathway from the substantia nigra to the caudate nucleus (cf. Frigyesi and Purpura, 1967; Feltz and MacKenzie, 1969). However, the possibility of polysynaptic components cannot be excluded.

In conclusion, there is good biochemical, histochemical, DA-release, and pharmacological evidence for an inhibitory dopaminergic nigro-caudate pathway in mammals. The synaptic *excitation* of caudate neurons evoked by stimulating the substantia nigra cannot be dopaminergic, because most caudate neurons excited by nigral stimuli were depressed or unaffected by DA and other monoamines (McLennan and York, 1967; Connor, 1970; Feltz, 1971a). Additional evidence for this conclusion was obtained by Feltz and de Champlain (1972), who depleted most of the endogenous DA in the caudate nucleus by chronic pretreatment

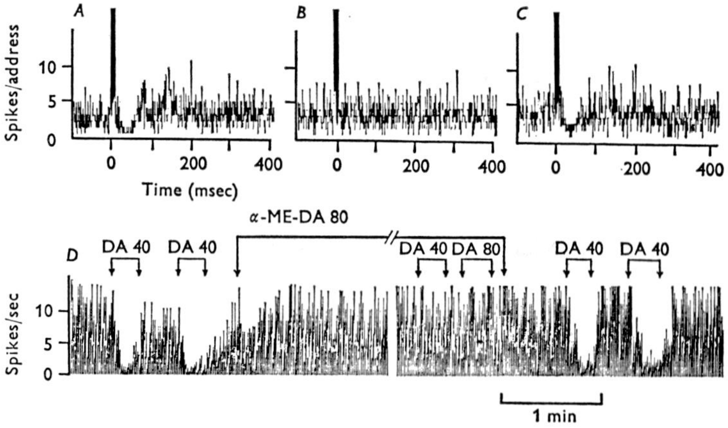

Fig. 64. Antagonism by alpha-methyldopamine of the inhibitory effects of nigral stimulation and DA on a caudate neuron. A–C are histograms (200 summations) showing the effects of stimulating the substantia nigra (10 msec. trains at 400/sec.) on the firing of the neuron. A, Before the electrophoretic ejection of alpha-methyldopamine. B, 6 min. after the beginning of the ejection of alpha-methyldopamine. C, After the ejection had been terminated. D, Chart recording of the firing frequency of the same cell, showing the depressant action of DA (40 nA) and the blocking action of alpha-methyldopamine (α-ME-DA, 80 nA) on this depression. The gap in the tracing represents 1·5 min. (Connor, 1970.)

with 6-hydroxydopamine. In cats so treated, caudate neurons were monosynaptically excited by nigral stimulation, as in normal animals. A precise understanding of the mechanisms operating in the dopaminergic pathway would be of considerable clinical importance, as incorrect functioning of this system is thought to result in Parkinsonism and other motor disturbances (Hornykiewicz, 1966; Gillingham and Donaldson, 1969; Barbeau and McDowell, 1970).

Noradrenaline and 5-Hydroxytryptamine

It is unlikely that NA is a transmitter in the mammalian caudate nucleus, as biochemical estimates have revealed very low quantities of this catecholamine (Vogt, 1954; Kuntzman and others 1961; McGeer and others, 1963; Hornykiewicz, 1966), and there is no histochemical evidence for the presence of NA in nerve terminals in the caudate nucleus (Fuxe, 1965; Ungerstedt, 1971).

It is therefore difficult to evaluate the significance of the observation that stimulation of the caudate nucleus resulted in a release of NA into the ventricles (von Voigtlander and Moore, 1971c); the efflux of NA may well have been from noradrenergic terminals localized outside the caudate nucleus. However, the actions of NA on caudate neurons did not always mimic those of DA (Bloom and others, 1965; Herz and Zieglgänsberger, 1968), indicating the presence of two different types of receptors for these catecholamines. Thus, although the available evidence suggests that NA cannot be a major transmitter in the caudate nucleus, the possibility of some noradrenergic synapses in this structure cannot be excluded.

The mammilian caudate nucleus contains relatively high concentrations of 5-HT (Bogdanski and others, 1957; Kuntzman and others, 1961; McGeer and others, 1963; Poirier and Sourkes, 1965), a proportion of which is localized in nerve-fibres and synaptosomes (Fuxe, 1965; Andén and others, 1966; Kim, Hassler, Kurokawa and Bak, 1970; Snyder and others, 1970). Lesions in the ventromedial tegmental area of the upper pons and midbrain sometimes resulted in a complete depletion of 5-HT and DA in the striatum, and pretreatment with precursors of these monoamines increased the concentrations of striatal 5-HT and DA in normal animals (Poirier and Sourkes, 1965; Poirier and others, 1966, 1967). Electrophoretically administered 5-HT depressed the firing of most caudate neurons sensitive to it (Bloom and others, 1965). However, only extremely small quantities of 5-HT were initially detected in perfusates of the anterior horn of the lateral ventricle (the borders of the caudate nucleus and septum), and this release was not increased by various forms of stimulation which increased the release of DA or ACh (Feldberg and Myers, 1966; Portig and Vogt, 1969). The clue as to the probable course of a 5-hydroxytryptaminergic pathway was the histochemical evidence for a projection from the raphe nuclei to the basal ganglia in rats (Dahlström and Fuxe, 1965a; Andén and others, 1966). Accordingly, it has recently been shown that stimulation of the most anterior raphe nuclei in cats *did* result in a significant release of 5-HT from the caudate nucleus (and septum), and such release of 5-HT was obtained using stimulus parameters which did not increase the release of ACh (Ashkenazi, Holman, and Vogt, 1972; Holman and Vogt, 1972). Anatomical evidence for a pathway from the anterior raphe nuclei to the basal ganglia in cats had been demonstrated earlier (Brodal, Taber, and Walberg, 1960). Various biochemical findings have led to the suggestion that the degree of Parkinsonism is related to changes in the balance of the normal DA content to the 5-HT plus ACh content of the striatum (Kim and others, 1970).

Acetylcholine

Electrophoretically administered ACh excited or depressed the firing of different populations of caudate neurons (Bloom and others, 1965; McLennan and York, 1966, 1967; Herz and Zieglgänsberger, 1968; Connor, 1970). In particular, caudate neurons, which were excited or depressed by stimulating nucleus ventralis anterior of the thalamus, were also excited and depressed, respectively, by ACh, and both the chemical and synaptic activation were blocked by atropine (McLennan and York, 1966). Results from one such neuron are illustrated in *Fig*. 65.

A pulse of DLH fired the caudate neuron (*Fig*. 65A), and this excitation was facilitated during the administration of ACh (*Fig*. 65C). Stimulation of nucleus ventralis anterior evoked a synaptic response from this neuron (*Fig*. 65, bottom left-hand record). After an intravenous injection of atropine the response to DLH was unaffected, but the facilitatory action of ACh was largely blocked, as was the synaptic excitation.

These and similar results suggested that ACh is a transmitter in the feline caudate nucleus, acting at both excitatory and inhibitory muscarinic synapses (McLennan and York, 1966). ACh-excited neurons appeared to be distributed both more superficially and deeper than ACh-depressed cells. The presence of relatively high amounts of ACh, choline acetyltransferase, and AChE in the mammalian caudate nucleus (MacIntosh, 1941; Feldberg and Vogt, 1948; Burgen and Chipman, 1951; Hebb and Silver, 1956), as well as the increase in resting release of ACh from this nucleus following stimulation of nucleus ventralis anterior (but not of nucleus centromedianus) of the thalamus (McLennan, 1964), is consistent with the concept of cholinergic transmission in the caudate nucleus.

However, it is not clear whether the cholinergic fibres project directly from the thalamus to the caudate nucleus as McLennan and York (1966) suggested, since their atropine-sensitive orthodromic responses were of polysynaptic latency (for example, *Fig*. 65), and the increased efflux of ACh evoked by thalamic stimulation (McLennan, 1964) might also have resulted indirectly. Subsequent studies involving 'superfusion' of the caudate nucleus have revealed a spontaneous release of ACh, which was increased by stimulating the substantia nigra, nucleus centralis lateralis of the thalamus, contralateral caudate nucleus, sciatic nerves, and paws, as well as by noise (Portig and Vogt, 1969), suggesting that the cholinergic afferents to the caudate nucleus are part of a 'reticular arousal system'. Histochemical studies of the rat brain

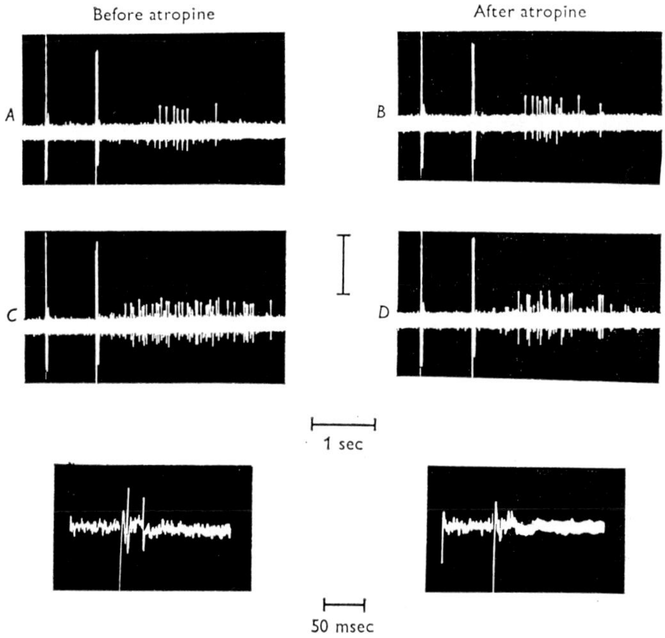

Fig. 65. The effect of intravenously administered atropine on the facilitation of firing by ACh and the synaptic excitation of a caudate neuron evoked by stimulating the nucleus ventralis anterior of the thalamus. Left-hand column before, and right-hand column 10 min. after an injection of atropine sulphate (0·1 mg./kg.). The upper four records show responses to pulses of DLH (20 nA for 800 msec.; the beginning and end of the pulses are marked by the large vertical deflections). A, B, Before; C, D, 30 sec. after the beginning of an electrophoretic application of ACh (75 nA). The two bottom records show the responses evoked by single stimuli (15 V, 0·1 msec. duration) delivered to nucleus ventralis anterior. *Voltage calibration*: 0·2 mV for the upper four and 0·25 mV for the lower two records. (McLennan and York, 1966.)

(Shute and Lewis, 1967) have indicated that the cholinesterase-containing fibres which terminate in the neostriatum arise in the globus pallidus, entopeduncular nucleus, lateral preoptic area, and possibly other regions in the brain-stem (*Fig.* 66). More recently, however, lesions in the cortex, thalamus, ventral tegmentum, and globus pallidus resulted in no significant changes in

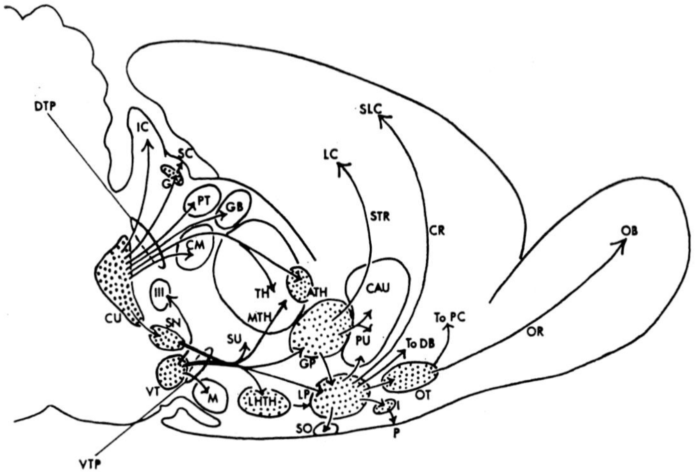

Fig. 66. Diagram showing the constituent nuclei (stippled) of the ascending cholinesterase-containing reticular system in the midbrain and forebrain of the rat, with projections to the cerebellum, tectum, thalamus, hypothalamus, striatum, lateral cortex, and olfactory bulb. *Abbreviations:* ATH, Anteroventral and anterodorsal thalamic nuclei; CAU, Caudate nucleus; CM, Centromedian (parafascicular) nucleus; CR, Cingulate radiation; CU, Cuneiform nucleus; DB, Diagonal band; DTP, Dorsal tegmental pathway; G, Stratum griseum intermediale of the superior colliculus; GB, Medial and lateral geniculate bodies; GP, Globus pallidus and entopeduncular nucleus; I, Islets of Calleja; IC, Inferior colliculus; III, Oculomotor nucleus; LC, Lateral cortex; LHTH, Lateral hypothalamic area; LP, Lateral preoptic area; M, Mammillary body; MTH, Mammillo-thalamic tract; OB, Olfactory bulb; OR, Olfactory radiation; OT, Olfactory tubercle; P, Plexiform layer of olfactory tubercle; PC, Precallosal cells; PT, Pretectal nuclei; PU, Putamen; SC, Superior colliculus; SLC, Supero-lateral cortex; SN, Substantia nigra pars compacta; SO, Supraoptic nucleus; STR, Striatal radiation; SU, Subthalamus; TH, Thalamus; TP, Nucleus reticularis tegmenti pontis (of Bechterew); VT, Ventral tegmental area and nucleus of basal optic root; VTP, Ventral tegmental pathway. (Shute and Lewis, 1967.)

the levels of AChE or choline acetyltransferase in the rat neostriatum, suggesting that cholinergic fibres of the caudate nucleus originate within the neostriatum itself (McGeer, McGeer, Fibiger, and Wickson, 1971). Thus, although ACh is likely to be a transmitter in the caudate nucleus, further neuropharmacological, histochemical, and biochemical studies of the one species are required to determine the origin and course of the cholinergic fibres in this region of the brain.

Amino-acids

It has been shown that glutamate, aspartate, and DLH excite, whereas GABA depresses, the firing of caudate neurons (Bloom and others, 1965; McLennan and York, 1966, 1967; York, 1967; Herz and von Freytag-Loringhoven, 1968; Herz and Zieglgänsberger, 1968; Connor, 1970; Feltz, 1971 a, b), but the significance of these actions is yet unknown. No studies have been made to relate the actions of the amino-acids to synaptic excitation or inhibition in the caudate nucleus. However, the relatively high concentrations of glutamate and GABA (a proportion of which is associated with synaptosomes), as well as the enzymes for the metabolism of these amino-acids (Salvador and Albers, 1959; Kržalić and others, 1962; Fahn and Côté, 1968a; De Feudis and others, 1970; Kuhar and others, 1970; Snyder and others, 1970; Hökfelt and others, 1970; Johnson and Aprison, 1971; Okada and others, 1971; see also Iversen and Bloom, 1972) suggest that glutamate and GABA may be transmitters in the caudate nucleus. This is supported by the finding that newly formed amino-acids (including glutamate and GABA) can be detected in perfusates of the monkey caudate nucleus, using a 'push-pull' cannula (De Feudis and others, 1970).

Putamen Neurons

Most of the biochemical and histochemical observations of the caudate nucleus mentioned in the preceding section also apply to the putamen, but fewer pharmacological studies have been made of neurons in the putamen. DA excited and depressed the firing of different proportions of neurons in the putamen of the cat and monkey (Herz and Zieglgänsberger, 1968; York, 1970, 1972). In contrast to the caudate nucleus, however, most of the cells in the putamen (cat) which were excited by DA were also excited by repetitive stimulation of the substantia nigra, and in a few cases it was possible to block the facilitatory effects of both DA and nigral stimulation by adrenergic antagonists (York, 1970). The results were interpreted as indicating that DA is an *excitatory* transmitter released on to putamen neurons from nigral fibres (York, 1970). No conclusions were made about the possible significance of the depressant actions of DA on putamen neurons (York, 1970, 1972). Previously, it had been shown that stimulation of the substantia nigra increased the release of DA from the putamen (McLennan, 1965), and histochemical and biochemical studies have established the presence of DA-containing fibres which project from the substantia nigra to the putamen (Andén and others, 1966; Moore and others, 1971). However, although it is likely that a dopaminergic pathway from the nigra to the putamen exists, conclusive neuropharmacological evidence for this is still needed. Repetitive stimulation of the substantia nigra (York, 1970) may cause complex phases of excitation and inhibition involving several pathways which are impossible to differentiate on a chart-recording (cf. reciprocal caudato-nigral relations; Frigyesi and Purpura, 1967; Feltz and MacKenzie, 1969). Furthermore, the adrenergic antagonists used to block the effects of nigral stimulation in the putamen were not specific for DA (York, 1970; cf. York, 1972).

Other amines, including 5-HT, NA, and ACh, excited or depressed the firing of a number of neurons in the putamen (Herz and Zieglgänsberger, 1968; York, 1970), but the significance of these actions is unknown. Similarly, the significance of the excitation by glutamate and the depression by GABA in the putamen (Herz and von Freytag-Loringhoven, 1968; Herz and Zieglgänsberger, 1968; York, 1970, 1972) is not yet clear.

Globus Pallidus Neurons

Even less is known about the possible identity of transmitters in the globus pallidus. Monoamines (including DA) affected pallidal neurons similarly to their actions on neurons of the putamen (York, 1970), but there is no direct evidence for any monoamine functioning as a transmitter in the globus pallidus.

The globus pallidus of the rat receives AChE-containing nerve-fibres which arise mainly from cells in the substantia nigra and the ventral tegmental area of the rostral mesencephalon (*Fig.* 66), but the role of ACh (if any) in the pallidus is unknown.

The highest levels of GABA and glutamic acid decarboxylase in the mammalian C.N.S. are present in the substantia nigra and globus pallidus (Albers and Brady, 1959; Fahn and Côté, 1968a; Chalmers, McGeer, Wickson, and McGeer, 1970; *see also* Hökfelt and Ljungdahl, 1971b), and a proportion of the GABA content of the globus pallidus is associated with synaptosomes (Kuhar and others, 1970; Iversen and Bloom, 1972). Slices of the rat globus pallidus take up GABA very readily (Hökfelt and others, 1970). Lesions in the globus pallidus led to a decrease of glutamic acid decarboxylase in the substantia nigra (McGeer, McGeer, Wada, and Jung, 1971). Since neurons of the globus pallidus are inhibited monosynaptically by axon collaterals of caudato-nigral fibres (Yoshida, Rabin, and Anderson, 1971), and since there is evidence that the caudato-nigral monosynaptic inhibition is mediated by GABA (*see below*), it is likely that GABA is the inhibitory transmitter released from caudate fibres on to pallidal neurons, a view supported by the biochemical findings. However, this has yet to be demonstrated directly.

Nigral Neurons

The substantia nigra is not considered part of the basal ganglia, but is discussed here rather than in Chapter 2 because of its direct relevance to the caudate nucleus, putamen, and globus pallidus.

GABA

The only example of an identified transmitter in the substantia nigra is GABA, which mediates a monosynaptic inhibition from the caudate nucleus (Yoshida and Precht, 1971; Precht and Yoshida, 1971; Feltz, 1971b). Stimulation of the caudate nucleus monosynaptically inhibited the firing of neurons in the substantia nigra (Yoshida and Precht, 1971), and this inhibition was blocked by intravenous picrotoxin, but unaffected by intravenous strychnine (Precht and Yoshida, 1971). Examples of this blocking action of picrotoxin are illustrated in *Fig.* 67.

Although the interactions of picrotoxin and strychnine with amino-acids were not determined, a comparison of the results with those obtained in other areas of the C.N.S. strongly suggests that the monosynaptic inhibition from the caudate nucleus to the substantia nigra is mediated by GABA (Precht and Yoshida, 1971). Subsequently, the existence of a powerful inhibitory pathway from the caudate nucleus to the substantia nigra was confirmed, and in

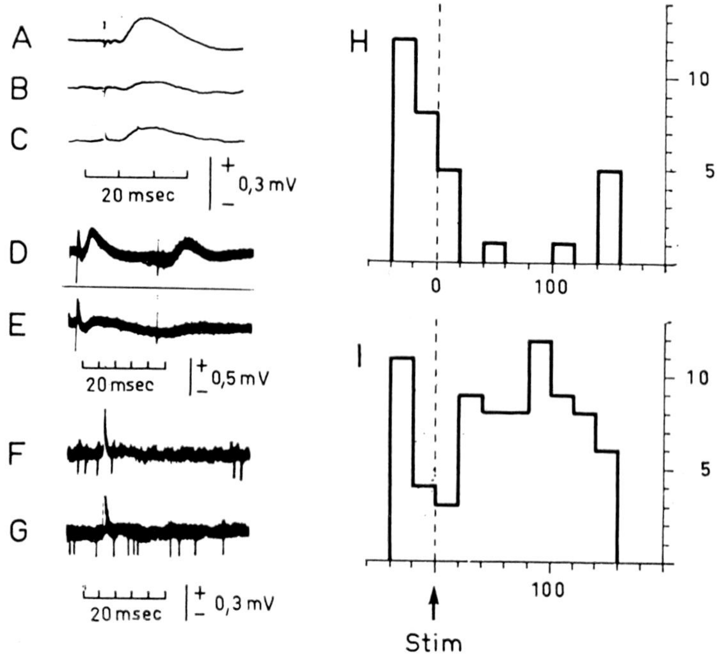

Fig. 67. Effects of picrotoxin on potentials recorded in the substantia nigra following stimulation of the caudate nucleus and cerebral peduncle. A, Control field potential evoked in the substantia nigra by stimulating the head of the caudate nucleus at a rate of 3/sec. (16 averaged traces). B, 20 min. after an intravenous injection of picrotoxin (2·5 mg./kg.), showing a marked decrease in the positive field potential. C, 75 min. after the injection, showing a partial recovery of the positive field potential. D, E, Field potentials recorded in the substantia nigra following stimulation of the ipsilateral cerebral peduncle (first stimulus artefact) and the head of the caudate nucleus (second stimulus artefact). D, A control record; E, 10 min. after an intravenous injection of picrotoxin (5 mg./kg.). F, Suppression of the spontaneous firing of a nigral neuron by stimulating the caudate nucleus; and G, Block of this inhibition 20 min. after picrotoxin (2·5 mg./kg.). H, I, Histograms of the firing of the same unit as in F, G, before (H) and after (I) picrotoxin (2·5 mg./kg.). The onset of caudate stimulation is indicated by an upward arrow (Stim) and vertical broken lines. About 45 sweeps were used to compute each histogram. *Abscissa*: time in msec.; *ordinate*: number of spikes. (Precht and Yoshida, 1971).

addition, a pronounced depressant action of GABA on nigral neurons was demonstrated (Feltz, 1971b). Some of these observations are illustrated in *Fig*. 68. Of particular interest was the finding that glycine was totally ineffective as a depressant, and both ACh and DA only weakly depressed nigral neurons which were markedly depressed by GABA.

A functional role of GABA in the substantia nigra is further supported by the very high concentrations of GABA and glutamic acid decarboxylase in this area (Albers and Brady, 1959; Fahn and Côté, 1968a; Chalmers and others, 1970; McGeer and others, 1971; Okada and others, 1971). Moreover, the GABA content of the substantia nigra decreased markedly following lesions of the caudate nucleus or its efferent fibres (Kim, Bak, Hassler, and Okada, 1971). GABA is taken up extremely readily by slices of the rat substantia nigra (Hökfelt and others, 1970). Finally, the terminals of striatonigral fibres contain elongated vesicles, and lesions in the striatum have been shown to lead to degeneration

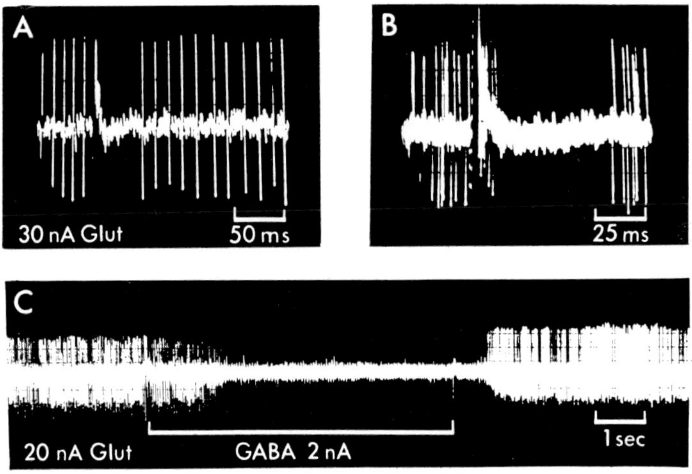

Fig. 68. Extracellular recordings from a nigral unit showing inhibition of firing by caudate stimulation, and depression by GABA. The firing was maintained by glutamate (Glut) administered continuously. A, Single stimulus (0·2 msec. duration, 1·5× threshold) to the caudate nucleus, evoking inhibition only. B, 2·0× threshold, showing an early excitation, but only slightly longer inhibition (four or five superimposed sweeps at 3/sec.). C Depression of glutamate-evoked firing of the same unit by GABA (2 nA, 6 sec. duration). (Feltz, 1971b.)

of some large axosomatic terminals containing elongated vesicles in the substantia nigra (Bak, Hassler, Parizek, and Wagner, 1970; Grofová and Rinvik, 1970; Kim and others, 1971). A point of interest is that this is another example of a long-axon neuron which releases GABA as a transmitter, the other being the Purkinje cell. Most other types of inhibitory cells which are thought to release GABA are interneurons.

Monoamines

Although the actions of the monoamines on nigral neurons have not been studied to any great extent (cf. Feltz, 1971b), investigations involving light and electron microscopic autoradiography of the substantia nigra following intraventricular administration of labelled DA, NA, and 5-HT have provided some information on the possible role of these monoamines in this region (Parizek, Hassler, and Bak, 1971). Labelled DA was limited largely to the cell bodies and their dendritic branches in the substantia nigra, although a small number of nerve terminals appeared to be labelled as well. This supports the concept of a nigro-neostriatal dopaminergic projection discussed earlier in this chapter. Labelled NA was distributed similarly, though more densely, but since the substantia nigra contains a high concentration of DA and little NA, it was thought that the labelled NA was taken up exclusively by dopaminergic neurons. The weak labelling of nerve terminals by DA and NA suggested that the *input* of catecholaminergic terminals to the substantia nigra is unlikely to be great.

In contrast, labelled 5-HT was associated with nerve terminals, not with perikarya, in the substantia nigra (Parizek and others, 1971). This observation, together with the high concentration of 5-HT in the substantia nigra, is in agreement with the concept of a 5-hydroxytryptaminergic projection to the nigra (*see* Andén and others, 1966; Ungerstedt, 1971). A large proportion of nigral 5-HT is present in the synaptosome fraction (Kim and others, 1970), supporting this view.

SUMMARY

Some of the transmitters which are thought to operate in the basal ganglia and substantia nigra are illustrated diagrammatically in *Fig.* 69.

Caudate Neurons

There is good evidence that DA is an inhibitory transmitter in the caudate nucleus, most probably released from fibres which originate in the substantia nigra. The caudate nucleus has the highest levels of DA in the C.N.S., and both biochemical and histochemical studies have established the presence of DA-containing nigral neurons which send their axons to the caudate nucleus. DA is released spontaneously from the caudate nucleus, and this release is increased by stimulating various areas, including the substantia nigra. In particular, caudate neurons which are inhibited by stimulating the substantia nigra are depressed by DA, and both effects can be reversibly blocked by alpha-methyl-dopamine. Although pharmacological observations have also raised the possibility that DA is an inhibitory transmitter released on to

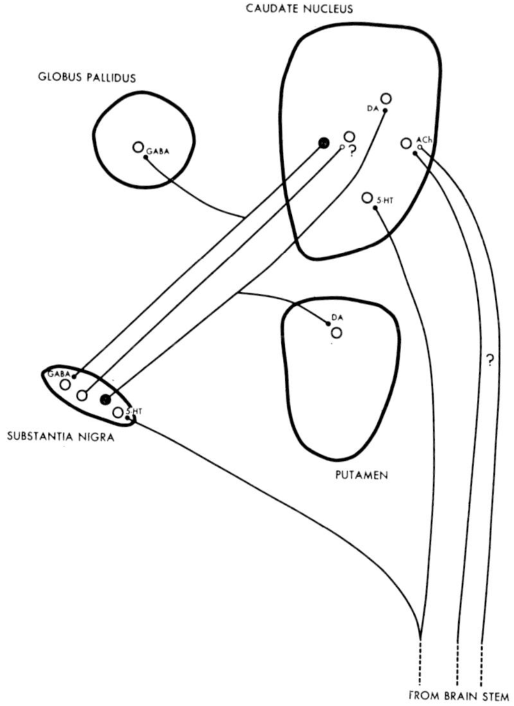

Fig. 69. Schematic diagram of the basal ganglia and substantia nigra, showing the transmitters likely to be involved in some of the neuronal pathways. Excitatory neurons and terminals are represented by O, inhibitory ones by ●. *See* text for details.

caudate neurons from fibres which originate in the thalamus, these results are best interpreted as having resulted from indirect activation of nigral fibres by thalamic stimulation, as neither biochemical nor histochemical evidence supports the postulate of dopaminergic thalamo-caudate fibres.

The excitatory pathway from the substantia nigra to the caudate nucleus is not dopaminergic, but the transmitter involved has not been identified.

NA is probably not a major transmitter in the caudate nucleus, but the presence of a few noradrenergic fibres in this structure cannot be excluded.

The caudate nucleus contains relatively high levels of 5-HT and 5-HT-containing nerve-fibres, which are thought to arise in the raphe nuclei of the brain-stem. Stimulation of the rostral raphe nuclei increases the release of 5-HT from the caudate nucleus. 5-HT depresses the firing of some caudate neurons. Thus, 5-HT is possibly an inhibitory transmitter in the caudate nucleus, but direct pharmacological evidence for the pathway involved has yet to be obtained.

High quantities of ACh and the enzymes for its synthesis and degradation are present in the mammalian caudate nucleus. ACh excites or depresses the firing of different populations of caudate neurons, both actions being mediated by muscarinic receptors. Some caudate neurons which are excited or inhibited by stimulation of nucleus ventralis anterior of the thalamus are excited or depressed by ACh, and both the chemical and the synaptic activation can be blocked by atropine. It seems, therefore, that ACh is both an excitatory and an inhibitory transmitter in the caudate nucleus, released from fibres which can be activated by thalamic stimulation. The presence of such cholinergic pathways is supported by the observation that the release of ACh from the caudate nucleus is increased following stimulation of the thalamus. However, results from the pharmacological, biochemical, and histochemical studies concerning cholinergic pathways to the caudate nucleus are conflicting. Thus, although the available data are consistent with the existence of cholinergic transmission in the mammalian caudate nucleus, the origins of these fibres are not at all certain.

The excitant action of glutamate and the depressant action of GABA on caudate neurons have not been related to synaptic activity in the caudate nucleus, although both amino-acids are possibly transmitters in this structure.

Putamen Neurons

As for the caudate nucleus, the results of biochemical and histochemical studies suggest the presence of a dopaminergic pathway from the substantia nigra to the putamen. Although DA excites and depresses the firing of different putamen neurons, and although it has been postulated that DA is an *excitatory* transmitter released from nigral fibres on to putamen neurons, the pharmacological evidence for this requires further investigation.

The significance of the actions of glutamate, GABA, 5-HT, NA, and ACh on putamen neurons has not been determined.

Globus Pallidus Neurons

Biochemical, histochemical, and other studies suggest that certain monoamines, amino-acids, and ACh have functions in the globus pallidus, but there is no direct pharmacological evidence that any of these substances is a transmitter in this structure. Indirect evidence, however, suggests that GABA is a transmitter in the globus pallidus, released from fibres which originate in the caudate nucleus.

Nigral Neurons

GABA is a powerful depressant of neurons in the substantia nigra, ACh and DA being weaker, and glycine being relatively ineffective. The monosynaptic inhibition of nigral neurons evoked by stimulating the caudate nucleus is blocked by picrotoxin, but unaffected by strychnine, suggesting that it is mediated by GABA. The concentration of GABA in the substantia nigra is the highest of any area in the C.N.S., and lesions of the caudate nucleus or its efferent fibres result in a decrease of this GABA content. These and other findings provide good evidence that GABA is an inhibitory transmitter released on to nigral neurons from fibres which originate in the caudate nucleus.

Histochemical and electron microscopy combined with autoradiography studies suggest that 5-HT, but not DA or NA, is likely to be a transmitter on many nigral neurons, but pharmacological studies have not yet been done to investigate this.

Chapter

6

LIMBIC SYSTEM

THE limbic system (rhinencephalon or 'nose-brain') consists of a large number of structurally diverse brain regions which are concerned with olfaction, emotion, short-term memory, sleep and arousal, and various other functions poorly understood. *Fig.* 70 illustrates the relationships between the main limbic structures of the mammalian brain. The neuroanatomy of the limbic system is too complex to describe here, and this chapter deals only with those limbic structures which have been studied pharmacologically at the neuronal level. The olfactory bulb is also included, although it does not strictly belong to the rhinencephalon.

HIPPOCAMPUS

The hippocampus is morphologically a relatively well-organized structure, the neurons being alined in well-defined layers. Nevertheless, the multiple connexions between the layers (at least eight have been described) and those between the hippocampus and other structures (for example, contralateral hippocampus, temporal and pyriform cortices, hypothalamus, olfactory tubercle, septum, anterior thalamus, medial forebrain bundle, mammillary body, midbrain tegmentum, amygdala, and others) have increased the complexity of interpreting the results of neurophysiological experiments on the hippocampus (*see* review by Green, 1964).

Hippocampal Pyramidal Cells

Hippocampal pyramidal cells may be identified electrophysiologically by their depth below the cortical surface (0·2–0·6 mm.), their antidromic responses to stimulation of the fimbria, and the characteristic inhibitory 'pause' which follows fimbrial, contralateral hippocampal, or 'local' cortical stimulation (Kandel, Spencer, and Brinley, 1961; Andersen, Eccles, and Løyning, 1964). The long-duration IPSPs evoked in pyramidal cells by these different modes of stimulation are thought to be mediated via pyramidal axon collaterals which synapse with basket cells, the inhibitory interneurons.

AMINO-ACIDS

Glutamate and DLH have been shown to excite hippocampal neurons, including pyramidal cells (Stefanis, 1964a; Herz and Nacimiento, 1965; Biscoe and Straughan, 1966; Steiner and Ruf, 1966; Steiner, 1969; Curtis, Felix, and McLennan, 1970; Curtis Duggan, Felix and others, 1971), but these effects have not been

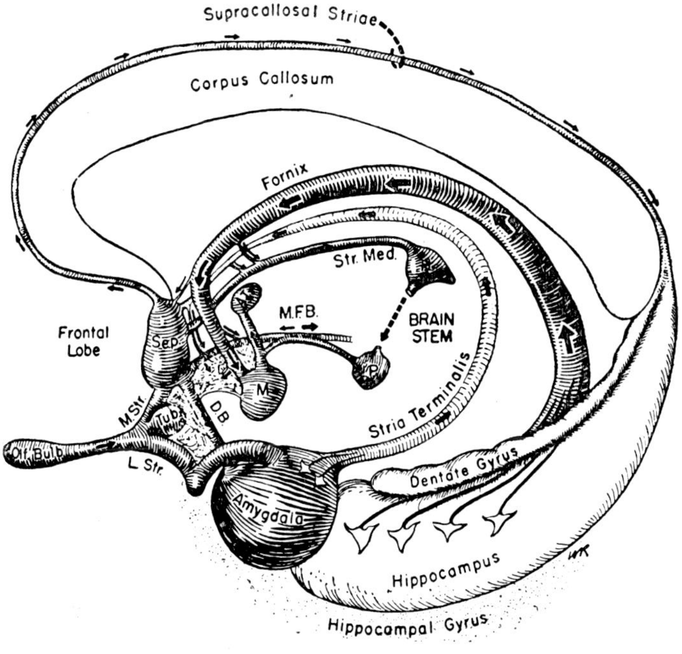

Fig. 70. Schematic representation of the relationships between the main subcortical structures and connexions of the rhinencephalon, drawn as though all of them could be seen from the medial aspect of the right hemisphere (*after* Krieg). Abbreviations: A.T., Anterior nucleus of the thalamus; D.B., Diagonal band of Broca; H., Habenula, a part of the epithalamus; I.P., Interpeduncular nucleus; L.Str., Lateral olfactory stria; M, Mammillary body, a part of the hypothalamus; M.F.B., Medial forebrain bundle; M.Str., Medial olfactory stria; Olf. Bulb, Olfactory bulb; Sep., Region of the septal nuclei; Str. Med., Stria medullaris; Tub. Olfactory tubercle. (MacLean, 1949.)

related to synaptic excitation in the hippocampus. Much more pharmacological and neurochemical data are required before it is permissible to suppose that an excitatory amino-acid functions as a transmitter in the hippocampus.

The available data on depressant amino-acids, however, provide relatively strong evidence that GABA is an inhibitory transmitter in the hippocampus. GABA is present in the mammalian hippocampus (Kržalić and others, 1962; Okada and others, 1971), concentrated especially in the pyramidal-cell layer of the cortex (Hökfelt and Ljungdahl, 1971b). The activity of the enzyme, glutamate decarboxylase, is distributed mainly in the molecular and pyramidal layers (Albers and Brady, 1959; Storm-Mathisen and Fonnum, 1971), which contain the terminals and cell bodies of the inhibitory basket cells. The pyramidal cells themselves do not exhibit much glutamate decarboxylase activity, nor does the fimbria, a major efferent and afferent tract (Storm-Mathisen and Fonnum, 1971). Moreover, hippocampal levels of glutamic acid decarboxylase remained unaltered after interruption of the fimbria and other afferent pathways, supporting the view that the decarboxylase activity is associated with interneurons (Storm-Mathisen, 1972). Similarly, the activity of GABA transaminase is high in the vicinity of pyramidal cell bodies (van Gelder, 1965b; Pohle and Matthies, 1970; cf. Knyihár and Csillik, 1970). In addition, GABA is readily taken up by slices of the rat hippocampus (Iversen and Johnston, 1971), and radioactive GABA can be accumulated by almost half of the nerve terminals in hippocampal homogenates (Iversen and Bloom, 1972).

GABA had a powerful depressant action on various types of hippocampal cortical neurons, including those identified as pyramidal cells (Stefanis, 1964a; Herz and Nacimiento, 1965; Biscoe and Straughan, 1966; Steiner and Ruf, 1966; Curtis, Felix and McLennan, 1970; Curtis, Duggan, Felix, and others, 1971). On the basis of a comparison of electrophoretic currents, GABA was a more effective depressant than glycine or β-alanine, and bicuculline reversibly reduced the depression by GABA and β-alanine in concentrations which had no effect on that by glycine (Curtis, Felix, and McLennan, 1970; Curtis, Duggan, Felix, and others, 1971). In particular, bicuculline diminished the basket cell inhibition of hippocampal pyramidal cells (Curtis, Felix, and McLennan, 1970). Such results are illustrated in *Fig.* 71.

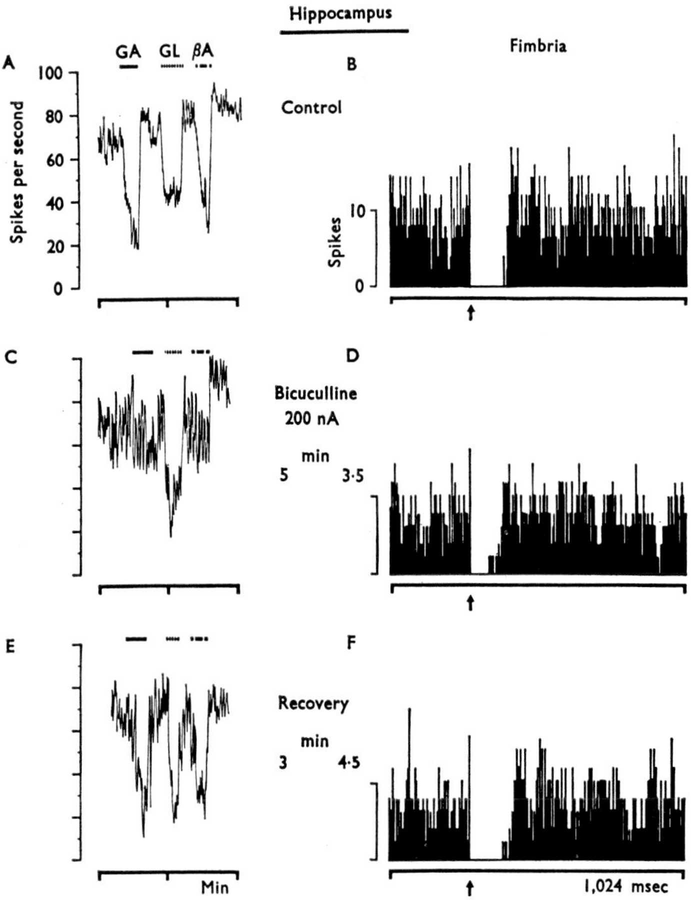

Fig. 71. Interaction of bicuculline with synaptic inhibition and depressant amino-acids on a hippocampal pyramidal cell. A, C, E, Chart recordings showing the depressant actions of GABA (GA, 20 nA, solid line), glycine (GL, 40 nA, dotted line) and β-alanine (βA, 40 nA, broken line) on the firing maintained by DLH. B, D, F, Inhibition of firing by electrical stimulation of the fimbria (3 V, 0·1 msec. duration, frequency 1/2 sec., onset marked by an arrow). The number of action potentials (ordinate) occurring in 256 intervals of 4 msec. duration (abscissa, 1024 msec. period of analysis) was analysed by a computer for a total of 30 sweeps. A, B, Control records. C, D, 5 and 3·5 min. respectively after the beginning of an ejection of bicuculline (5 m*M* in 165 m*M* NaCl; 200 nA for 6 min.). E, F, 3 and 4·5 min. respectively, after the bicuculline ejection was terminated. (Curtis, Felix, and McLennan, 1970.)

These findings, in conjunction with the neurochemical data, suggest that GABA is an inhibitory transmitter in the hippocampal cortex, most probably released from basket axons on to pyramidal cells. Unfortunately, some unusual pharmacological observations are inconsistent with this interpretation. In contrast to other structures of the C.N.S. studied, electrophoretically administered strychnine apparently reduced the depressant actions of glycine, β-alanine, and GABA on hippocampal neurons with no obvious selectivity towards glycine (Curtis, Felix, and McLennan, 1970; Curtis, Duggan, Felix, and others, 1971). If this is true, strychnine should reduce the inhibition of pyramidal cells by impulses in basket axons. However, although electrophoretic strychnine occasionally reduced such inhibition (Curtis, Felix, and McLennan, 1970), intravenous strychnine had no effect on this inhibition in earlier studies (Andersen and others, 1963; *see also* Andersen, Holmqvist, and Voorhoeve, 1966). These findings are difficult to interpret at present, but may indicate that either the inhibition studied is more complex (possibly involving a link with glycine as a transmitter), or that GABA receptors in the hippocampus are readily affected by strychnine compared with other central structures. Still another possibility is that the transmitter is another amino-acid (taurine ?), although the neurochemical findings favour GABA as the most likely transmitter. Clearly, these problems require further investigation.

Acetylcholine

ACh is likely to be a transmitter in the mammalian hippocampus. ACh excited or depressed the firing of different proportions of unidentified neurons in the hippocampal cortex (Stefanis, 1964a; Herz and Nacimiento, 1965; Biscoe and Straughan, 1966; Steiner, 1968). Most cholinoceptive neurons were located in the superficial layers of the cortex, corresponding to the hippocampal pyramidal cell bodies and their main dendritic processes (Biscoe and Straughan, 1966); and in unanaesthetized cats, ACh consistently excited hippocampal neurons identified as pyramidal cells (Stefanis, 1964a; Salmoiraghi and Stefanis, 1965). The excitation of pyramidal cells by ACh was of a slow time course and mediated mainly by muscarinic receptors (*Fig.* 72).

The distribution of cholinoceptive cells in the hippocampal cortex of the cat corresponded with the distribution of AChE (Biscoe and Straughan, 1966), and in various mammals there is a

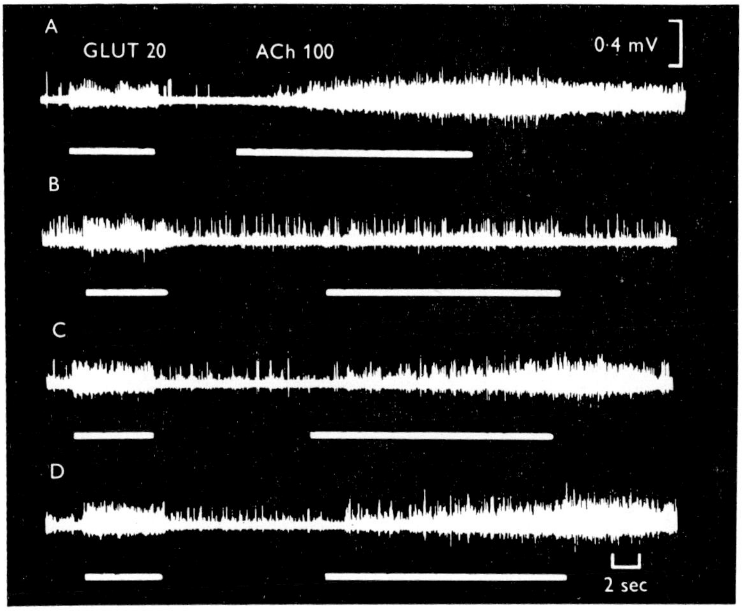

Fig. 72. Extracellularly recorded spike discharges evoked by glutamate and ACh from a group of neurons located at a depth of 1000 μm. from the surface of the hippocampus (region CA 2–3). The periods of administration of L-glutamate (GLUT, 20 nA) and ACh (100 nA) are indicated by the short and long bars, respectively. Atropine (40 nA for 3 min.) was administered between A and B. B, 3·5 min. after atropine, showing the unchanged response to glutamate, and the lack of effect of ACh. The spontaneous firing which had been initiated by the first application of ACh was unaffected by atropine. C, D, 3 and 4 min. respectively, after the atropine ejection had been terminated (Biscoe and Straughan, 1966.)

close relation between the distribution of AChE and choline acetyltransferase (Lewis, Shute, and Silver, 1967; Storm-Mathisen, 1970; Fonnum, 1970). *Fig.* 73 is a micrograph of a section through the guinea-pig hippocampus, stained for AChE. The differential distribution of AChE in well-defined layers is typical of the hippocampus. Additional evidence for cholinergic transmission in the hippocampus is that the choline acetyltransferase content of hippocampal homogenates is confined mainly to the synaptosome fraction (Fonnum, 1970).

Various observations suggest that cholinergic fibres to the hippocampal cortex arise in the septum. Physiological evidence for the presence of a septo-hippocampal pathway was first obtained by Andersen, Bruland, and Kaada (1961), who evoked potentials in the hippocampus by stimulating the septum. Of particular relevance to the present discussion is that intravenous eserine potentiated the synaptic responses of pyramidal cells to septal

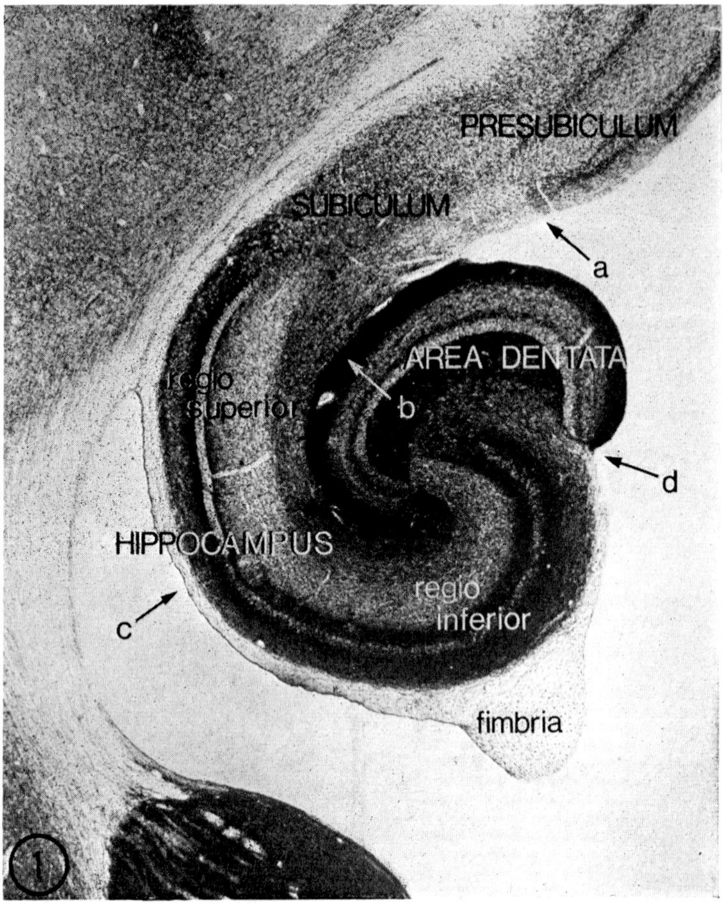

Fig. 73. Photomicrograph of a horizontal section (50 μm.) through the cerebral hemisphere of an adult guinea-pig, showing the distribution of AChE. The general relationships between the presubiculum, subiculum, hippocampus, and dentate area are labelled. Subdivisions of the hippocampus are indicated by arrows labelled a to d. (× 27). (Geneser-Jensen, 1972.)

stimulation, and this effect was antagonized by scopolamine, a muscarinic antagonist of ACh (Brücke, Gogolák, and Stumpf, 1963). Moreover, the cholinesterase-containing fibres to the rat hippocampus mainly make up the septal radiation, the fibres of which arise in the medial septal nucleus and the nucleus of the diagonal band (Lewis and Shute, 1967; Storm-Mathisen, 1970).

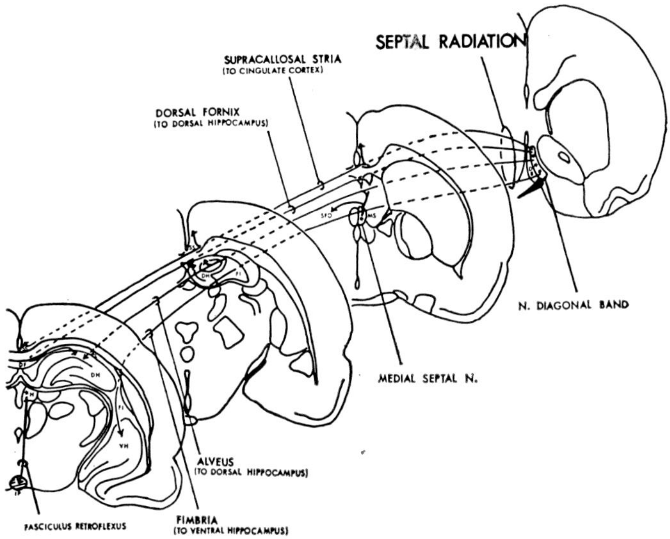

Fig. 74. Expanded diagram (natural spacing of transverse sections increased five times) of the rat brain, showing the septal radiation of cholinesterase-containing fibres which arise from the medial septal nucleus (MS) and the nucleus of the diagonal band (DB). These fibres supply the subfornical organ (SFO), cingulate cortex via the supra-callosal stria (SS), dorsal hippocampus (DH) via the dorsal fornix (DF) and alveus (AL), and ventral hippocampus (VH) via the fimbria (FI). The fasciculus retroflexus, with cholinesterase-containing fibres running from the habenular nuclei (H) to the interpeduncular nucleus (IP) is also included. (Lewis and Shute, 1967.)

This AChE-containing projection is illustrated diagrammatically in *Fig*. 74. Hippocampal afferents other than those which originate in the septal region (for example, temporo-ammonic, commissural, and mossy fibres) contain no cholinesterase, and are thought to be non-cholinergic (Lewis and Shute, 1967). Investigations involving choline acetyltransferase analyses support the view that the hippocampus receives a massive cholinergic innervation from the septum via the fimbria, particularly as lesions in the septum or in hippocampal afferents from the septum led to a loss of choline acetyltransferase in the hippocampus (Lewis and others, 1967; McGeer, Wada, Terao, and Jung, 1969; Storm-Mathisen, 1972). Finally, it has recently been demonstrated that the release of ACh from the dorsal hippocampus was increased by stimulating the surface of the septum (Smith, 1972). Despite these studies, however, it is still necessary to demonstrate that the synaptic excitation of a single pyramidal cell, evoked by stimulation of the septum, as well as the excitant action of ACh are reversibly and specifically blocked by antagonists of ACh, and potentiated by inhibitors of cholinesterase.

Monoamines

The firing of hippocampal pyramidal cells was consistently depressed by 5-HT and NA when these cells were sensitive to the monoamines (Stefanis, 1964a; Salmoiraghi and Stefanis, 1965). Similarly, 5-HT, NA, and DA mainly depressed the spontaneous or glutamate-evoked firing of unidentified neurons in the hippocampal cortex, most of which were recorded at depths corresponding to the pyramidal-cell layer (Stefanis, 1964a; Herz and Nacimiento, 1965; Biscoe and Straughan, 1966). In a few cases, 5-HT excited hippocampal neurons (Biscoe and Straughan, 1966). The relatively high concentrations of 5-HT in limbic structures, including the hippocampus (Paasonen, MacLean, and Giarman, 1957; Kuntzman and others, 1961), and the presence of 5-HT and NA in nerve-fibres (Fuxe, 1965; Blackstad, Fuxe, and Hökfelt, 1967), raise the possibility that these monoamines are transmitters in the hippocampus. Moreover, the distribution of monoamine fluorescence in the hippocampal cortex corresponds reasonably well with that of monoamine oxidase (Manocha and others, 1967; Pohle and Matthies, 1970; Mellgren and Geneser-Jensen, 1972). Eidelberg, Goldstein, and Deza (1967) have described what appear to be axosomatic boutons containing 5-HT on hippocampal neurons. However, there is yet no direct pharmacological evidence for a transmitter role in the hippocampus for any of the monoamines.

Amygdala

Glutamate excited and GABA depressed all cells, whereas ACh excited a small proportion and the monoamines (5-HT, NA, and

DA) depressed the firing of most neurons in the amygdala (Straughan and Legge, 1965). Intravenously administered 5-HTP decreased, whereas L-DOPA increased the rate of spontaneous firing of neurons in the amygdaloid complex, and histochemical studies suggested that amygdaloid neurons appear to be covered with spots containing 5-HT (Eidelberg and others, 1967). However, these studies are too limited to draw conclusions about possible transmitters in the amygdala.

Newly formed amino-acids (including glutamate and GABA) were detected in perfusates of the monkey amygdala (De Feudis and others, 1970), but little else is known about amino-acids and amygdaloid neurons.

Septum

Neurons in the diagonal band of Broca (septum) were insensitive to ACh, but were fired by glutamate, depressed by GABA, and in most cases also depressed by 5-HT and NA (Herz and Gogolák, 1965). The sample of neurons tested was small, however, and little can be said about the significance of these actions in the septum. The conclusions about transmitter release drawn from perfusion studies of the anterior horn of the lateral ventricle (Chapter 5; caudate neurons) apply equally well to the septum.

Olfactory Bulb

The olfactory bulb is a relatively simple structure consisting of nerve-cells arranged in concentric layers (*see* references by Andres, 1965). Each layer is composed largely of a particular nerve-cell type which may be identified according to its depth and response to electrical stimulation of efferent and afferent fibres to the olfactory bulb. Tufted cells are 500–700 μm. deep and have long axons which form the rostral part of the anterior commissure and terminate in the granule-cell layer of the contralateral olfactory bulb. Mitral cells are located at a depth of 700–1100 μm. and their axons form the lateral olfactory tract which terminates in the prepyriform cortex. The deepest layer (greater than 1200 μm.) consists of granule cells. Some of these features are represented diagrammatically in *Fig.* 75.

Mitral Cells

Mitral cells are inhibited for long periods (several tens or hundreds of msec.) following stimulation of their axons which make up the lateral olfactory tract, and although it is agreed that this inhibition involves an initial excitation of granule cells, opinion is divided as to whether this excitation is mediated predominantly by recurrent collaterals of mitral axons or by a dendro-dendritic mechanism (Green, Mancia, and von Baumgarten, 1962; Phillips, Powell, and Shepherd, 1963; Yamamoto,

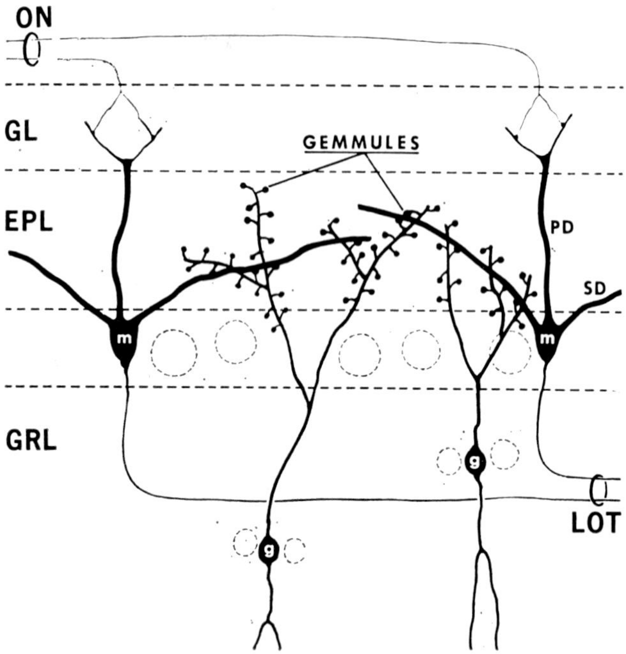

Fig. 75. Layers and connexions in the olfactory bulb (*after Cajal*). Abbreviations: EPL, External plexiform layer; GRL, Granular layer; ON, Olfactory nerve; LOT, Lateral olfactory tract; g, Granule cell; m, Mitral cell; PD, Primary mitral dendrite; SD, Secondary mitral dendrite. Tufted cells (in the EPL) are not shown. (Rall, Shepherd, Reese, and Brightman, 1966.)

Yamamoto, and Iwama, 1963; Ochi, 1963; Rall, Shepherd, Reese, and Brightman, 1966; Nicoll, 1969, 1972; Westecker, 1970). The more recent evidence favours the existence of a dendro-dendritic mechanism, and is the view adopted here. In either case there is relatively good pharmacological evidence that the transmitter responsible for the inhibition is GABA (Felix and McLennan, 1971; McLennan, 1971; Nicoll, 1971).

AMINO-ACIDS

The olfactory bulb contains large quantities of GABA (Popov, Pohle, Rösler, and Matthies, 1967). GABA was usually more effective than glycine as a depressant of mitral cells, and the depressant action of GABA, but not that of glycine, was reversibly blocked by bicuculline and picrotoxin (Felix and McLennan, 1971; McLennan, 1971; Nicoll, 1971). Although glutamate, asparate, and DLH excited some mitral cells, in contrast to all other structures of the mammalian C.N.S. studied, these 'excitant' amino-acids depressed the firing of most mitral cells, this depression sometimes being associated with a simultaneous increase in the rate of firing of neighbouring neurons having small action potentials (von Baumgarten, Bloom, Oliver, and Salmoiraghi, 1963; Felix and McLennan, 1971; McLennan, 1971; Nicoll, 1971). Since the depression by DLH, glutamate, and aspartate could be reversed into an excitation by moving the electrode tip or by administering either bicuculline or picrotoxin, the depression was interpreted as resulting from the chemical excitation of granule cells, which in turn inhibited the mitral cells (Felix and McLennan, 1971; McLennan, 1971; Nicoll, 1971). A rather elegant confirmation of this view was provided by the observation that electrophoretically administered Mg^{2+} ions (which, amongst other actions, can block the release of synaptic transmitters) antagonized the depression of mitral cells by DLH, glutamate, and aspartate (Nicoll, 1971). The results with convulsants suggested that the granule cell to mitral cell inhibition is mediated by GABA. However, attempts to shorten the inhibition of mitral cell firing evoked by stimulation of the lateral olfactory tract or olfactory mucosa with bicuculline or picrotoxin have failed; on the contrary, bicuculline often increased the duration of the inhibitory 'pause' (Felix and McLennan, 1971; McLennan, 1971). Such results from a mitral cell are illustrated in *Fig.* 76.

The spontaneous firing of the mitral cell was depressed to approximately equal extents by GABA (20 nA), DLH (10 nA), and glycine (10 nA), and stimulation of the lateral olfactory tract inhibited the firing for several hundred msec. (*Fig.* 76A, D). During the electrophoretic ejection of bicuculline, the depression by GABA was blocked, that by DLH was reversed to an excitation, whereas that by glycine was unaffected (*Fig.* 76B), and the inhibition by lateral olfactory tract stimulation was increased in duration (*Fig.* 76E). Recovery occurred within a few minutes after the current ejecting bicuculline had been terminated.

The increase in duration of the inhibition of mitral cell firing evoked by a volley to the lateral olfactory tract during the administration of bicuculline was interpreted as a block of a tonic GABA mediated inhibition of granule cells (Felix and McLennan, 1971;

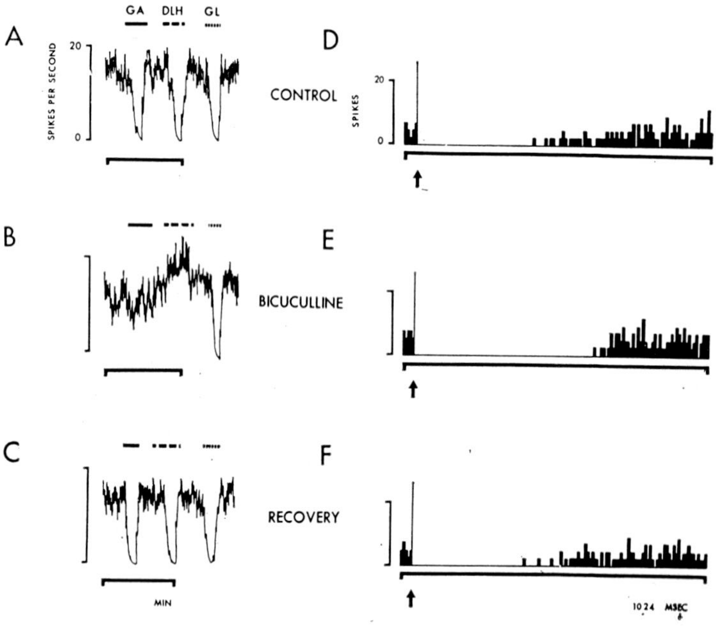

Fig. 76. Effects of bicuculline on the actions of amino-acids and synaptic inhibition of a mitral cell. A, B, C, Chart recordings of the firing frequency of the cell, showing the actions of GABA (GA, solid line, 20 nA; retaining current 17 nA), DLH (dashed line, 10 nA; retaining current 3·5 nA), and glycine (GL, dotted line, 10 nA; retaining current 7 nA) before (A), during (B), and 1 min. after (C) an ejection of bicuculline (5 m*M* in 165 m*M* NaCl; 80 nA for 3 min.; retaining current 9 nA). The histograms in D, E, F, show the inhibition of spontaneous firing of the same cell produced by stimulation of the lateral olfactory tract (40 V, 0·1 msec. duration, frequency 0·5/sec., marked with an arrow) before (D), during (E), and 6 min. after (F) an ejection of bicuculline (80 nA for 8·5 min.). The number of action potentials (ordinate) occurring in 128 intervals each of 8 msec. duration (abscissa, 1024 msec. period of analysis) was counted for a total of 40 sweeps by means of a computer. (Felix and McLennan, 1971.)

McLennan, 1971). In other words, it was assumed that GABA is released synaptically on to granule cells from inhibitory fibres of an unknown origin, as well as on to mitral dendrites from granule processes, and that the inhibitory synapses on the granule cells are more sensitive to bicuculline. However, no lengthening of the inhibitory pause was observed by Nicoll (1971), using either bicuculline or picrotoxin. Furthermore, there is no evidence for a disinhibition of mitral cell firing by stimulation of inputs to the olfactory bulb, although the possibility of such a pathway nevertheless exists.

If paired stimuli are delivered to the lateral olfactory tract, the second response is inhibited and the degree of inhibition declines when the interval between the stimuli is increased (Westecker, 1969; Nicoll, 1970). Bicuculline and picrotoxin reduced or blocked this inhibition, supporting the hypothesis that GABA is the transmitter mediating granule cell to mitral cell inhibition (McLennan, 1971; Nicoll, 1971). Nicoll (1971) has interpreted this in terms of dendro-dendritic inhibition, as shown in *Fig.* 77. The dendrites of mitral cells make extensive contact with the processes of granule cells (*Fig.* 77A), and reciprocal synapses occur between the mitral dendrites and the gemmules of granule cell processes (*Fig.* 77C). Field potentials recorded from the external plexiform layer are composed of two negative waves following a single stimulus to the lateral olfactory tract (*Fig.* 77B). The first negative wave (N_1) reflects the depolarization of mitral cell dendrites which activate excitatory synapses (arrow labelled E in *Fig.* 77C) on granule cell dendrites, and the depolarization of granule cell dendrites which produces the second negative wave (N_2) in turn activate inhibitory synapses oriented in the opposite direction (arrow labelled I in *Fig.* 77C). Superimposed field potentials recorded in the external plexiform layer following paired stimuli to the lateral olfactory tract are shown in *Fig.* 77D–G. Intravenous picrotoxin reduced the inhibition in one animal (*Fig.* 77D, E) and bicuculline blocked the inhibition in another preparation (*Fig.* 77F, G).

The inhibition of firing by stimulation of the lateral olfactory tract has been shown to be resistant to strychnine (Green and others, 1962; Salmoiraghi, Bloom, and Costa, 1964; Felix and McLennan, 1971; McLennan, 1971; Nicoll, 1971), which blocked the depressant action of glycine without affecting that of GABA, DLH, glutamate, or aspartate (Felix and McLennan,

1971; McLennan, 1971; Nicoll, 1971). This inhibition is also resistant to antagonists of ACh (Salmoiraghi and others, 1964). Thus, although interpretation differs as to the complete sequence of events leading to the inhibition of mitral cell firing, it is agreed that the final inhibitory synapse involves GABA as the transmitter.

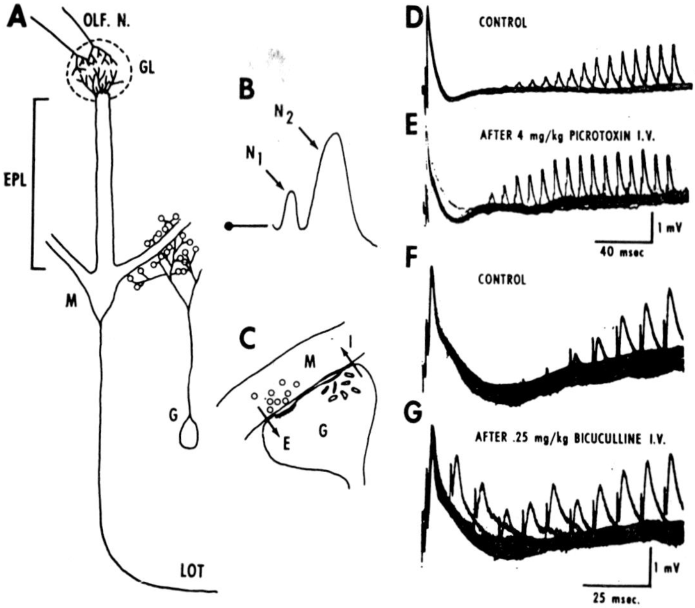

Fig. 77. A, Diagram of a mitral cell and its connexions in the olfactory bulb. Olfactory nerves (OLF.N) synapse with mitral cells (M) in the glomerulus (GL). Mitral axons form the lateral olfactory tract (LOT). Mitral cell dendrites make contacts with granule cell (G) processes in the external plexiform layer (EPL). B, Tracing of extracellular field potential recorded in the external plexiform layer (negativity upwards). The time between the two negative peaks is 3·5 msec. C, Diagram of reciprocal synapses between mitral (M) and granule (G) cell processes. D, Superimposed extracellular field potentials evoked in the EPL by pairing stimuli to the LOT (negativity upwards). E, As in D, but after an intravenous injection of picrotoxin (4 mg./kg.). F, Before, and G, After intravenous bicuculline (0·25 mg./kg.) from another preparation. (Nicoll, 1971.)

Monoamines

NA depressed the firing of the majority of mitral cells tested (von Baumgarten and others, 1963; Bloom and others, 1964;

Salmoiraghi and others, 1964; McLennan, 1971). The alpha-adrenergic blocking agent, dibenamine (*N*-(2-chloroethyl)-dibenzylamine), reduced the inhibition of mitral cells produced by stimulation of the lateral olfactory tract, as well as the depressant action of NA (but not that of 5-HT or ACh) (Salmoiraghi and others, 1964). Reserpine and α-methyl-*m*-tyrosine, which deplete central stores of monoamines, also diminished this inhibition. These results suggested the participation of an adrenergic mechanism in the mediation of inhibition following stimulation of the lateral olfactory tract (Salmoiraghi and others, 1964). Recent studies by McLennan (1971), however, suggest that the depressant effect of NA on mitral cells is mediated indirectly, as is the depressant action of DLH. The depressant actions of both GABA and NA on mitral cells were blocked by bicuculline, whereas that of DLH was reversed to an excitation by the convulsant. Moreover, the depression by DLH was associated with firing of neighbouring neurons with small action potentials, and in some cases, when the electrode was moved slightly, DLH and NA excited neurons. It was postulated that DLH and NA excite granule cells, which in turn, inhibit the firing of mitral cells (McLennan, 1971). If this explains all the depressant actions of NA observed on mitral cells, it is not clear whether NA affects mitral cells directly at all.

Similarly, although mainly depressant actions of 5-HT have been observed on mitral cells (von Baumgarten and others, 1963; Bloom and others, 1964; Salmoiraghi and others, 1964), it is not known whether these are direct actions.

Acetylcholine

In contrast to neurons in all other regions studied in the C.N.S., most olfactory bulb neurons were depressed by ACh, excitation being rare (von Baumgarten and others, 1963; Bloom and others, 1964; Salmoiraghi and others, 1964). Physostigmine also depressed the firing of olfactory bulb neurons and prolonged the depressant action of ACh. A number of cholinergic antagonists blocked the depression by ACh, but had no effect on the inhibition of firing of mitral and tufted cells evoked by stimulating the lateral olfactory tract (Salmoiraghi and others, 1964). However, the functional significance of the action of ACh in the olfactory bulb is obscure. The depression of mitral cells by ACh may be largely or wholly indirect, as occurs with DLH, glutamate, aspartate, and NA, and the olfactory bulb may not be very different from other structures of the C.N.S. where ACh has predominantly excitant actions on neurons sensitive to it. It is clear that many aspects of the pharmacology of different types of olfactory bulb neurons require reinvestigation.

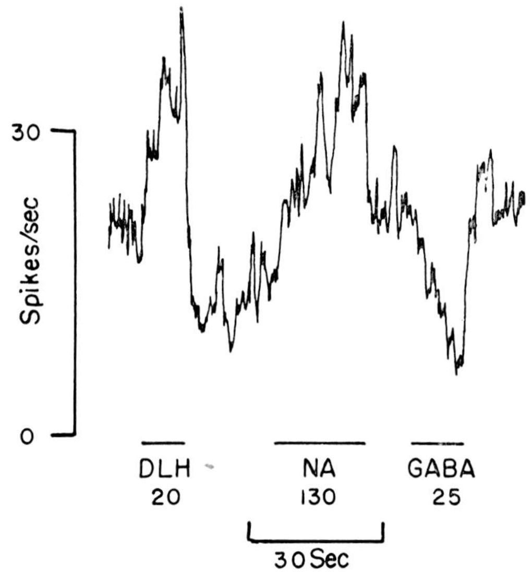

Fig. 78. Ratemeter record of the firing frequency of a neuron (granule cell ?) 50 μm distant from an identified mitral cell. DLH (20 nA), NA (130 nA), and GABA (25 nA) were administered electrophoretically during the periods indicated by the horizontal bars. (McLennan, 1971.)

Granule Cells

Neurons in the vicinity of mitral cells, but which were not mitral cells themselves and were most likely to have been granule cells, were excited by NA (McLennan, 1971), as shown in *Fig.* 78. The vast majority of catecholamine-containing nerve-fibres in the olfactory bulb are in the granular layer, and this fluorescence is predominantly from NA, as little DA is present in the olfactory bulb (Dahlström, Fuxe, Olson, and Ungerstedt, 1965; Ungerstedt, 1971). This raises the possibility that NA is an excitatory transmitter released on to granule cells (McLennan, 1971). The activation of such a noradrenergic excitatory pathway during stimulation of the lateral olfactory tract would account for the reduction of the inhibition of mitral cells by dibenamine (Salmoiraghi and others, 1964). This is supported by the observation that a neuron in the granular layer (presumably a granule cell) was fired by stimulating the lateral olfactory tract, the time course of the excitation paralleling that of the inhibition of mitral cells (Salmoiraghi and others, 1964).

GABA depressed the firing of granule cells (*Fig.* 78), and it has been proposed that GABA is an inhibitory transmitter released on to these cells (McLennan, 1971). However, evidence for this postulate is lacking.

The excitant and depressant actions of ACh (von Baumgarten and others, 1963) and the excitant actions of glutamate and aspartate (von Baumgarten and others, 1963; Nicoll, 1971; McLennan, 1971) on granule cells have not been related to synaptic pathways terminating on these neurons. The relatively high concentrations of glutamate and aspartate in the olfactory bulb (Popov and others, 1967) raise the possibility that both of these amino-acids are excitatory transmitters in this structure.

Pyriform Cortex

Only one study has been made of the sensitivity of neurons in the pyriform cortex to electrophoretically administered transmitter candidates (Legge, Randić, and Straughan, 1966). ACh excited less than one-third of the palaeocortical neurons tested, and in contrast to neocortical neurons (*see* Chapter 7) the cholinoceptive cells in the pyriform cortex appeared to be uniformly distributed in depth. The excitation of these cells by ACh had mainly muscarinic properties, although some evidence for nicotinic excitation was also observed. A small proportion of pyriform cortical neurons was depressed by ACh. ACh, choline acetyltransferase, and AChE are present in the pyriform cortex (MacIntosh, 1941; Feldberg and Vogt, 1948; Burgen and Chipman, 1951; Hebb and Silver, 1956; Lewis and Shute, 1967), but the role of ACh in this structure remains to be elucidated.

The monoamines, 5-HT, NA, DA, depressed the firing of approximately one-half of the palaeocortical neurons tested, no excitant actions being observed (Legge and others, 1966). 5-HT also blocked the synaptic excitation evoked by stimulating the lateral olfactory tract. However, although the pyriform cortex contains large quantities of 5-HT and moderate quantities of NA compared with the neocortex (Amin and others, 1954; Vogt,

1954; Bogdanski and others, 1957; Paasonen and others, 1957; Kuntzman and others, 1961), there is no direct evidence that either of these monoamines is a transmitter in the pyriform cortex.

Glutamate excited, whereas GABA depressed, the firing of palaeocortical neurons (Legge and others, 1966), but the significance of this is not clear. The transmitter mediating inhibition of palaeocortical neurons following stimulation of the lateral olfactory tract (Legge and others, 1966) has not been identified.

SUMMARY

Hippocampal Neurons

Glutamate excites pyramidal and other neurons in the hippocampal cortex, but this excitation has not been related to synaptic firing in the hippocampus.

GABA and glycine depress the firing of hippocampal pyramidal cells, GABA being more effective. Bicuculline reduces the inhibition of pyramidal cells following basket cell activation, and blocks the depressant action of GABA without affecting that of glycine. Moreover, the inhibition of pyramidal cells produced by basket-cell axons is relatively resistant to strychnine. This suggests that GABA is the inhibitory transmitter released on to hippocampal pyramidal cells by basket axon terminals, a hypothesis supported by the regional distribution of GABA, glutamic acid decarboxylase, and GABA transaminase in the hippocampal cortex, as well as by other neurochemical findings. The unusual observation that strychnine reduces the depressant actions of both glycine and GABA to a comparable extent, although it usually has little effect on the synaptic inhibition of pyramidal cells, cannot be readily explained at present.

ACh excites hippocampal pyramidal cells, this excitation being mediated mainly by muscarinic receptors. The distribution of ACh-excited cells in the hippocampal cortex corresponds well with the distribution of both AChE and choline acetyltransferase. Biochemical, histochemical, subcellular fractionation, and some pharmacological studies using intravenous eserine and scopolamine suggest that the main cholinergic innervation of the hippocampus arises in the septum, but this has yet to be unequivocally demonstrated by physiological experiments involving single-unit recording.

5-HT and NA depress the firing of hippocampal pyramidal cells sensitive to these substances, and biochemical and histochemical studies suggest that the monoamines have a synaptic function in the hippocampus. However, direct pharmacological evidence for this has not been obtained.

Amygdaloid and Septal Neurons

Amino-acids (glutamate, GABA) and endogenous amines (ACh, 5-HT, NA, DA) excite or depress the firing of different proportions of neurons in the amygdala and septum, but the significance of these actions has not been investigated.

Olfactory Bulb Neurons

The olfactory bulb contains relatively high GABA levels. GABA is usually more effective than glycine as a depressant of mitral cells, and the depression by GABA is blocked by bicuculline and picrotoxin, which do not affect the depression by glycine. Glutamate, aspartate, DLH, and NA also often depress the firing of mitral cells, and this depression is reversed to an excitation by bicuculline and/or picrotoxin, suggesting that the inhibition of mitral cells by granule-cell processes is mediated by GABA. The inhibition of firing of mitral cells following stimulation of the lateral olfactory tract or olfactory mucosa is unaffected by strychnine and antagonists of ACh, but is sometimes prolonged by bicuculline. However, the inhibition of mitral cells demonstrated by the effects of paired stimuli to the lateral olfactory tract on field potentials recorded in the external plexiform layer can be blocked by bicuculline and picrotoxin. Thus, although the complete sequence of events leading to mitral-cell inhibition by volleys in mitral axons is not entirely clear, there is good evidence that the inhibitory transmitter released on to mitral cells from granule processes is GABA.

The amines, ACh, 5-HT, and NA mainly depress the firing of mitral cells sensitive to them, but since the action of NA has been shown to be indirect, this may also be true for ACh and 5-HT.

Glutamate, aspartate, and NA excite granule cells, whereas GABA depresses their firing. Other evidence raises the possibility that these substances are transmitters on granule cells, but there are insufficient data to make definitive conclusions at this stage. In the case of NA, however, a noradrenergic excitatory input to granule cells (supported by fluorescence histochemistry) could account for the early observations that adrenergic antagonists can reduce the inhibition of mitral cells evoked by stimulating the lateral olfactory tract.

Palaeocortical Neurons

Certain neurons of the pyriform cortex can be excited by ACh and glutamate, and depressed by the monoamines (5-HT, NA, DA) and GABA, but whether any of these substances is a transmitter in the pyriform cortex remains to be investigated.

Chapter

7

CEREBRAL CORTEX

THE cellular architecture of the cerebral cortex is particularly complex, and is the reason why most pharmacological experiments have been done on unidentified neurons in this structure. To limit this chapter to a discussion of Betz cells would be a pity, as great progress has been made in understanding the pharmacology of synaptic transmission from studies of unidentified cerebral cortical neurons. This chapter is therefore subdivided into sections dealing with the three main groups of transmitters—ACh, amino-acids, and monoamines—and their actions on various types of cortical neurons. Only the neocortex is discussed, and for the sake of simplicity the different areas (sensorimotor cortex, visual cortex, and others) are treated as a single group, although it is understood that many subtle differences may exist.

ACETYLCHOLINE

Most of the recent information on transmitter mechanisms in the cortex indicates that ACh is not the principal transmitter mediating rapid synaptic excitation or inhibition, but may be responsible for the relatively slow, prolonged excitation and inhibition involving only certain populations of cortical cells. Three main lines of evidence suggest that ACh is a transmitter in the cerebral cortex: biochemical and histochemical evidence for the presence of ACh and the enzymes which synthesize and destroy it; pharmacological studies of cortical neurons; and experiments which have demonstrated the release of ACh from the surface of the cortex in amounts which correlate with the state of cortical activity.

It has long been known that ACh, choline acetyltransferase, and AChE are present in the cerebral cortex (MacIntosh, 1941; Feldberg, 1945; Feldberg and Vogt, 1948; Burgen and Chipman, 1951; Hebb and Silver, 1956; Hebb, 1957; MacIntosh, 1963; *see also* Bartolini, Weisenthal, and Domino, 1972). A proportion of the ACh content of the cerebral cortex is known to be present in synaptic vesicles (Whittaker and Sheridan, 1965). *Fig.* 79 is a diagram illustrating the distribution of AChE-containing fibres in the cerebral cortex of the cat (Krnjević and Silver, 1965). In the convexity of a gyrus, cholinesterase fibres travel subcortically on either side of the cholinesterase-free core of white matter. As they enter the cortex the fibres form a network in the deepest

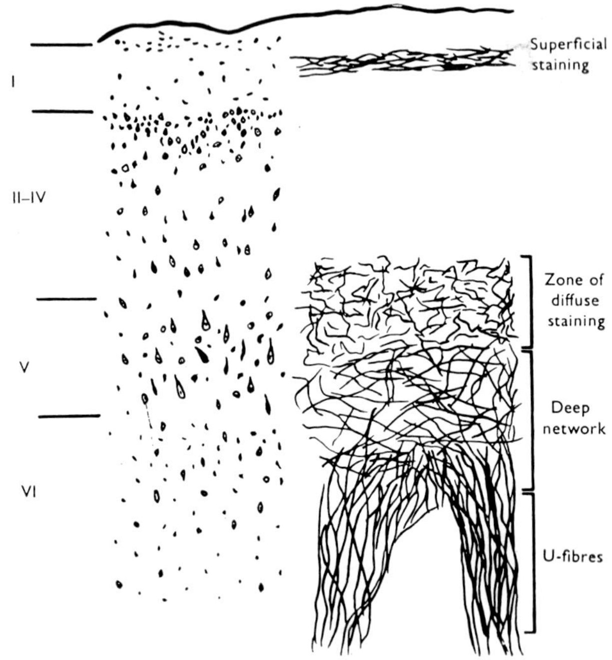

Fig. 79. Diagram illustrating the distribution of AChE-containing fibres in a gyrus (right) in relation to the different cell layers (left) of the cat cerebral cortex. (Krnjević and Silver, 1965.)

layer; more superficially, they pass into a zone of more diffuse fibrillar staining which extends through layers V and IV. Since the superficial cell layers contain few cholinesterase fibres, the outermost edge of this zone is seen as a relatively sharp line at a depth of 0·8–1·0 mm. (which includes the pyramidal-cell layer). Cellular staining is confined mainly to pyramidal cells of layer V and polymorph cells of layer VI. However, on the basis of these histochemical studies, the pyramidal cells are unlikely to be cholinergic because the cholinesterase stain appears to be mainly at the surface of the soma, intracellular staining being weak. Polymorph cells, on the other hand, show denser staining which is clearly intracellular and extends into the various neuronal processes. Histochemical studies of various areas of the cat and monkey brain indicated that the AChE-containing fibres of the cerebral cortex appear to arise mainly from the phylogenetically older regions of the forebrain (including the deep layers of the cerebral cortex) and subcortical areas (lenticular nucleus, septum, and other regions of the brain-stem) (Krnjević and Silver, 1965, 1966). Such projections are illustrated diagrammatically in *Fig.* 80. These conclusions are similar to those of Shute and Lewis (1967) for the rat cerebral cortex (cf. *Fig.* 66).

Cholinergic Excitation

The first direct indication that ACh might be an excitatory transmitter in the cerebral cortex was the observation that it

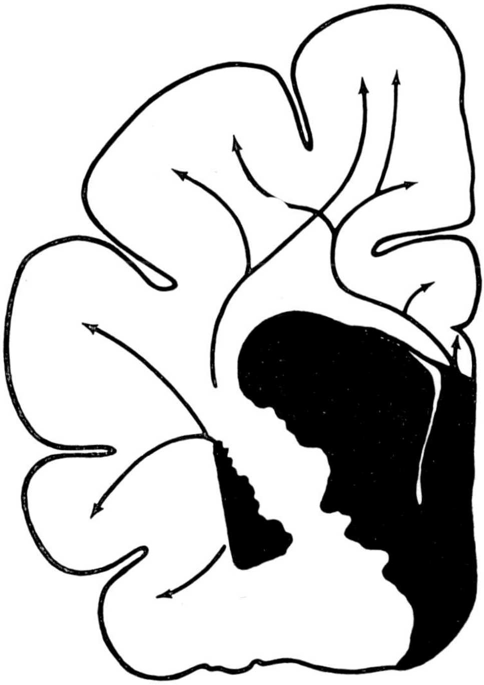

Fig. 80. Diagram showing the origins (shaded area) and projections (arrows) of AChE-containing fibres which terminate in the cat cerebral cortex. (Krnjević and Silver, 1965.)

excited certain cortical neurons when administered electrophoretically (Krnjević and Phillis, 1963a, b, c; Spehlmann, 1963; Stefanis, 1964b; Crawford and Curtis, 1966). The majority of ACh-excited neurons were located in the primary sensory areas at depths of 0·8–1·3 mm. (layers IV and V). In particular, the large majority of the neurons identified by antidromic invasion as

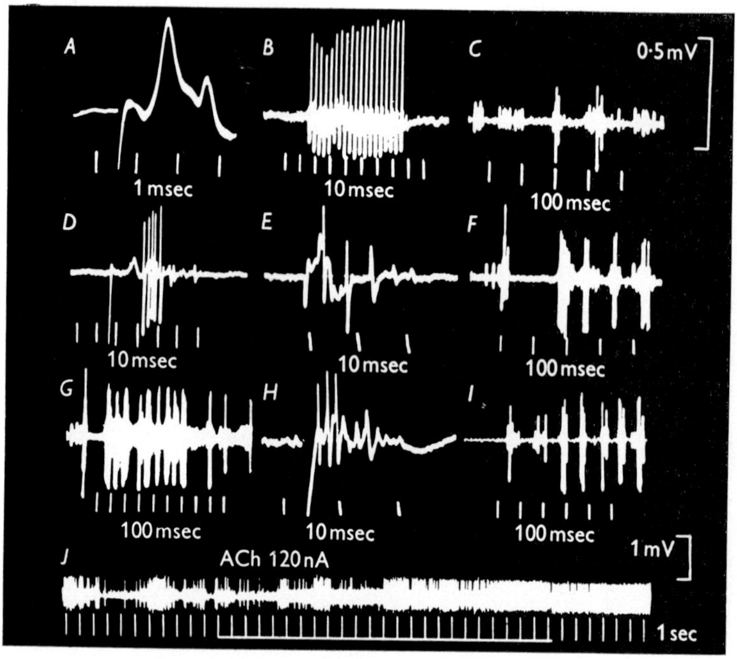

Fig. 81. Simultaneous extracellular recordings of two different Betz cells located at a depth of 1·1 mm. in the lateral precruciate cortex of a cat anaesthetized with Dial compound. A, Antidromic action potentials (large and small) of 0·7 and 2·0 msec. latency respectively, evoked by stimulating the medullary pyramid. B, Both could be driven repetitively at 300/sec. C, Spontaneous firing of both cells. D, Early responses to peripheral (contralateral fore-paw) stimulation. E, Responses to stimulating a specific afferent nucleus in the thalamus. F, G, Additional late repetitive discharges following peripheral and specific thalamic excitation, respectively, with single shocks. Similar early firing and late repetitive responses were evoked by transcallosal stimulation in H and I. Both units were excited by ACh (120 nA) administered electrophoretically during the period indicated by the white line in J. (Krnjević and Phillis, 1963b.)

pyramidal tract cells were excited by ACh, this excitation being of slow time course. The actions of ACh on two such neurons are illustrated in *Fig*. 81J. The excitant action of ACh on pyramidal tract cells was potentiated by anticholinesterases, and studies with various cholinomimetics and antagonists of ACh suggested that this excitation is mediated by muscarinic receptors (Krnjević and Phillis, 1963a, b, c; Spehlmann, 1963; Crawford and Curtis, 1966; Spehlmann, Daniels, and Smathers, 1971). In more recent investigations of the somatosensory area of the rat cerebral cortex, Stone (1972a, b) reported the existence of nicotinic excitation by ACh of some non-pyramidal cells located at depths of less than 600 μm., in addition to the muscarinic excitation of pyramidal cells located deeper.

Most of the pathways responsible for the synaptic excitation studied in the cerebral cortex are likely to be non-cholinergic. Although it had been postulated that the callosal pathway is exclusively or largely cholinergic (Marrazzi, 1953), subsequent studies did not confirm this, even though many cholinoceptive cortical neurons were excited by transcallosal stimulation (Krnjević and Phillis, 1963b; Spehlmann, 1971a; Stone, 1972b). It has also been concluded that the direct pathway from the specific nuclei of the thalamus to the somatosensory cortex, as well as the non-specific, short-latency thalamo-cortical pathways are non-cholinergic, as there was little correlation between the ability of cortical neurons to respond synaptically (with short latency) to electrical stimulation and their response to ACh (Krnjević and Phillis, 1963b). Similarly, the sensitivity of neurons in the visual cortex to ACh did not correlate with their responsiveness to stimulation of the optic radiation or flashes of light, and atropine did not block either form of synaptic excitation in concentrations greater than those required to block the excitant action of ACh (Spehlmann and others, 1971). Thus, the transmission of 'specific' impulses to neurons of the primary visual cortex is also thought to be mediated by non-cholinergic fibres. In contrast, Stone (1972b) has recently reported that almost all the pyramidal tract cells (of the rat cortex) that were excited by ACh responded to specific thalamic stimulation (nucleus ventralis posterolateralis) with a short-latency burst of action potentials, and that this synaptic excitation could occasionally be reduced by electrophoretically administered atropine. However, the pharmacological specificity of the action of atropine was not demonstrated in these cases. It is possible that the depression by atropine may have resulted from a 'local anaesthetic' type of action, especially in view of the brevity of the depression. It is therefore not fair to conclude that the cholinergic fibres terminating on pyramidal tract cells arise from specific thalamic nuclei (Mitchell, 1963; Stone, 1972b).

Krnjević and Phillis (1963b) did observe a strong correlation between the excitation of sensorimotor cortical neurons (cat) and the *long-latency* repetitive firing evoked by stimulating specific thalamic nuclei, the internal capsule, and the fore-paws. Stimulation of the fore-paws and thalamus sometimes evoked relatively early discharges (as in *Fig.* 81D, E), but the late repetitive firing (*Fig.* 81F, G) was typical of ACh-excited neurons. Similar early and late repetitive responses were sometimes evoked by trans-callosal stimulation (*Fig.* 81H, I). The cholinoceptive neurons fired spontaneously in a characteristic, irregular manner, and both the spontaneous activity and the late repetitive firing to peripheral nerve or specific thalamic stimulation were potentiated by cholinesterase inhibitors and reduced by atropine, suggesting that this type of synaptic excitation is mediated by cholinergic fibres (Krnjević and Phillis, 1963b, c). These findings are consistent with the early observations on the effects of systemically administered cholinesterase inhibitors and atropine on the electrocorticogram (Chatfield and Dempsey, 1942; Morison and Dempsey, 1943). In addition, the late repetitive firing evoked from cortical neurons was depressed by choralose (Chatfield and Dempsey, 1942; Krnjević and Phillis, 1963c), which has been shown to depress the release of ACh in the C.N.S. (Mitchell, 1963). However, the full significance of this observation is not clear, as such depression of release may be neither specific for ACh nor a reliable action of chloralose (*see* Biscoe and Krnjević, 1963).

In the visual cortex ACh facilitated the synaptic excitation of approximately one-third of the neurons evoked by stimulating the mesencephalic reticular formation, and atropine specifically reduced the excitant action of ACh, the spontaneous firing and the responses to reticular stimulation in some of these cases (Spehlmann, 1969, 1971a; Spehlmann and others, 1971). Some of these observations are illustrated in *Fig.* 82. The spontaneous firing rate of this visual cortical neuron was increased by ACh and reduced by both electrophoretic and intravenous atropine. Reticular stimulation also increased the firing of the neuron, and this increase was facilitated by ACh. In particular, both actions were depressed by atropine. Light flashes to the eyes also increased the firing, but this was unaffected by ACh and electrophoretic atropine, although reduced by intravenous atropine. Similar

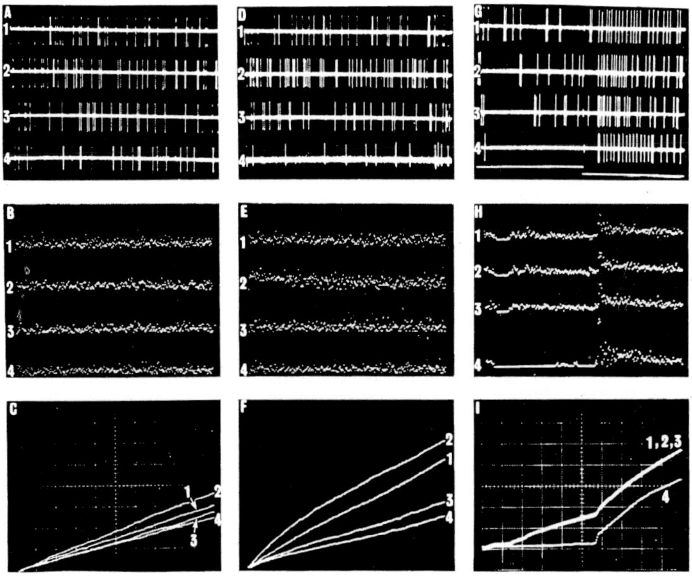

Fig. 82. Neuron in the visual cortex of the cat firing spontaneously (A–C), immediately after 30 stimuli at 300 per sec. to the mesencephalic reticular formation (D–F), and during intermittent retinal illumination (G–I; the upward deflection of the bottom trace in G indicates the period of illumination). 1, Before drug application; 2, During the electrophoretic ejection of ACh (100 nA); 3, During atropine (50 nA); 4, After intravenous atropine (0·5 mg./kg.). A, D, G, Playback of original data from a magnetic tape. B, E, H, Histograms of neuron discharges, each histogram representing 64 sec. of firing accumulated in 256 addresses of a Fabri-Tek 1060 Instrument Computer. Sweep length is 2 sec. for A–F, 1 sec. for G–I. Gain is 16 counts for graticule division for B and E, 32 counts per division for H. C, F, I, Integrals of histograms at a gain of 256 counts per division. The ordinate of each point represents the total number of counts accumulated to the abscissa of that point. For instance, spontaneous firing before drug application (C, I) reaches 3 x 256 discharges for 64 sec. which indicates a mean firing rate of 12 discharges per sec. (Spehlmann, 1969.)

observations have been made recently in the motor cortex (Spehlmann, 1971b).

A mechanism has recently been postulated for the relatively slow excitation of cerebral cortical neurons by ACh (Godfraind, Krnjević, and Pumain, 1970; Godfraind, Kawamura, Krnjević, and Pumain, 1971; Krnjević, Pumain, and Renaud, 1971a, b). In an earlier study involving intracellular recording from cortical neurons, it was suggested that the primary action of ACh may be on a metabolic process which controls the activity of an electrogenic ion pump, as no clear changes in membrane resistance were detected during the depolarizing action of ACh (Krnjević and Schwartz, 1967a). Further studies, however, suggested that the slow 'muscarinic' excitation of cortical cells by ACh was similar to the situation in sympathetic ganglia, where the slow EPSP generated by muscarinic cholinergic activity is associated with an increase in membrane resistance, thought to be due to a decrease in permeability to K^+ (Kobayashi and Libet, 1968, 1970; Nishi, Soeda, and Koketsu, 1969; Weight and Votava, 1970).

The depolarization of cerebral cortical neurons by ACh was often accompanied by an increase in membrane resistance (Krnjević and others, 1971b). An example of this is illustrated in *Fig.* 83. The precruciate neuron, which was firing spontaneously (*Fig.* 83A), was inhibited for several tens of msec. following stimulation of the cortical surface (*Fig.* 83B), and during this IPSP there was an increase in membrane conductance (*Fig.* 83C). ACh depolarized the membrane and increased its resistance (*Fig.* 83D), these effects still being evident several seconds after the current ejecting ACh had been terminated (*Fig.* 83E).

DNP and other inhibitors of oxidative phosphorylation blocked the excitant action of ACh, but not that of glutamate, on cortical neurons (indicative of a metabolic role for ACh), and DNP hyperpolarized the membrane and increased its conductance (Godfraind and others, 1971). Intracellular recordings showing the interaction between ACh and DNP on a cortical neuron are illustrated in *Fig.* 84. The mean reversal potential for DNP was nearly 30 mV more negative than the resting potential, and DNP hyperpolarized the membrane after IPSPs had been reversed by injection of chloride ions. It was concluded that DNP lowered excitability by increasing the membrane conductance to potassium ions, and that it selectively blocked the excitation by ACh because ACh had the contrary action. This was supported by the observation that the

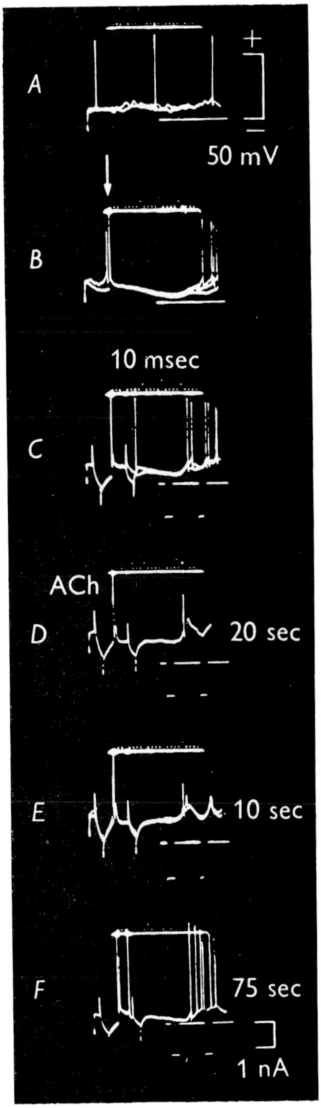

Fig. 83. Effect of ACh on the membrane potential and resistance of a precruciate cortical neuron, recorded intracellularly with an electrode containing potassium acetate (resting potential −60 mV). A, Spontaneous firing. B, IPSP evoked by a surface shock (onset indicated by the arrow). C, Two equal pulses of hyperpolarizing current (monitored on the trace below, displaced to the right), showing a decrease in amplitude of the pulse during the IPSP (indicative of increased conductance). D, 20 sec. after the beginning of an ejection of ACh (140 nA) extracellularly; note the depolarizing shift upwards, the greater negativity of the IPSP, the increase in resistance and the delayed repolarization after the spike. E, 10 sec. after terminating the ACh application. F, About 1 min. later. (Krnjević, and others, 1971b).

mean reversal potential for the depolarizing action of ACh was identical with the reversal potential for the hyperpolarizing action of DNP (Krnjević and others, 1971b). Furthermore, barium ions, which can reduce membrane permeability to potassium, resembled ACh in their action on cortical neurons (Krnjević and others, 1971a). Further experiments suggested that ACh acts by both reducing the resting permeability to K^+ and the delayed K^+ current of action potentials (Krnjević and others, 1971b). The immediate cause of this effect may be a decreased amount of free calcium ions inside the neuron (*see* Feltz, Krnjević, and Lisiewicz, 1972; spinal

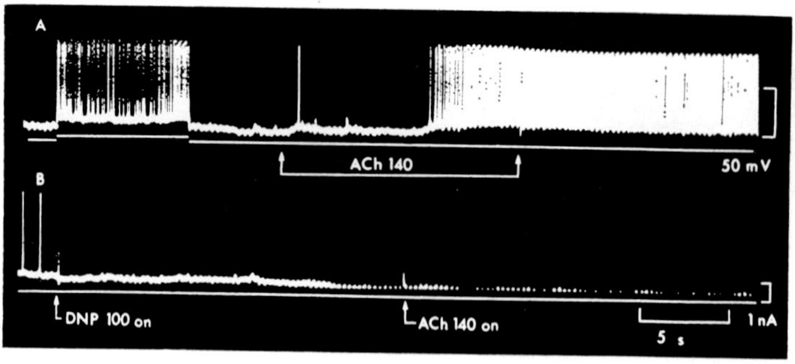

Fig. 84. Intracellular recordings from a precruciate cortical neuron, showing the interaction between ACh and DNP. A, Firing evoked first by an intracellular injection of depolarizing current (magnitude monitored on the trace below, in nA), and subsequently by the electrophoretic administration of ACh (140 nA) extracellularly. B, DNP (100 nA) was ejected extracellularly, beginning at the arrow and maintained until the end of the trace. The beginning of another ejection of ACh (140 nA) is indicated by the second arrow. (Godfraind and others, 1970.)

motoneurons). Sophisticated experiments of this kind have contributed greatly towards the understanding of the basic mechanisms underlying synaptic transmission, although, of course, it is not known whether the mechanisms studied operate throughout the C.N.S. (cf. Renshaw cells; Curtis and Felix, 1971c).

Cholinergic Inhibition

The possibility that ACh is also an inhibitory transmitter in the cerebral cortex has been put forward by Phillis and York (1967a, b; 1968a, b; *see also* Jordan and Phillis, 1972), who observed that ACh depressed the firing of a large proportion (one-third to one-half) of cortical neurons tested in all but the most superficial layers, and most frequently in layers II, III, and IV. Examples of this are illustrated in *Fig.* 85. In particular, the depression by ACh and the long-duration inhibition produced by repetitive stimulation of the cortical surface, pyramidal tract, mesencephalic reticular formation, and lateral hypothalamus were blocked by strychnine and a variety of cholinergic blocking agents (atropine, hyoscine, d-tubocurarine, dihydro-β-erythroidine, and gallamine). Anticholinesterases also depressed the firing of cortical neurons and sometimes potentiated both the depression by ACh and the

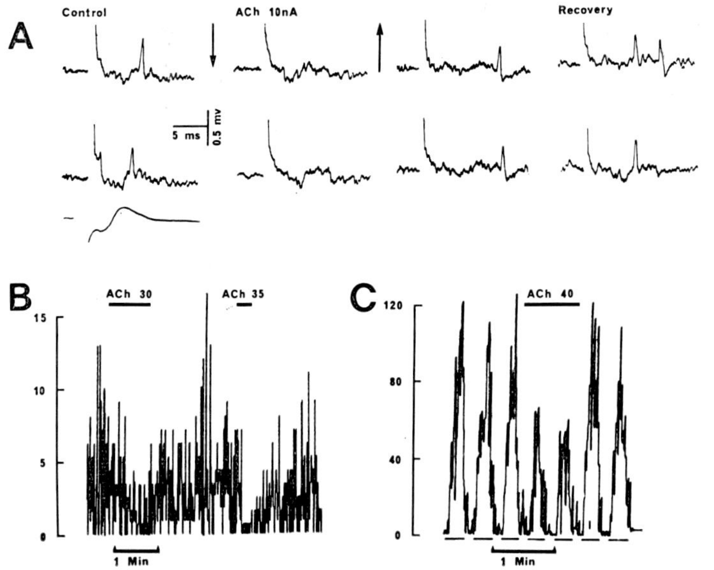

Fig. 85. Depressant action of ACh on synaptic, spontaneous, and glutamate-evoked firing of a single cortical neuron. A, Synaptic responses were evoked by electrical stimulation of the surface of the cerebral cortex, at a short distance from the tip of the recording electrode. The records show a reversible block of this firing during two separate applications of ACh (10 nA; administered between the arrows). The third trace in the control column is the N-wave direct cortical response recorded from the surface of the cortex. B, C. The depressant action of ACh on spontaneous and glutamate-evoked firing, respectively. Glutamate (40 nA) pulses were applied during the periods indicated by the horizontal bars below the trace in C. *Ordinate*: spikes per sec. (Jordan and Phillis, 1972.)

inhibition evoked synaptically. Some of these observations are illustrated in *Fig.* 86. Since stimulation of several different areas of the brain inhibited the firing of cortical neurons for long periods, and since the depressant action of ACh and the inhibition evoked by stimulation of the cortical surface were preserved in isolated cortical slabs, the presence of cholinergic inhibitory interneurons in the cerebral cortex was postulated (Phillis and York, 1967a, b; 1968a, b; Jordan and Phillis, 1972). Studies with antagonists of ACh and various cholinomimetics suggested that the cholinoceptive receptors are intermediate in type between the classic muscarinic and nicotinic receptors. This evidence is compatible with previous findings suggesting cortical cholinergic inhibition (Chatfield and Purpura, 1954; cf. MacIntosh, 1963; Szerb, 1965). The observation that some AChE and choline acetyltransferase remains in the undercut cerebral cortex (Hebb, Krnjević, and Silver, 1963; Green, Halpern, and Van Niel, 1970;

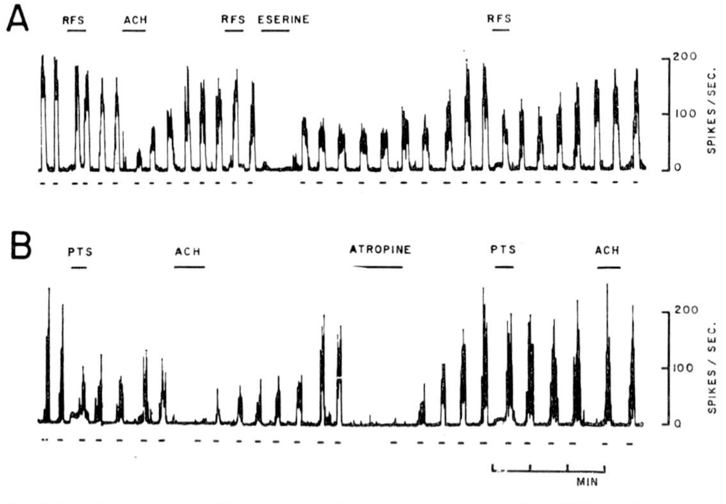

Fig. 86. Chart recordings, showing the change in firing frequency of two different neurons (A and B) during the application of ACh and electrical stimulation. A, The excitation evoked by pulses of L-glutamic acid (20 nA, horizontal bars below tracing) was initially only slightly inhibited by repetitively stimulating the mesencephalic reticular formation (RFS; 5·5 V, 10/sec.), but markedly depressed by ACh (60 nA). After an ejection of eserine (100 nA), which itself had a depressant action, RFS clearly inhibited the firing. Depth 300 µm. B, Records from another cortical neuron which was inhibited by stimulation of the pyramidal tract (PTS; 4 V, 10/sec.) and depressed by ACh (60 nA). After atropine (80 nA) had been administered electrophoretically, the effects of PTS and ACh were abolished. Depth 870 µm. (Phillis and York, 1967b.)

but cf. Krnjević, Reiffenstein, and Silver, 1970) was regarded as supporting the concept of local cholinergic circuits in the cerebral cortex.

However, the evidence for cholinergic inhibition in the cerebral cortex is controversial. A number of investigators have explored the different layers of the cat cerebral cortex using various types of preparations (including unanaesthetized animals), but have observed depressant actions of ACh on only a very small proportion of cells (Krnjević and Phillis, 1963a; Randić, Siminoff, and Straughan, 1964; Crawford and Curtis, 1966; Crawford, 1970; McLennan, 1970b; Krnjević and others, 1971b; Spehlmann and others, 1971). More recently, Stone (1972a, b) has obtained evidence *for* muscarinic depression by ACh of cells other than pyramidal tract cells in the rat cerebral cortex, and has suggested that such depression was not observed by many earlier workers because of the high concentrations of ACh and cholinergic antagonists used. Similarly, studies of cortical inhibitions are at variance. For example, stimulation of the cortical surface with single shocks evoked a postsynaptic inhibition of about 100–300 msec. duration in all layers of the cortex, but predominantly in the superficial half of the grey matter (Krnjević, Randić, and Straughan, 1966a, b, c). Repetitive stimulation evoked a much longer inhibition. This inhibition was fully preserved in isolated cortical slabs, suggesting a widespread system of inhibitory intracortical interneurons. However, this cortical inhibition was unaffected by a wide variety of substances, including strychnine and cholinergic antagonists, and subsequent studies suggested that the inhibitory transmitter involved is GABA (*see* next section in this chapter on Amino-acids). Similarly, although McLennan (1970b) observed an inhibition which lasted for many seconds following repetitive stimulation of the cerebral cortex, there was no evidence that the inhibition was related to cells depressed by ACh, and neither atropine nor strychnine (administered both electrophoretically and intravenously) affected the inhibition. At present it is not clear why such conflicting results have been obtained from different laboratories, but they may possibly be due to the different methods of producing and studying cortical inhibition, and/or to the different populations of neurons sampled.

Studies which have demonstrated the spontaneous and evoked release of ACh from the cerebral cortex support the evidence for cholinergic synapses in this structure. The first demonstration of this was by MacIntosh and Oborin (1953), who showed that the release of ACh from the cortex correlated well with its electrical activity, and decreased with greater depth of anaesthesia. A number of subsequent investigators have also shown that the rate of liberation of ACh from the cortical surface increases during spontaneous or induced periods of alertness and E.E.G. desynchronization (Mitchell, 1963; Beleslin, Polak, and Sproull, 1965; Phillis and Chong, 1965; Celesia and Jasper, 1966; Collier and Mitchell, 1966, 1967; Bartolini and Pepeu, 1967; Szerb, 1967; Beani, Bianchi, Santinoceto, and Marchetti, 1968; Phillis, 1968b; Dudar and Szerb, 1969; Bartolini and others, 1972). These studies support other findings which collectively suggest that cortical arousal and consciousness may be mediated by cholinergic mechanisms (*see* Funderburk and Case, 1951; Krnjević, 1967).

Amino-acids

Acidic Amino-acids and Synaptic Excitation

Hayashi (1954, 1956) was the first to show the excitant action of glutamate when applied topically to the cerebral cortex. The possibility that this amino-acid is an excitatory transmitter in the cerebral cortex received support when Krnjević and Phillis (1963a) demonstrated the excitant action of electrophoretically administered glutamate on single cortical neurons. The powerful, rapid, and readily reversible excitation of neurons by glutamate was considered indicative of its likely transmitter role. Similar observations were made soon afterwards by Crawford and Curtis (1964), and since then by many others. Other acidic amino-acids, including aspartate, have also been shown to excite cortical neurons, similar to glutamate, but the available evidence favours glutamate as the major amino-acid operating as an excitatory transmitter in the cerebral cortex (Krnjević, 1970; Curtis and Johnston, 1970; Johnson, 1972).

The mammalian cerebral cortex contains relatively high levels of glutamate (Tallan, 1962; Johnson and Aprison, 1971). Subcellular fractionation studies of the rat cerebral cortex have revealed that a proportion of this glutamate content is localized in synaptosomes (Kuhar and Snyder, 1970). In particular, such studies have indicated further the existence of unique synaptosome fractions that selectively accumulate glutamic and aspartic acids (Wofsey, Kuhar, and Snyder, 1971). These particles are distinctly different from those that store other amino-acids, including GABA. Various amino-acids (including glutamic and aspartic acids) can be taken up by slices of the cerebral cortex, and released by solutions containing high potassium concentrations (Tsukada, Nagata, Hirano, and Matsutani, 1963; Arnfred and Hertz, 1971). A more efficient uptake mechanism for the naturally occurring amino-acids than for DLH may explain why synaptic inhibition evoked by stimulating the cortical surface was considerably longer when the neurons were excited by glutamate than when excited by DLH (McLennan, 1970b). Glutamate has been shown to depolarize and increase the conductance of the membrane of cortical neurons (Krnjević and Schwartz, 1967a). The rate of release of glutamate from the cerebral cortex was lower in 'sleeping' than in 'aroused' cats, whereas the rate of release of aspartate and glutamine remained constant (Jasper, Khan, and Elliott, 1965). In addition, stimulation of the midbrain reticular formation (which produced a desynchronized cortical E.E.G.) greatly increased the liberation of glutamate from the cerebral cortex without changing that of glutamine, valine, and leucine (Jasper and Koyama, 1969). The synaptic release of glutamate is thought to be the main process involved in spreading depression in the cerebral cortex (van Harreveld, 1959, 1966; Phillis and Ochs, 1971).

Other evidence for glutamate as a transmitter in the cerebral cortex and elsewhere in the C.N.S. has recently been reviewed (Krnjević, 1970; Johnson, 1972). Despite these studies, however, the action of glutamate (or any other acidic amino-acid) has not yet been related to an excitatory pathway in the cerebral cortex. This evidence may become available when a specific antagonist of glutamate is discovered (*see* Haldeman and others, 1972; Curtis and others, 1972; Davies and Watkins, 1972).

GABA and Cortical Inhibition

Much stronger evidence is available for GABA as an inhibitory transmitter in the cerebral cortex (Krnjević, 1970). It has been known for more than twenty years that GABA is present in the cerebral cortex (Awapara, Landua, Fuerst, and Seale, 1950; cf. Roberts and Frankel, 1950). Subsequent analyses of different C.N.S. regions have revealed that the GABA levels of the cerebral cortex are relatively high, particularly in the more superficial layers (Hirsch and Robins, 1962; Singh and Malhotra, 1962;

Okada and others, 1971). GABA is readily taken up by slices of the rat cerebral cortex (Tsukada and others, 1963; Iversen and Neal, 1968; Hökfelt and others, 1970), and similar proportions of endogenous and labelled GABA are associated with synaptosomes, which also contain glutamic acid decarboxylase (Neal and Iversen, 1969; Hökfelt and others, 1970). Moreover, the uptake of GABA by slices of cerebral cortex is much greater than that of glycine (Iversen and Johnston, 1971; Johnston and Iversen, 1971), which is unlikely to be a transmitter in this structure (*see below*). Furthermore, a combination of electron microscopic and autoradiographic studies of cortical slices and homogenates has revealed that radioactive GABA is accumulated predominantly by nerve terminals, and that these total approximately 30 per cent of the terminals in the cerebral cortex (Bloom and Iversen, 1971; Iversen and Bloom, 1972). Autoradiographic studies of the distribution of labelled GABA in the cerebral cortex showed a high diffuse radioactivity in layer I, and strong accumulations of grains over several cell bodies, mainly in layers II and III, but sometimes also in deeper layers (Hökfelt and Ljungdahl, 1971b, 1972). *Fig.* 87 is an autoradiograph showing the relatively intense radioactivity (labelled GABA) over cell bodies in the superficial layers of the rat cerebral cortex. The studies indicated that stellate cells in the cerebral cortex strongly accumulate GABA, whereas pyramidal cells show only low activity.

Hayashi (1956) first observed the depressant actions of GABA and some other omega monocarboxylic amino-acids, applied topically to the cerebral cortex. The possibility that GABA may be an inhibitory transmitter was raised by Elliott and his collaborators (Bazemore, Elliott, and Florey, 1957; Levin, Lovell, Elliott, and Elliott, 1961), who identified GABA as the active agent in brain extracts which had a unique inhibitory action on crayfish stretch receptors (Florey, 1954; Elliott and Florey, 1956). Krnjević and Phillis (1963a) first demonstrated the depressant action of electrophoretically administered GABA on the firing of single cortical neurons (including pyramidal tract cells), and were impressed by the powerful, rapid, and readily reversible action of this amino-acid. Since then, a number of investigators have confirmed these observations (*see* review by Curtis and Crawford, 1969), and abundant pharmacological evidence has been obtained that GABA is the transmitter mediating the synaptic inhibition produced by a certain population of interneurons in the mammalian cerebral cortex.

GABA has been shown to be more effective than glycine as a depressant of cortical neurons, and investigations involving intracellular recording have demonstrated that GABA, but not glycine,

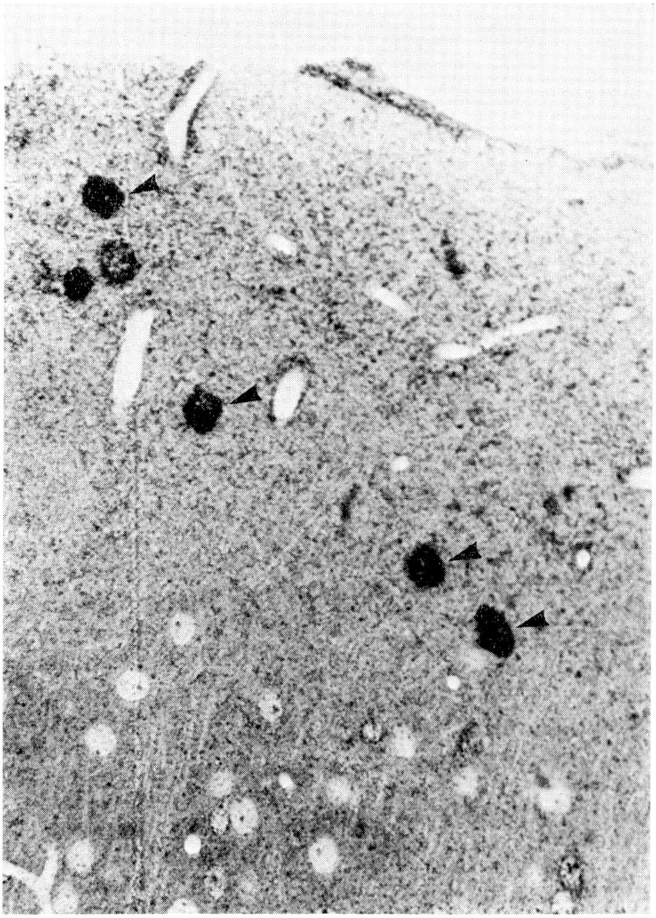

Fig. 87. Light microscope autoradiograph of the superficial layers of the rat cerebral cortex following intracerebral injection of labelled GABA. Several cell bodies (indicated by arrow heads) are labelled in the upper part, whereas only a low activity is found over most cell bodies in the lower part of the figure. Magnification × 375. (Hökfelt and Ljungdahl, 1972.)

mimics the action of an inhibitory transmitter in the cerebral cortex (Krnjević and Schwartz, 1967b; Dreifuss, Kelly, and Krnjević, 1969; Kelly and Krnjević, 1969). GABA usually hyperpolarized and increased the conductance of the membrane of cortical neurons. Both the IPSP evoked by stimulating the cortical surface and the hyperpolarization by GABA could be reversed by increasing the membrane potential and by injecting chloride ions, and the reversal potentials for GABA and the IPSP

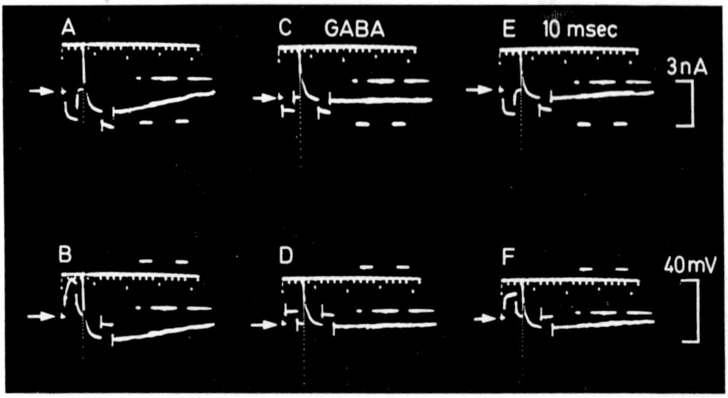

Fig. 88. Intracellular recordings from a cortical neuron (potassium citrate electrode, resting potential initially —45 mV) showing the action of GABA on the membrane potential and conductance. Each record indicates the level of the resting potential (arrow), and the transmembrane potential developed by a 20-msec. pulse of constant intracellular current, at rest and during an IPSP evoked by stimulating the cortical surface. Negative pulses were used in the upper traces, and positive ones in the lower traces, the current pulses being monitored on a second beam, displaced to the right. A, B, Controls. C, D, 30 and 50 sec. after beginning an ejection of GABA (140 nA) extracellularly. E, F, 30 and 40 sec. after the end of the GABA release. Time marks, 10 and 100 msec. (Dreifuss and others, 1969.)

were similar, both as a result of current and chloride injection. *Fig.* 88 illustrates some properties of this cortical inhibition and the action of GABA. Stimulation of the cortical surface evoked a characteristic long-duration IPSP in a cortical neuron, and an examination of the potential changes induced by pulses of positive and negative current passed across the membrane at regular intervals revealed an increase in membrane conductance during this IPSP (*Fig.* 88A, B). During the electrophoretic release of

GABA, the membrane was slightly hyperpolarized and its conductance was increased (*Fig.* 88C, D).

In contrast, the depressant action of glycine did not mimic synaptic inhibition in the cerebral cortex (Kelly and Krnjević, 1969). Glycine usually had no effect on the membrane potential or slightly depolarized cortical neurons, and produced only slight changes in membrane conductance. Although there was a positive correlation between the reversal potential for the IPSP and that for glycine, the latter was usually much more positive.

Studies using relatively selective antagonists of these two amino-acids have been particularly rewarding in providing pharmacological evidence for GABA synapses in the cerebral cortex. The depressant action of GABA on cortical neurons was blocked by bicuculline and picrotoxin, which had no effect on the depression by glycine (Curtis, Duggan, Felix and others, 1971; Curtis and Felix, 1971b; Straughan, Neal, Simmonds, Collins, and Hill, 1971; Hill, Simmonds, and Straughan, 1972). Conversely, the action of glycine on cortical neurons was blocked by strychnine in concentrations which had no effect on the depression by GABA (Curtis, Hösli, and Johnston, 1968; Johnson, Roberts, and Straughan, 1970; Curtis, Duggan, Felix and others, 1971). The GABA-mediated cortical inhibition is thought to be due to a widespread system of intracortical interneurons, as it can be evoked by stimulating a number of different sites, including the cortical surface, and it is fully preserved in isolated cortical slabs (Krnjević and others, 1966a, b). Various methods of studying such inhibition of pyramidal tract neurons are illustrated diagrammatically in *Fig.* 89. This inhibition has been shown to be resistant to a number of diverse pharmacological agents, including strychnine, tetanus toxin, and antagonists of ACh and monoamines (Crawford and others, 1963a; Krnjević and others, 1966c; Biscoe and Curtis, 1967). In particular, the synaptic inhibition produced by stellate and basket cells in the cerebral cortex can be blocked by bicuculline, which also blocks the action of GABA (Curtis, Duggan, Felix and others, 1971; Curtis and Felix, 1971b). *Fig.* 90 illustrates the antagonistic effects of bicuculline on the depression of a pyramidal tract cell by GABA, as well as on the synaptic inhibition evoked both recurrently and indirectly by chemical stimulation of a nearby stellate cell.

The results of more recent experiments with AOAA, which is known to increase the GABA content of brain by inhibiting

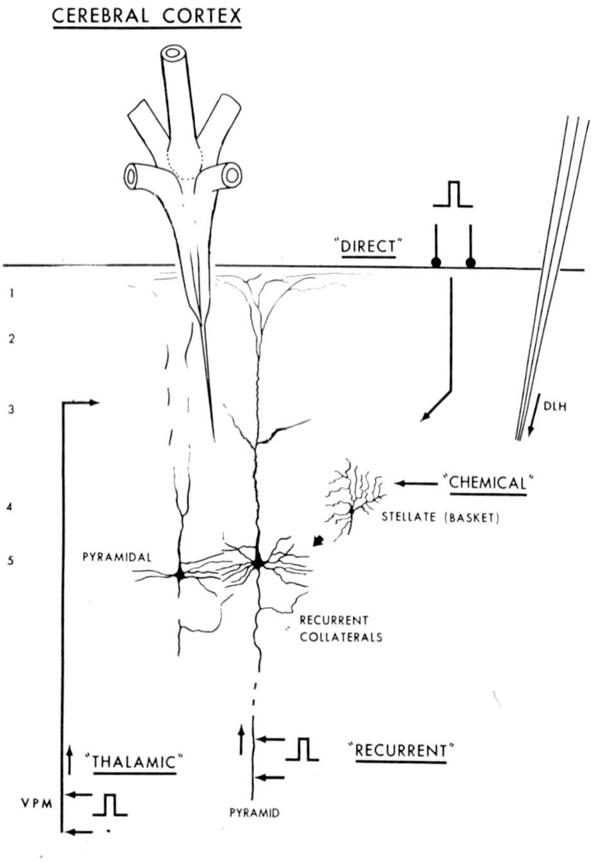

Fig. 89. Diagram illustrating various methods of synaptically inhibiting the firing of pyramidal tract neurons in the cerebral cortex. Action potentials of pyramidal tract neurons may be recorded through one barrel of a multibarrel micropipette, the other barrels being used for administering substances electrophoretically. These cells may be identified antidromically by electrically stimulating the exposed pyramid, which also results in recurrent inhibition via collaterals of pyramidal axons and stellate or basket cells. Electrical stimulation of the thalamus (nucleus ventralis posteromedialis; VPM) or the cortical surface within a few mm. of the site of recording also inhibits the firing of pyramidal tract cells. In addition, these cells may be inhibited following chemical excitation of neighbouring stellate or basket cells produced by administering excitant substances (for example, DLH) from another electrode assembly (one barrel of which is used for recording). (Curtis and Felix, 1971b).

GABA α-ketoglutaric transaminase (Baxter and Roberts, 1961; Wallach, 1961; Kuriyama, Roberts, and Rubinstein, 1966), support the view that stellate and basket-cell inhibition of pyramidal cells is mediated by GABA (Kelly, Renaud, and Gottesfeld, 1971; Gottesfeld, Kelly, and Renaud, 1972). Intraperitoneal injections of AOAA significantly increased the GABA levels of the rat and cat cerebral cortex. Following such administration of AOAA in the cat, pyramidal tract cells became progressively less sensitive to glutamate, but progressively more sensitive to GABA. In particular, the synaptic inhibition of pyramidal cells evoked by

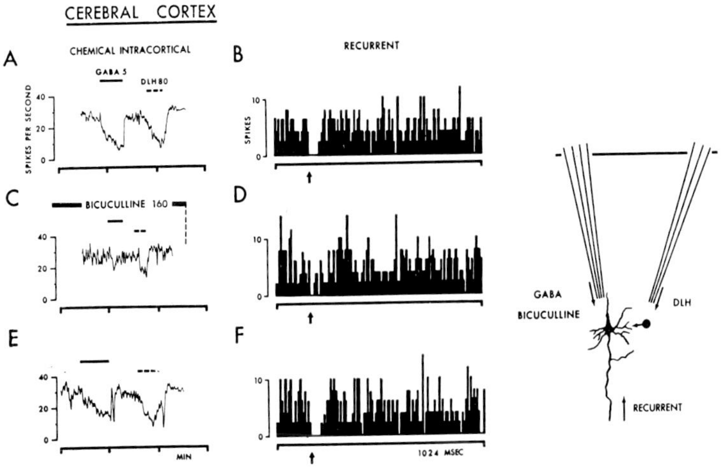

Fig. 90. Effects of bicuculline on synaptic inhibition and the depression by GABA of a pyramidal tract neuron. A, C, E, Chart recordings of the firing frequency maintained by DLH ejected continuously. Before bicuculline was administered, GABA (5 nA) depressed the firing, as did DLH (—80 nA) ejected from another micropipette situated in the same cortical layer, but 210 μm. rostral to the pyramidal tract neuron (as illustrated in the diagram on the right). DLH excited neurons in the immediate vicinity of the micropipette tip from which it was expelled. B, D, F, Histograms of the probability of firing in 128 consecutive 8 msec. periods, counted over 30 sweeps at intervals of 1·8 sec. The pyramid was stimulated with a single pulse (25 V, 0·2 msec.) at the time indicated by the arrow. A, B, Controls. C, D, During the electrophoretic administration of bicuculline (160 nA), which was terminated after 5 min. (vertical broken line). E, F, Approximately 3 min. after terminating the bicullinee jection. (Curtis and Felix, 1971b.)

stimulating the cortical surface and pyramidal tract increased in duration and intensity with time. *Electrophoretically* administered AOAA had no effect on the synaptic inhibition. *Fig.* 91 illustrates the progressive reduction in sensitivity to glutamate, accompanied by a progressive increase in the inhibition evoked by surface shocks of a pyramidal tract neuron in the cortex, after an intraperitoneal injection of AOAA. Although various explanations are possible for these results, the most likely possibility is that AOAA had impaired the GABA uptake mechanism in the cerebral cortex (Gottesfeld and others, 1972).

Studies of the release of GABA support the view that this amino-acid is a transmitter in the cerebral cortex. The rate of

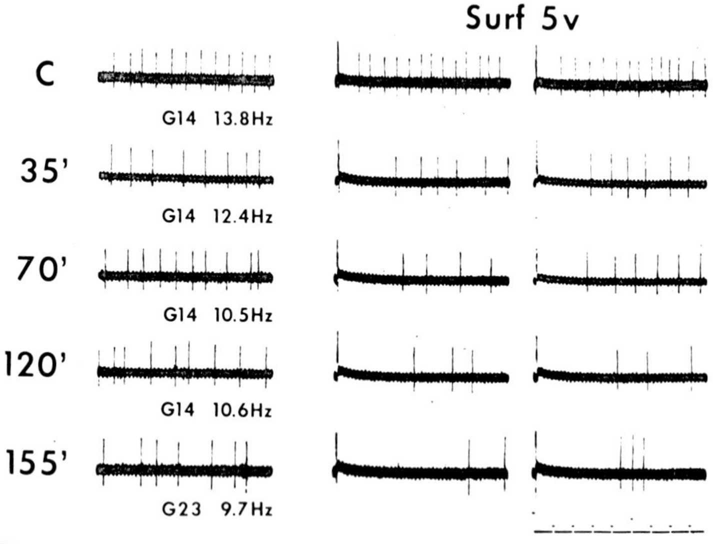

Fig. 91. Effects of amino-oxyacetic acid on glutamate-evoked firing and synaptic inhibition of a pyramidal tract neuron. The first column of records shows the spike discharges evoked by glutamate (G, ejecting currents, in nA, indicated alongside), the average firing frequency (Hz) having been calculated from post-stimulus latency histograms. The first record (C) is a control; the others represent records obtained at various times (in min.) after the intraperitoneal injection of amino-oxyacetic acid (30 mg./kg.). For each time selected, two further records show the 'inhibitory pause' evoked by a monopolar cathodal shock to the surface of the cortex (5 V, 0·1 msec. duration) applied 10 msec. after the start of each sweep. Time calibration units of 100 msec. (Gottesfeld and others, 1972.)

release of GABA from the cortex was tripled during 'sleep', but not during 'arousal' (Jasper and others, 1965; Jasper and Koyama, 1969; *see also* Crowshaw, Jessup, and Ramwell, 1967). Similarly, the release of GABA was increased several-fold by stimulation of the cortical surface and lateral geniculate nucleus, using stimulus parameters which inhibited the firing of cortical neurons (Iversen, Mitchell, and Srinivasan, 1971). Removal of calcium ions from the collection fluid did not affect the resting release of GABA, but prevented both the increase in GABA release and the inhibition of cortical neurons produced by stimulating the surface of the cortex and the lateral geniculate nucleus. There were no significant differences between the rates of release of a number of other amino-acids, during or in the absence of electrical stimulation. In addition, the spontaneous release of GABA from the cortex was increased following treatment with AOAA. Studies of cortical slices have also demonstrated a release of GABA by stimulation with electric currents and potassium ions (Srinivasan, Neal, and Mitchell, 1969; Arnfred and Hertz, 1971).

Taurine

Taurine is present in relatively high concentrations in the cerebral (and cerebellar) cortex compared with a number of other structures of the mammalian (including human) brain (Okumura, Otsuki, and Kameyama, 1960; Tallan, 1962; Shaw and Heine, 1965; Battistin, Grynbaum, and Lajtha, 1969; Shank and Aprison, 1970; Agrawal, Davison, and Kaczmarek, 1971; Guidotti and others, 1972). Approximately half of the taurine and cysteine-sulphinate decarboxylase content of the rat cortex is present in the nerve-ending fraction, similar to GABA and glutamic acid decarboxylase (Agrawal and others, 1971). Taurine is released from cortical slices by electrical stimulation, and is also taken up by these slices, the kinetics of the uptake process being similar to those of GABA (Davison and Kaczmarek, 1971). *In vivo* studies of the cat have shown that taurine is released spontaneously from the cerebral cortex, and this release is higher during 'arousal' than during 'sleep' (Jasper and Koyama, 1969). Similarly, the taurine content of some areas of the cerebral cortex was higher in cats which exhibited a synchronized E.E.G. pattern (indicative of 'sleep') than in animals showing an activated pattern (Guidotti and others, 1972). Taurine depresses the firing of cortical neurons

(Crawford and Curtis, 1964), the depression being blocked by both bicuculline and strychnine (Curtis, Duggan, Felix and others, 1971). Thus, it is possible that taurine is an inhibitory transmitter in the mammalian cerebral cortex. More direct evidence for this would be the demonstration of a synaptic inhibition sensitive to both bicuculline and strychnine in the cortex.

Monoamines

Reports on the effects of monoamines on cortical neurons are conflicting. Originally it was observed that electrophoretically applied NA and DA depressed the firing of most cortical neurons in anaesthetized animals, whereas 5-HT depressed in low concentrations and occasionally excited in higher concentrations (Krnjević and Phillis, 1963a, d; Crawford and Curtis, 1966). Although depression was the predominant action of these monoamines on cortical neurons of cats anaesthetized with nitrous oxide and halothane or methoxyflurane (Phillis and York, 1967a; Phillis and others, 1968c), it was subsequently reported that 5-HT and NA usually excited cortical neurons of unanaesthetized or halothane-anaesthetized cats, and produced mainly depression in animals anaesthetized with barbiturate (Roberts and Straughan, 1967; Johnson, Roberts, and Straughan, 1969; Johnson, Roberts, Sobieszek, and Straughan, 1969; Bradshaw, Roberts, and Szabadi, 1971). The infrequency of excitation by the monoamines in the early studies was attributed to a number of different reasons, including the effects of different anaesthetics, pontine decerebration, the sampling of cells fired by glutamate rather than spontaneously firing cells, and (in the case of 5-HT) the depressant actions of creatinine ejected from solutions of 5-HT creatinine sulphate. However, it now appears that the discrepancies may have resulted predominantly from the different *p*H of the monoamine solutions used. It has recently been shown that the main action of both NA and 5-HT ejected from solutions at *p*H 4 or higher was a depression of cortical neurons, whereas at lower *p*H values excitation by the monoamines was observed more frequently (Frederickson and others, 1971; Jordan, Frederickson, Phillis, and Lake, 1972). *Fig.* 92 shows the depressant action of 5-HT administered from a solution at *p*H 4, and its excitant action (following a depression) when ejected from a solution at *p*H 3, on the spontaneous firing of a cortical neuron in an unanaesthetized cat. When solutions at a *p*H above 4 were used, both 5-HT and NA depressed the firing of large proportions of cortical neurons. In some cases strong depression was observed using relatively low ejecting currents (*Fig.* 93, for example). In addition, the actions of the monoamines on blood-vessels in the cerebral cortex (Stone, 1971, 1972c), and the different criteria used for depression and excitation have probably also contributed to

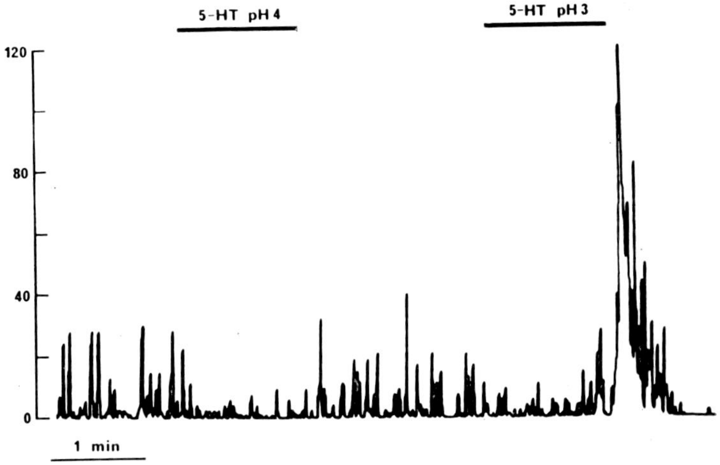

Fig. 92. Responses of a spontaneously firing neuron, located approximately 700 μm. below the surface of the postcruciate cortex of a cerveau isolé cat, to the electrophoretic application of 5-HT (60 nA) administered from solutions containing 5-HT creatinine sulphate at *p*H 4·0 and *p*H 3·0. The firing frequency of the neuron (ordinate, in spikes/sec.) is plotted against time. (Jordan, Frederickson, Phillis, and Lake, 1972.)

the differences of opinion regarding the actions of monoamines in the cortex. These conclusions may, of course, apply to all regions of the C.N.S. to a greater or lesser extent.

Thus 5-HT and NA depress the firing of cortical neurons sensitive to them. Only depressant actions of these monoamines have been observed on cortical cells identified as pyramidal tract neurons (Salmoiraghi and Stefanis, 1965; Phillis and others, 1968c). Histamine usually weakly depressed the firing of cortical cells (including pyramidal tract cells), whereas the related amino-acid, imidazole acetic acid, had a strong depressant action (Krnjević and Phillis, 1963d; Phillis and others, 1968c). Examples of the actions of histamine on a pyramidal tract cell and other cortical neurons are shown in *Fig.* 94.

Fluorescence histochemical and biochemical studies have demonstrated the presence of both 5-HT- and NA-containing nerve-fibres which originate in the brain-stem and project to the cerebral cortex (Fuxe, 1965; Andén and others, 1966; Heller,

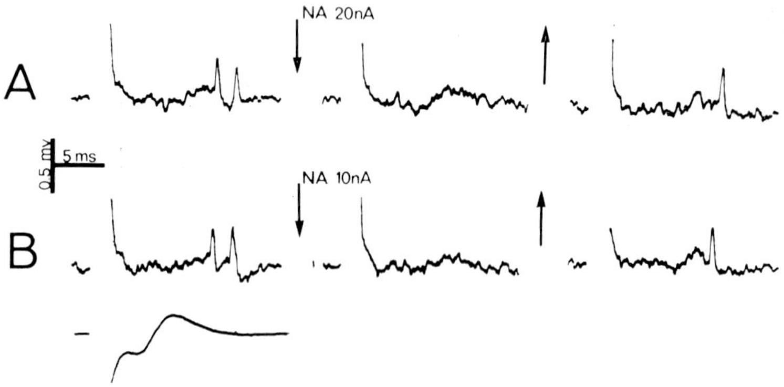

Fig. 93. Depression of synaptic excitation in the cerebral cortex by NA. Stimulation of the surface of the cortex with a bipolar electrode placed within 2 mm. of the recording electrode evoked two action potentials from a neuron located 500 μm. beneath the surface (first record in A and B). This synaptic response was blocked by NA (20 nA and subsequently, by 10 nA; administered electrophoretically between the arrows). Some depression was still evident shortly after the current ejecting NA had been terminated (third records). The bottom trace in B is a simultaneous recording of the direct cortical response. (Frederickson and others 1971.)

Seiden, and Moore, 1966; Andén, Fuxe, and Ungerstedt, 1967; Fuxe, Hamberger, and Hökfelt, 1968; Fuxe and others, 1970; Moore, 1970; Ungerstedt, 1971). A proportion of the 5-HT and NA content of the cerebral cortex is associated with synaptosomes (Kuhar and Snyder, 1970; Snyder and others, 1970; Kuhar, Shaskan, and Snyder, 1971). Stimulation of the midbrain raphe nuclei enhanced the release of 5-HT and 5-hydroxyindol-3-ylacetic acid in the forebrain, including the cerebral cortex (Aghajanian, and others, 1967; Padjen and Randić, 1970). The cerebral cortex also contains histamine, a proportion of which is associated with synaptosomes similar to those that store NA and 5-HT (Kataoka and De Robertis, 1967; Kuhar,

Shaskan, and Snyder, 1971; Kuhar, Taylor, and Snyder, 1971; Taylor and others, 1972). Thus, it is likely that these monoamines (at least 5-HT and NA) function as transmitters in the mammalian cerebral cortex, but direct pharmacological evidence for this has yet to be obtained.

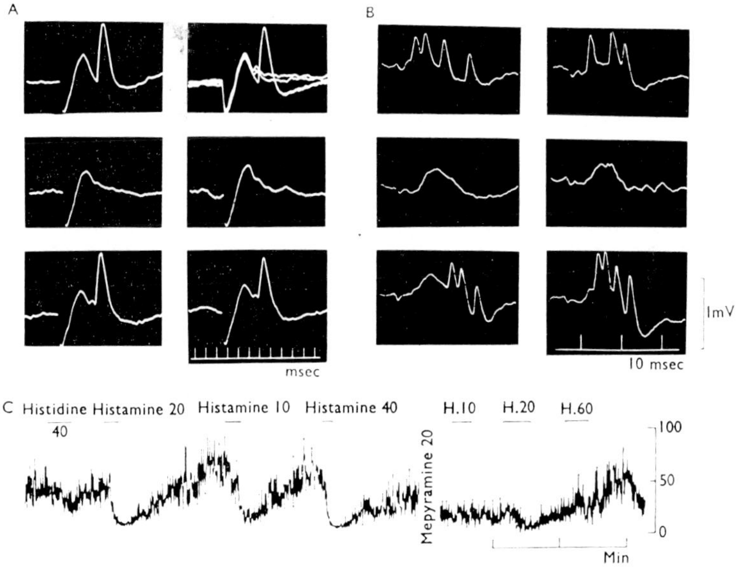

Fig. 94. Depression by histamine of antidromic (A), orthodromic (B), and spontaneous (C) firing of three different cerebral cortical neurons. A, Pairs of consecutive records of antidromic spikes evoked by stimulating the medullary pyramid. At threshold stimulus intensity the spike appeared in an all-or-none manner (top records). Ten sec. after the application of histamine (80 nA) antidromic invasion failed (middle records), and recovery occurred within 12 sec. of terminating the histamine ejection (bottom records). B, A similar series of records showing the synaptic responses evoked by stimulation of the lateral hypothalamus before, during, and 15 sec. after the application of histamine (30 nA). C, Chart recording of the spontaneous firing frequency of another neuron. Histamine (10, 20, and 40 nA) markedly depressed the firing, whereas DL-histidine (40 nA) had only a weak depressant action. After the application of mepyramine (20 nA), the depression by histamine was virtually blocked, and higher ejecting currents of this monoamine (H. 60 nA) had an excitant action. Ordinate, spikes/sec. (Phillis and others, 1968c).

SUMMARY

Acetylcholine

Various lines of evidence suggest that ACh is a transmitter mediating relatively slow effects on only certain populations of neurons in the mammalian cerebral cortex. ACh, choline acetyltransferase, and AChE are present in the cerebral cortex. Cholinesterase-containing fibres project from the mesencephalic reticular formation, certain other subcortical areas (including the lenticular nucleus, septum), and the phylogenetically older regions of the forebrain, and terminate mainly in layers IV and V of the cortex (which include the pyramidal cells).

Pyramidal tract cells are excited by ACh, the excitation being of relatively slow time course and mediated predominantly by muscarinic receptors. The 'muscarinic' depolarization of cortical neurons by ACh is accompanied by an increase in membrane resistance, thought to be due to a decrease in permeability to potassium ions. The *long-latency* repetitive firing of cortical cells evoked by stimulating the mesencephalic reticular formation, specific thalamic nuclei, the internal capsule, fore-paws, and other subcortical areas can be sometimes potentiated by cholinesterase inhibitors and specifically blocked by atropine, suggesting that the terminations of the pathways stimulated form cholinergic synapses on ACh-excited cortical neurons. The *short-latency* excitatory responses evoked by impulses in callosal afferents and fibres from specific and non-specific thalamic nuclei are probably mediated predominantly by non-cholinergic fibres.

It has been postulated that ACh is also an inhibitory transmitter in the cerebral cortex, released from intracortical interneurons located more superficially than the pyramidal cells. The depressant action of ACh and the long-duration inhibition of these neurons evoked by repetitive stimulation of the cortical surface, pyramidal tract, mesencephalic reticular formation, and lateral hypothalamus can be blocked by strychnine and antagonists of ACh, and both effects can be potentiated by anticholinesterases. However, the existence of cholinergic inhibition in the cerebral cortex is controversial, as most investigators have reported that ACh depresses the firing of only a small proportion of cortical cells, and cortical inhibitions have in most cases proved to be resistant to strychnine and antagonists of ACh. The reasons for the discrepancies are not clear.

The efflux of ACh from the cerebral cortex correlates well with the state of arousal of the animal. This and some histochemical and pharmacological observations suggest that cortical arousal and consciousness may be mediated by cholinergic mechanisms.

AMINO-ACIDS

Relatively high levels of glutamate are present in the cerebral cortex. Some of the cortical glutamate content is associated with a certain population of synaptosomes, which differ from those that store other amino-acids. Glutamate and related acidic amino-acids excite cortical neurons, including pyramidal tract cells. The depolarization of cortical neurons by glutamate is associated with an increase in membrane conductance. The release of glutamate from the cerebral cortex correlates with the state of arousal. *In vitro* studies of cortical tissue have revealed an efficient uptake mechanism for glutamate, as well as a release of this amino-acid following electrical or chemical stimulation. Thus it is likely that glutamate is an excitatory transmitter in the cerebral cortex, but direct pharmacological evidence for this has not yet been obtained.

Good evidence has been obtained that GABA is an inhibitory transmitter released from stellate and basket cell axons predominantly in the superficial layers of the cerebral cortex. The cerebral cortex contains relatively high concentrations of GABA (a proportion of which is present in the nerve-ending fraction) and the enzymes for its metabolism. Labelled GABA is accumulated mainly by stellate cells of layers II and III, pyramidal cells showing weak activity. GABA is more effective than glycine as a depressant of cortical neurons, including pyramidal tract cells. GABA hyperpolarizes and increases the conductance of the membrane, and both the IPSP evoked by electrical stimulation of the cortical surface and the hyperpolarization by GABA can be reversed by intracellular injection of current and chloride ions. In particular, the reversal potentials for the IPSP and GABA are similar in both cases. Glycine does not mimic the action of the inhibitory transmitter. The inhibition of pyramidal tract cells by basket and stellate cell axons, as well as the action of GABA, are blocked by bicuculline, which has no effect on the depression by glycine. Conversely, the depression by glycine is blocked by strychnine in concentrations which do not affect the action of GABA or synaptic inhibition. AOAA increases cortical GABA levels and the release of GABA from the cortex. Concomitantly, AOAA decreases the sensitivity of pyramidal tract cells to glutamate, but increases their sensitivity to GABA and increases the intensity and duration of the inhibition produced by stellate cells. The spontaneous release of GABA from the cerebral cortex correlates with the state of arousal of the animal, and is enhanced several times by stimulating the cortical surface and lateral geniculate nucleus, using stimulus parameters which inhibit the firing of cortical neurons and do not significantly alter the release of many other amino-acids.

Various biochemical, release, and pharmacological studies raise the possibility that taurine is also an inhibitory transmitter in the cerebral cortex, but direct evidence for this has not been demonstrated.

Monoamines

The monoamines, 5-HT, NA, DA, and histamine, depress the firing of many cortical neurons, including pyramidal tract cells. Histochemical, neurochemical, and release studies suggest that 5-HT and NA (and possibly histamine) are involved in transmitter functions in the cerebral cortex, but there is no direct evidence for this.

Chapter

8

RETINA

From embryological, morphological, and functional points of view, the mammalian retina may be considered to be part of the C.N.S. It is a relatively well-organized entity, consisting of three main layers of densely packed cell bodies and two intervening 'synaptic layers' of intertwining dendrites and axons. A diagram of the primate retina is shown in *Fig.* 95. The layer nearest the choroid and pigment epithelium consists of the rods and cones (the photoreceptors of the retina), the nuclear part of their cell bodies forming the outer nuclear layer. The axons of the rod and cone cells terminate on the dendrites of the bipolar cells. The region of these synaptic contacts between photoreceptor axons and bipolar dendrites is known as the outer plexiform layer, in which horizontal cells are found. The bipolar cell bodies form the inner nuclear layer, and their axons terminate on the dendrites of ganglion cells in the inner plexiform layer, in which amacrine cells are located. The axons of the ganglion cells make up the optic nerve. This, understandably, is a grossly over-simplified description of the neuroanatomy of the retina, but will serve as a basis for discussing the histochemistry and pharmacology of retinal neurons. As in the preceding chapter, this chapter is divided into sections on monoamines, amino-acids, and ACh, rather than into the different types of retinal neurons, as it has not yet been possible to speculate about the possible transmitters that operate on the various types of cells in the retina. In fact, little is known about the possible transmitters that operate in the retina as a whole, the strongest evidence being histochemical, rather than pharmacological.

Monoamines

Dopamine

The first suggestion that a monoamine may be involved in synaptic transmission in the retina arose from the observation that reserpine, which is known to deplete tissue stores of monoamines, produced an extreme constriction of the pupils in rats and increased their sensitivity to light, manifested as a photophobia (Malmfors, 1963). This hypersensitivity to light was interpreted as having resulted from a block of monoaminergic inhibition in the retina. Histochemical studies of the retina of rats, rabbits, and guinea-pigs revealed a specific catecholamine fluorescence in cell bodies and fibres at the junction of the inner nuclear and inner plexiform layers, and biochemical analyses indicated further that almost all of the catecholamine content in the retina was DA (Malmfors, 1963; Häggendal and Malmfors, 1965). In addition, a rich plexus of catecholamine-containing nerve-fibres was observed to surround the blood-vessels of the choroid, but in

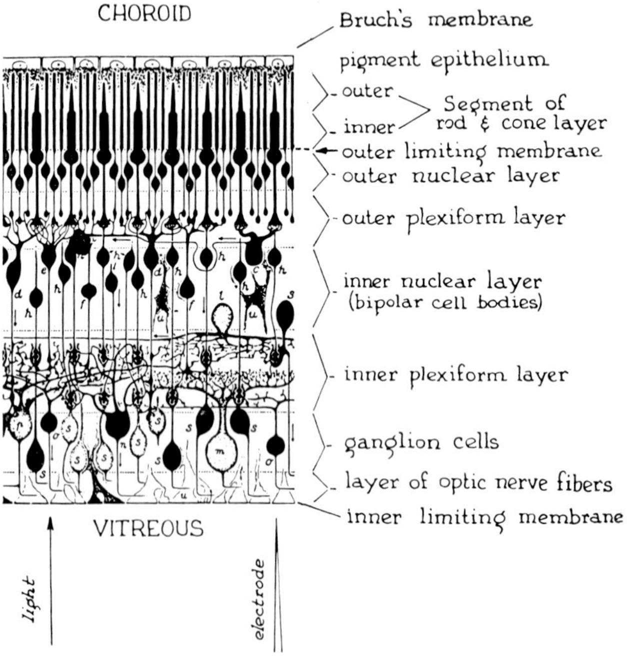

Fig. 95. Diagram of the primate retina (based on studies involving Golgi impregnation) showing the principal cell types and their synaptic connexions. The cell types are: *c*, Horizontal cells; *d*, *e*, *f*, Diffuse or polysynaptic bipolar cells; *h*, Individual cone (midget) bipolar cell; *i*, *l*, 'Amacrine cells'; *m*, *n*, *o*, *p*, *r*, *s*, Ganglion cells, of which *s* is the individual or monosynaptic ganglion cell. (Polyak, 1941).

contrast to the retina, the choroid contained mainly NA and very little DA. Moreover, sympathetic denervation led to a decrease in the NA content of the choroid, but had no effect on the DA content of the retina. These results first suggested the presence of a dopaminergic neuron in the mammalian retina (Häggendal and Malmfors, 1965).

Since then, these findings have been confirmed and extended in a number of different mammals (including the cat), using both fluorescence histochemical (Ehinger, 1966a, b; Laties and Jacobowitz, 1966) and autoradiographic (Ehinger and Falck, 1971; Kramer, Potts, and Mangnall, 1971) techniques. The band of DA fluorescence in the retina has been subdivided into three layers consisting of fine nerve terminals, and three different types of DA-containing cells—'outer adrenergic, eremite, and alloganglionic cells' (Ehinger, 1966b). The reason for the new terminology was that none of the fluorescent cell types fitted very easily into any single 'classical' type described previously. For the sake of simplicity, however, it is convenient to regard the most common type of DA-containing cell (the 'adrenergic junctional cell') as a type of amacrine cell, which it resembles morphologically, although it comprises only about 10 per cent of the amacrine cells.

Autoradiographic studies of the mammalian retina following either systemic or topical application of labelled DA revealed a distribution of radioactivity which corresponded extremely well with the distribution of DA fluorescence described earlier (Ehinger and Falck, 1971; Kramer and others, 1971). Thus, radioactivity was concentrated mainly at the junction of the inner nuclear and inner plexiform layers, as illustrated in *Fig.* 96. This is an autoradiograph of the cat retina which had been perfused with labelled DA.

An unusual finding from both the fluorescence histochemical and autoradiographic studies was the occasional presence of a DA-containing cell in the ganglion-cell layer (Ehinger, 1966b; Laties and Jacobowitz, 1966; Kramer and others, 1971). An example of this is illustrated in *Fig.* 97. Retinal ganglion axons are generally believed to project almost exclusively to the lateral geniculate nucleus and superior colliculus, where they exert excitatory effects. The connexions of the DA-containing 'alloganglionic' cells are unknown, but if they are similar to those of other ganglion cells, it is possible that DA is released synaptically from some optic nerve-fibres on to lateral geniculate and/or superior collicular neurons. The proportion of such fibres (if they exist) must be small, as few ganglion cells fluoresce for DA, and the optic nerve has a low catecholamine content (Vogt, 1954; Cobbin

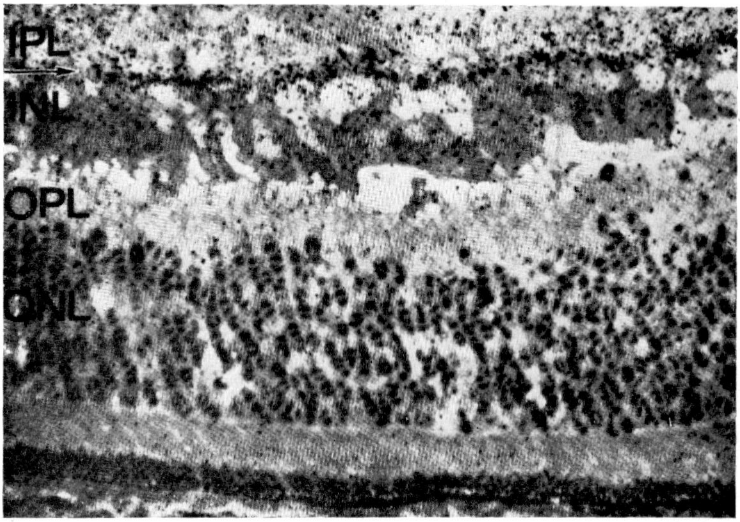

Fig. 96. Autoradiograph of the cat retina showing the differential distribution of radioactivity following preretinal perfusion with labelled DA (3-week exposure). The arrow indicates the band of silver grains at the junction of the inner nuclear and inner plexiform layers. Abbreviations: IPL, Inner plexiform layer; INL, Inner nuclear layer; OPL, Outer plexiform layer; ONL, Outer nuclear layer. Magnification ×80. (Kramer and others, 1971.)

and others, 1965). Nevertheless, this is an interesting problem to solve, as it raises the possibility of dopaminergic inhibition in the lateral geniculate nucleus (*see* Chapter 4). Most of the other DA-containing neurons in the retina are without doubt interneurons (Ehinger, 1966b).

Further evidence for the involvement of DA in retinal function was obtained by demonstrating the presence of DA metabolites in radiochromatographic analyses of retinal extracts following intracarotid injections of labelled DA into cats (Kramer, 1971). Furthermore, the labelled DA was retained by the retina after the cervical ganglion had been removed, confirming the earlier conclusion (Häggendal and Malmfors, 1965) that retinal DA is independent of the ocular sympathetic nerve-supply. Labelled DA was taken up by the retina *in vivo*, and in light-adapted cats was released spontaneously (Kramer, 1971). In the dark-adapted state DA was not released spontaneously, but could be released

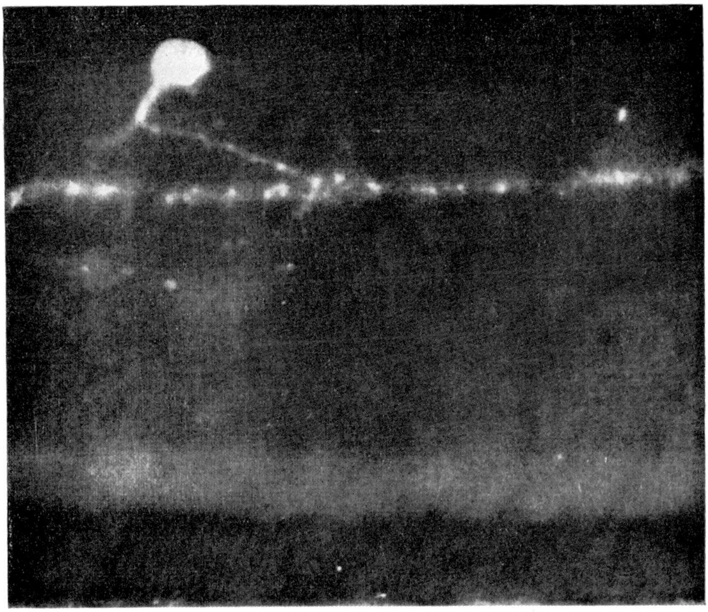

Fig. 97. Fluorescence micrograph of the rat retina, showing the presence of DA fluorescence in a neuron of the ganglion cell layer ('alloganglionic' cell). A process of this cell can be seen to branch in the inner plexiform layer, which also contains a band of DA fluorescence. The outer adrenergic layer demarcates the border between the inner plexiform and inner nuclear layers. (×460.) (Ehinger, 1966b.)

by light flashes to the eye, the rate of release increasing with higher frequencies of stimulation.

Few pharmacological studies have been made of the retina, and these have been mainly of unidentified neurons. However, since only ganglion cells have been shown to fire action potentials in the mammalian retina, it is reasonably safe to conclude that the pharmacological experiments have demonstrated actions of drugs on ganglion cells, in the electrophoretic studies, at least. Intravenous L-DOPA depressed the amplitude of the potential evoked in the optic tract by light flashes, and electrophoretically administered DA depressed the spontaneous firing of retinal ganglion cells, as well as the excitation evoked by light, glutamate, and stimulation of the optic nerve (Straschill and Perwein, 1969). *Fig.* 98A illustrates that increasing concentrations of DA

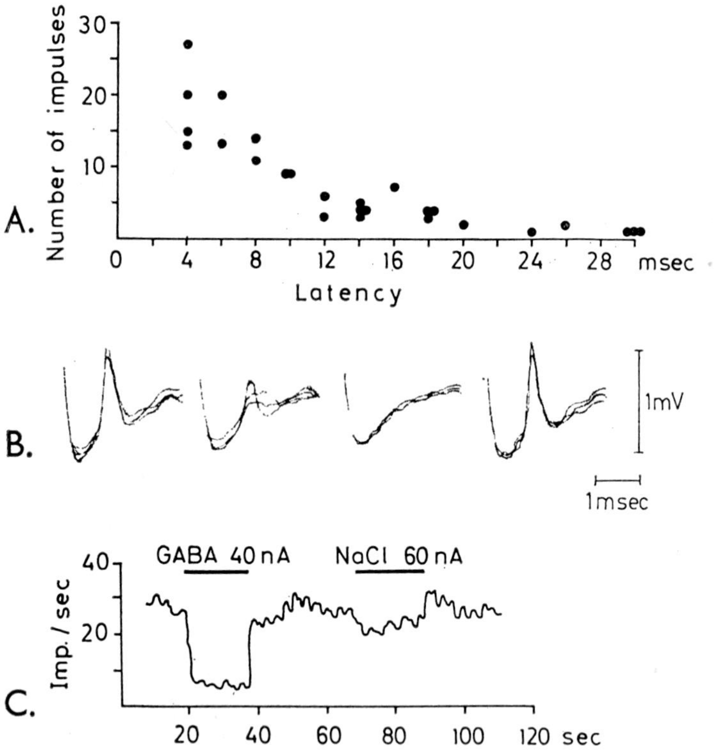

Fig. 98. Effects of electrophoretically administered DA (A, B) and GABA (C) on retinal ganglion cells of the cat. A, A plot showing the decrease in the number of discharges (ordinate) and increase in the latency (abscissa) of the 'off' responses of a retinal 'off' neuron with increasing concentrations of DA. B, Depressant action of DA on the antidromic invasion of a retinal ganglion cell. Each record consists of seven superimposed responses to supramaximal stimulation (1/sec.) of the optic nerve head. The first record is a control. The second and third records were obtained 10–17 sec. and 20–27 sec. respectively, after the beginning of an ejection of DA (150 nA); the last record, 20–27 sec. after the DA ejection was terminated. C, Actions of GABA (40 nA) and Na^+ (60 nA) on the spontaneous firing frequency of a retinal ganglion cell. (Straschill and Perwein, 1969.)

(administered electrophoretically) progressively decreased the number of discharges and increased the latency of discharges of an 'off' retinal ganglion cell to photic stimulation. The reversible block of antidromic invasion of a ganglion cell by DA is shown in *Fig.* 98B. In general, DA enhanced inhibitory activity and depressed excitatory activity of retinal ganglion cells (Straschill and Perwein, 1969).

Studies involving topical application of DA to the isolated rabbit retina indicated that the catecholamine usually increased the firing of 'off' cells, and decreased that of 'on' and 'on-off' cells (Ames and Pollen, 1969). However, the enhancement of firing may not necessarily indicate an excitant action of DA, as the topical method of drug administration precludes any definitive conclusions on the precise site of action. A depression of the firing of inhibitory neurons by DA (disinhibition) may have facilitated firing. Thus, although only limited studies have been made of the actions of DA in the retina, it seems that this substance has predominantly or wholly a depressant action on retinal neurons.

The function of the dopaminergic neurons in the retina is not known, but may be related to dark and light adaptation. The different rates of release of DA from the retina of cats in the dark- and light-adapted states (Kramer, 1971) was mentioned above. Nichols, Jacobowitz, and Hottenstein (1967) found that the DA levels of the retina, as well as the intensity of DA fluorescence in 'amacrine' cells, increased consistently in rats, rabbits, and guinea-pigs which had been light-adapted. This is in contrast to an earlier report by Drujan, Díaz Borges, and Alvarez (1965), who observed a fall in the catecholamine content of the frog retina with light adaptation. Thus, although the precise function of DA in the retina is not clear, it appears to be involved in light and dark adaptation.

5-Hydroxytryptamine, Noradrenaline, and Histamine

It is doubtful whether 5-HT and NA have transmitter functions in the mammalian retina, as neither monoamine has been detected in neurons or nerve-fibres within the retina (Häggendal and Malmfors, 1965; Ehinger, 1966b; Laties and Jacobowitz, 1966). Both substances have been shown to alter the firing of retinal ganglion cells, but the different methods of drug administration used (systemic, Straschill, 1968; electrophoretic, Straschill and Perwein, 1969; topical, Ames and Pollen, 1969) do not allow an adequate comparison of the results. Electrophoretically administered 5-HT had no action on the large majority, and weakly depressed the firing of the remainder of the retinal ganglion cells tested (Straschill and Perwein, 1969), whereas 5-HT applied topically to the whole retina usually increased the firing of ganglion cells (Ames and Pollen, 1969). The actions of NA resembled those of DA on the same cell, suggesting that NA probably acts on DA receptors (Ames and Pollen, 1969; Straschill and Perwein, 1969).

Autoradiographic studies of the rabbit retina following intravitreal administration of labelled histamine revealed a diffusely distributed radioactivity in the retina, the pigment epithelium showing a higher activity than the retina proper (Ehinger and Falck, 1971). The significance of this is not known, however, as no other studies of the retina involving histamine have been made.

Amino-acids

Excitatory Amino-acids

The results of various studies suggest that glutamate is an excitatory transmitter in the mammalian retina, although definitive pharmacological evidence for this is lacking (as elsewhere in the C.N.S.). Glutamate and the enzymes involved in its metabolism have been found in the retina of the rabbit and monkey, although the levels are low compared with other areas of the brain (Lowry, Roberts, and Lewis, 1956). Autoradiographic studies of the rabbit retina performed after the intravitreal injection of labelled glutamic and aspartic acids revealed a similar distribution of radioactivity for the two amino-acids (Ehinger and Falck, 1971). Activity was distributed non-uniformly in all layers of the retina; the inner nuclear layer usually showed a moderate-to-weak activity, the outer plexiform layer, the outer nuclear layer, and the photoreceptor layer revealed only a low activity, whereas ganglion cells and the bundles of nerve-fibres had no or very little radioactivity. Recently, a high affinity and relatively specific uptake mechanism for L-glutamate has been demonstrated in the rat retina (Neal and White, 1971).

A wide variety of methods has shown that glutamate excites retinal neurons. In immature chickens (which have a deficient blood-brain barrier) systemically administered glutamate produced behavioural depression and altered the electroretinogram in a different manner from that of GABA (Kramer, Sherman, and Seifter, 1967). Systemic injections of glutamate have also been shown to increase the firing rate of retinal ganglion cells in the adult cat (Straschill, 1968). More directly, glutamate administered electrophoretically excited ganglion cells (Noell, 1959; Straschill and Perwein, 1969), as has been observed in all other central areas studied. In addition, topical application of either glutamate or aspartate to the isolated bullfrog (Kishida and Naka, 1967) or rabbit (Ames and Pollen, 1969) retina resulted in increased firing of ganglion cells.

Studies of the isolated chicken retina have led to the proposal that a major component of spreading depression results from a release of the transmitter, glutamate (van Harreveld and Fifková, 1970, 1971). Stimulation of the retina with glutamate (or aspartate in higher concentrations), potassium ions, or direct current increased the transparency of the retina (indicative of spreading depression), and all forms of stimulation resulted in a release of labelled glutamate (but not glutamine) from the retina. This release of glutamate is thought to be the primary process in the development of spreading depression (van Harreveld and Fifková, 1970, 1971).

Depressant Amino-acids

As for glutamate, the results of a number of different types of investigations suggest that both GABA and glycine are transmitters in the retina, but definitive pharmacological evidence for this has yet to be obtained. GABA has been shown to depress the firing of retinal ganglion cells, whether administered systemically to whole animals (Straschill, 1968; cf. Scholes and Roberts, 1964; Kramer and others, 1967), electrophoretically to neurons (Noell, 1959; Straschill and Perwein, 1969), or topically to the isolated retina (Kishida and Naka, 1967; Ames and Pollen, 1969). An example of the depressant action of electrophoretically administered GABA on a ganglion cell in the cat retina is shown in *Fig.* 98C. Only on the isolated retina of the bullfrog and rabbit have depressant actions of glycine been demonstrated (Kishida and Naka, 1967; Ames and Pollen, 1969).

Strychnine (administered systemically or topically) usually increased the firing of retinal ganglion cells, and produced rhythmical bursts of excitation, but did not reduce most types of postsynaptic inhibition evoked by flashes of light to the eye (Heiss and Bornschein, 1965; Bornschein and Heiss, 1966; Straschill, 1968; Ames and Pollen, 1969). *Fig.* 99 illustrates examples of inhibitions recorded from single units in the optic nerve of cats in which convulsive doses of strychnine had been injected intravenously. However, more detailed analyses have revealed the existence of a strychnine-sensitive inhibition in the cat retina, in addition to the strychnine-insensitive inhibitions (Heiss, 1967; Chu, 1968). In particular, intravenous strychnine blocked the 'on' inhibition produced by diffuse illumination of the retina of 'off' cells, but not of 'on' cells (Chu, 1968). Picrotoxin blocked certain retinal inhibitions which were insensitive to strychnine (Heiss, 1967; Chu, 1968). In the isolated rabbit retina, topically applied

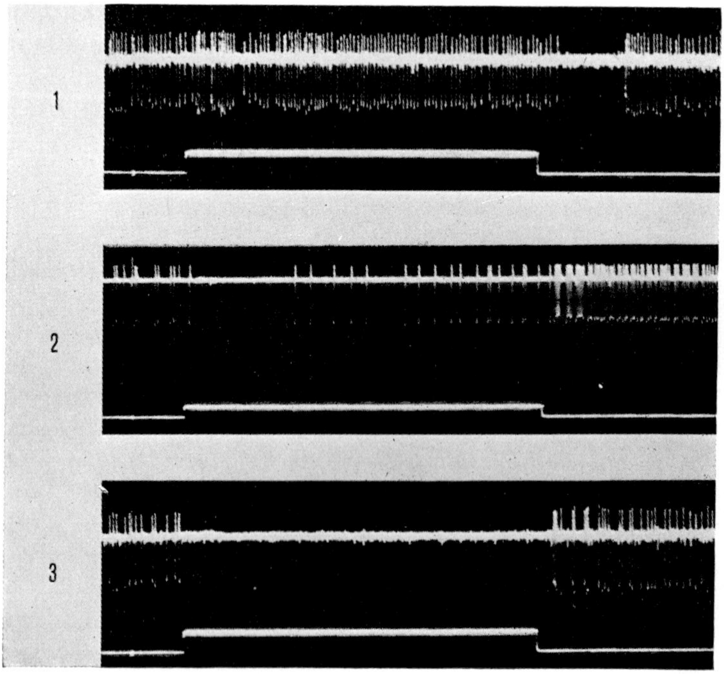

Fig. 99. Light responses of three different single units recorded in the optic nerve of the cat after strychnine (0·4 mg./kg.) had been injected intravenously. 1, 'On-centre' unit (dark discharge 89/sec.). 2, 'Off-centre' unit (dark discharge 46/sec.). 3, 'Off-centre' unit (dark discharge 58/sec.). The duration of the light stimulus (1 sec.) is indicated in the lower trace. Periods of inhibition are apparent in all three records. (Bornschein and Heiss, 1966.)

picrotoxin had actions similar to those of strychnine on the same preparation, but no firm conclusions could be made about the sites of action of these convulsants (Ames and Pollen, 1969). In addition, some evidence was obtained for a retinal inhibition that was resistant to both strychnine and picrotoxin (Heiss, 1967; Chu, 1968). This type of inhibition may be dopaminergic (p. 237), but no experiments have been done to investigate this possibility. Thus, although the pharmacological results raise the possibility that both GABA and glycine are inhibitory transmitters in the mammalian retina, this has yet to be demonstrated directly. The strongest available evidence for a functional role of GABA and glycine in the mammalian retina has been obtained from neurochemical and autoradiographic studies.

Since the first report on the presence of GABA in the retina (Kojima, Mizuno, and Miyazaki, 1958), several papers have appeared describing the presence of this amino-acid in all layers of the retina in a number of different vertebrates. A biochemical analysis of the rabbit retina revealed the highest levels of GABA and glutamic acid decarboxylase in the ganglion-cell layer, somewhat lower levels in the photoreceptor layer, and even lower levels in the bipolar-cell layer (Kuriyama, Sisken, Haber, and Roberts, 1968). Since the receptor, bipolar, and ganglion cells are excitatory neurons, and since there was little or no GABA or glutamic acid decarboxylase in the optic nerve (which, interestingly, contains relatively large amounts of GABA-transaminase), it was argued that the GABA content of the retinal layers was in retinal interneurons other than receptor, bipolar, and ganglion cells (Kuriyama and others, 1968). In the dark-adapted frog retina, the highest concentrations of GABA and glutamic acid decarboxylase were found to be in the inner synaptic layer with somewhat lower levels in the amacrine-bipolar, horizontal-bipolar, and ganglion-fibre layers (Graham, 1972). An evaluation of the data suggested that GABA is stored mainly within amacrine cells, but also within horizontal cells to a lesser extent.

Autoradiographic studies of the rabbit retina following the administration of labelled GABA into the eye indicated that radioactivity was accumulated principally in the inner plexiform, inner nuclear, and ganglion-cell and nerve-fibre layers, with label occupying the positions of amacrine cells and some neurons in the ganglion-cell layer (Ehinger, 1970; Ehinger and Falck, 1971). An autoradiograph of the rabbit retina is shown in *Fig.* 100, which shows radioactivity associated with some amacrine cells (arrows) and two ganglion cells (g).

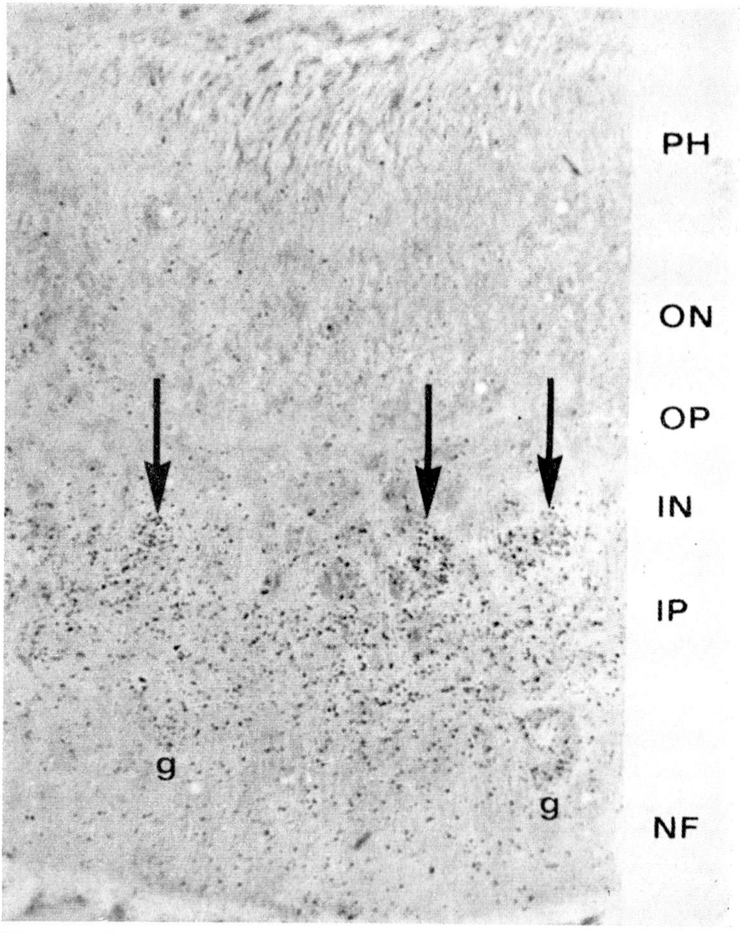

Fig. 100. Autoradiograph of the rabbit retina following intravitreal application of (^{3}H) GABA (50 μc., 4 hours). Radioactivity is present in some cells with the position of amacrines (arrows) and in two ganglion cells (g). The radioactivity of the third ganglion cell does not differ much from that of the surrounding tissue. Activity is also present in the inner plexiform layer, and more sparsely, in the nerve-fibre layer and in the innermost part of the inner nuclear layer. Abbreviations: PH, Photoreceptors; ON, Outer nuclear layer; OP, Outer plexiform layer; IN, Inner nuclear layer; IP, Inner plexiform layer; NF, Nerve-fibre layer. (×900.) (Ehinger and Falck, 1971.)

A subsequent autoradiographic study of the rat retina also revealed activity in all layers (except the outer segments of the rods), although radioactivity was concentrated in the ganglion-cell and nerve-fibre layer, inner nuclear, and outer plexiform layers, and at the outer limiting membrane (Neal and Iversen, 1972). Within the ganglion-cell layer, labelled GABA was conspicuously absent from the ganglion-cell bodies, an observation consistent with the fact that ganglion cells are excitatory neurons. The radioactivity associated with some ganglion cells observed in the rabbit retina (Ehinger, 1970; Ehinger and Falck, 1971) are therefore likely to represent GABA-containing nerve terminals *on* ganglion cells, rather than the presence of GABA *within* ganglion cell bodies. An unusual observation on the rat retina, however, was the finding that unlike other areas of the C.N.S. (including the retina of rabbits), where labelled GABA accumulates predominantly on neurons, labelled GABA in the rat retina was largely taken up by glial cells, the Müller fibres which extend through the retina as column-like structures at right angles to the retinal layers (Neal and Iversen, 1972). It is clear that autoradiographic studies must be interpreted with caution, especially when attempting to assign a transmitter function to the substance studied.

A highly efficient and specific uptake mechanism for GABA has been demonstrated in the rat retina (Goodchild and Neal, 1970; Starr and Voaden, 1972). The spontaneous release of GABA *in vitro* was higher from the dark-adapted retina, and this release could be increased by stimulating the retina with electric currents and solutions containing high potassium concentrations (Voaden and Starr, 1972).

It is useful to mention studies involving GABA and the amphibian retina, as there are likely to be similarities in retinal function between different vertebrates. In the retina of frogs GABA is concentrated primarily within layers which are rich in amacrine cells (Graham, Baxter, and Lolley, 1970; Graham, 1972). In particular, the GABA and glutamic acid decarboxylase levels decreased in the bipolar and ganglion-cell layers of the dark-adapted frog retina, whereas adaptation to light and darkness had little or no effect on either the content or localization of retinal glutamate. This temporal decrease of retinal GABA, and the decrease in threshold excitability during dark adaptation described earlier (Donner and Reuter, 1965), suggested a correlation between these two phenomena (Graham and others, 1970).

An evaluation of the findings discussed above suggests that despite some interspecies variability, GABA-containing neurons of the retina are largely or wholly interneurons. A proportion of these appear to be amacrine cells, and possibly also horizontal cells. Like the dopaminergic neurons of the retina (*see* beginning of this chapter), the GABA-containing neurons appear to be involved in light and dark adaptation mechanisms. It is therefore interesting that the bipolar layer of the retina (in which amacrine cells are located) is the site of non-photochemical visual adaptation, and it has been suggested that the amacrine cells could provide the mechanism for this adaptation because of their reciprocal synaptic arrangements with the bipolar cells and their lateral connexions with other amacrine cells (Dowling, 1967).

Autoradiographic studies of the rabbit retina have revealed that exogenous glycine is also unevenly distributed in different layers, being particularly associated with amacrine cells and, to a lesser

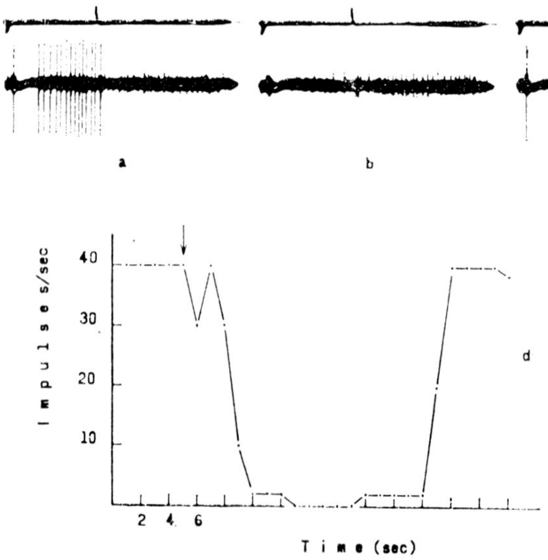

Fig. 101. Action of ACh on the responses of a retinal 'on-centre' unit to intermittent diffuse illumination of its receptive field. The stimulus artefacts indicate the onset (downward deflection) and termination of the light stimulus (intensity 6×10^{-5} lm./mm.^{2}). a, Before injection. b, 8 sec. after the onset of an injection of ACh (6 μg) into the external carotid artery. c, 26 sec. after the injection. (Straschill, 1968.)

extent, ganglion cells (Ehinger and Falck, 1971). In continuation of these studies Bruun and Ehinger (1972) showed that the rabbit retina is able to take up and retain glycine from an external medium over a wide range of concentrations, very similar to the situation in the spinal cord. The high-affinity uptake of glycine by retinal neurons was unaffected by a number of other amino-acids (including GABA, L-α-alanine, β-alanine, glutamate, aspartate), indicating a high specificity for glycine. Pharmacological observations relevant to glycine were mentioned earlier (p. 245).

In conclusion, although the available evidence suggests that both GABA and glycine are inhibitory transmitters in the mammalian retina, this has yet to be firmly established. In particular, the actions of electrophoretically administered GABA and glycine and their interactions with bicuculline, picrotoxin, and strychnine on retinal neurons warrant investigation.

The relatively high levels of taurine in the retina (Kubíček and Dolének, 1958) and lateral geniculate nucleus (Guidotti and others, 1972) raise the possibility that this amino-acid is involved in synaptic transmission in visual pathways, but little more can be said at this stage.

Acetylcholine

Noell and Lasansky (1959) were the first to report (briefly) that retinal ganglion cells of the cat and rabbit can be excited by electrophoretically administered ACh, nicotine, and physostigmine. A subsequent study involving systemic administration of substances (cat) indicated that ACh usually excited retinal neurons, although sometimes it depressed their firing, and that physostigmine increased the 'spontaneous' firing and the proportion of ACh-sensitive units recorded (Straschill, 1968). *Fig.* 101 illustrates the depressant action of ACh (injected into the carotid artery) on the firing of a retinal 'on-centre' neuron. Subsequently, Perwein and Straschill (1972) briefly reported that all retinal ganglion cells that were excited by intra-arterially injected ACh proved to be 'off-centre' neurons, whereas all ganglion cells depressed by ACh had 'on-centre' properties. In other words, ACh behaved like a dark stimulus. The excitation, but not the depression by ACh, could be blocked by atropine, suggesting the presence of muscarinic receptors on 'off-centre' ganglion cells. Neither atropine (2–4 mg./kg.) nor dihydro-β-erythroidine (0·1–0·3 mg./kg.) affected light-driven or spontaneous activity, although less ACh-sensitive cells were encountered (Straschill, 1968; Perwein and Straschill, 1972). This indicates that the excitatory transmitter released from axon terminals of bipolar cells on to ganglion cells in the cat is unlikely to be ACh, although ACh may well be a transmitter elsewhere in the retina.

In the isolated rabbit retina, physostigmine potentiated, whereas *d*-tubocurarine blocked, various types of responses of ganglion cells, also suggesting the presence of cholinergic synapses in the retina (Ames and Pollen, 1969). However, it was concluded that both excitatory and inhibitory cholinergic pathways to the 'off-type' of ganglion cell exist, the receptors being mainly nicotinic. The different methods of drug application used by Straschill (1968) and Ames and Pollen (1969) may account for the different observations reported. It is therefore still necessary to determine the effects of ACh and its antagonists, as well as cholinesterase inhibitors, on orthodromic firing of retinal neurons by the method of microelectrophoresis, which localizes the site of action of the administered drug to the vicinity of the neuron recorded.

Unfortunately, histochemical studies of the distribution of AChE in the mammalian retina have contributed little towards the identification of cholinergic pathways in this structure. For example, since AChE was present mainly or wholly in the inner synaptic layer between the bipolar cells and the ganglion cells of the retina of various mammals, it has been concluded that bipolar cells with their nerve-fibres are cholinergic neurons (Koelle and Friedenwald, 1950; Francis, 1953). Subsequently, however, intense staining for AChE was observed in amacrine cells and most ganglion cells (Nichols and Koelle, 1968). Similarly, although the results of early studies suggested that little or no AChE is present in the outer plexiform layer, the reverse was reported recently (Dickson, Flumerfelt, Hollenberg, and Gwyn, 1971). Considerable interspecies differences are probably the main reasons for the different conclusions reached (*see* Silver, 1967). At present, therefore, the significance of the differential distribution of AChE in the mammalian retina is not clear.

SUMMARY

Dopamine

A large body of different types of evidence suggests that DA is an inhibitory transmitter in the mammalian retina. The results of fluorescence histochemical and autoradiographic studies indicate that DA is distributed within nerve-fibres mainly at the junction of the inner nuclear and inner plexiform layers, in three distinct bands, as well as in a small proportion of amacrine-like cell bodies in this region. DA also appears to be present in an even smaller proportion of ganglion cells. DA is taken up by the retina and is released either spontaneously or as a result of light flashes to the eye. DA has mainly depressant actions on retinal ganglion cells. Some studies suggest that DA is involved in light and dark adaptation. Despite these studies, however, direct pharmacological evidence for an inhibitory pathway in the retina that releases DA has yet to be obtained.

5-Hydroxytryptamine, Noradrenaline, and Histamine

Although 5-HT and NA have actions on some retinal ganglion cells, there is no other evidence to suggest that either monoamine is a transmitter in the retina. Histamine is present in the rabbit retina, but the significance of this is not clear.

Excitatory Amino-acids

Glutamate and its metabolizing enzymes are present in the mammalian retina, and autoradiographic studies have shown further that glutamate is differentially distributed in the rabbit retina. Glutamate excites retinal neurons, and various lines of evidence suggest that a major component of spreading depression in the retina results from the release of glutamate as a transmitter.

Depressant Amino-acids

Both GABA and glycine depress the firing of retinal ganglion cells, and both amino-acids appear to be unevenly distributed in the retina, associated mainly with amacrine cells. The results of pharmacological, biochemical, and autoradiographic studies suggest that these two amino-acids are transmitters in the retina, probably released from inhibitory interneurons. However, pharmacological experiments are needed to investigate this further.

Acetylcholine

The actions of ACh and related substances on retinal neurons have been insufficiently studied, but the available evidence raises the possibility of cholinergic pathways in the mammalian retina.

Chapter

9

GENERAL CONCLUSIONS

An evaluation of the data obtained from the studies reviewed in the preceding chapters permits some generalizations to be made about transmitter substances in the mammalian C.N.S. Probably the most obvious observation is that virtually all the naturally occurring substances which have hitherto been postulated as central transmitters in mammals fall into one of three categories—choline esters, monoamines, or amino-acids. Of course, other types of substances, such as certain polypeptides, ATP (cf. purinergic transmission in the gut; Burnstock, 1972), or unidentified compounds from brain extracts, may well prove to be mediators of the nerve impulse at some central synapses, but, as a first approximation, it appears that the major transmitters of the mammalian C.N.S. belong to these three groups. Although ACh, and subsequently the biogenic monoamines, have been regarded as likely transmitters for far longer than the amino-acids, it is the latter group which has been established as transmitters with the most certainty. Recent studies of glycine and GABA demonstrating their transmitter-like actions, which may be blocked by specific antagonists in the same manner as particular synaptic inhibitions, as well as their presence, specific uptake, and release from nervous tissue (both *in vitro* and *in vivo*), leave little doubt that these two amino-acids are inhibitory transmitters. Unfortunately, the complexity of the C.N.S. has precluded the demonstration of certain transmitter criteria at the synapse. For example, with the methods currently available it is impossible to demonstrate that stimulation of a central neuron results in a release of the postulated transmitter from the axon terminal of that neuron. Despite these limitations, however, some useful generalizations may be made if it is assumed that certain amino-acids, ACh, and monoamines are transmitters in the mammalian C.N.S.

Amino-acids

Glycine is an inhibitory transmitter in particular areas of the spinal cord and medulla, and neurochemical findings suggest that this is also true for the pons. There is no evidence that glycine functions as a transmitter in any central structures rostral to the brain-stem. In fact, glycine has relatively little or no action on many neurons of the cerebral, cerebellar, and hippocampal cortices, thalamus, and other higher structures, although the firing of the same neurons can be markedly depressed by GABA. It thus seems possible that the transmitter function of glycine, the simplest amino-acid, is confined to the phylogenetically older structures of the C.N.S.

GABA is also an inhibitory transmitter at certain sites in the spinal cord and brain-stem, but in addition mediates various inhibitions in all the more rostral brain structures (cerebral cortex, cerebellum, hippocampus, thalamus, hypothalamus, olfactory bulb, basal ganglia, and others). In these areas GABA is usually considerably more effective than glycine as a depressant, whereas in the spinal cord and medulla the reverse is often true. In different nuclei of the brain-stem (which may be regarded as a 'transition' between the spinal cord and the upper brain) it is difficult to distinguish between glycine and GABA on the basis of relative potency. Thus, it appears that GABA is involved in more recently developed inhibitions than is glycine, when considered on an evolutionary time-scale.

A point of interest in this regard is that the inhibitions thought to be mediated by glycine are usually of very short latency and duration (of the order of a few msec., as first described for spinal motoneurons by Eccles in 1964), whereas the inhibitions mediated by GABA are of relatively short latency, but of longer duration (several tens or hundreds of msec.). Of course, exceptions may be found, as for example the relatively long-duration inhibition of motoneurons mediated by Renshaw cells (glycine) and the relatively short-duration inhibition mediated by Purkinje axons in Deiters' nucleus (GABA). Nevertheless, the generalization that glycine mediates shorter inhibitions than GABA may have some predictive value as to the possible identity of the inhibitory transmitter involved in areas not yet studied. Since the qualitative actions of GABA and glycine are indistinguishable when administered electrophoretically to the same neuron, the longer duration of GABA-mediated synaptic inhibitions most likely results from repetitive firing of, or prolonged release of, transmitter from the inhibitory interneurons involved.

Little direct pharmacological evidence is available that the naturally occurring excitatory amino-acids are transmitters in the mammalian C.N.S., although various other types of studies suggest that they are. The fact that the synaptic excitation of most neural pathways in the C.N.S. cannot be reduced by antagonists of ACh and monoamines is indirect evidence that excitatory amino-acids could be the transmitters involved, in view of what is known about them. For example, glutamate and aspartate have powerful excitant actions on neurons in all areas of the C.N.S. studied, and there is some evidence for a differential sensitivity of various types of neurons to these amino-acids. Their actions on neurons as revealed by intracellular recording are not incompatible with an excitatory transmitter role, but mimicry of synaptic excitation has yet to be directly demonstrated. Glutamate and aspartate are differentially distributed in the C.N.S.; they can be released by stimulation of particular sites; they are present in the synaptosome fraction of nervous tissue; and there are specific uptake mechanisms to inactivate them. Thus, it is likely that both glutamate and aspartate are excitatory transmitters in the mammalian C.N.S. The greatest obstacle to obtaining clear evidence that glutamate and aspartate *are* central transmitters is the lack of specific antagonists of these amino-acids (*see* Curtis and others, 1972; Davies and Watkins, 1972; Haldeman and others, 1972). The discovery of such antagonists should lead to a breakthrough in the identification of excitatory transmitters in the C.N.S.

An evaluation of the actions and distribution of amino-acids in the mammalian C.N.S. provides a clue as to their possible functions. For example, properties common to both the excitant and depressant naturally occurring amino-acids are that they have a rapid time course of action and are consistent on repeated electrophoretic applications. Similarly, the inhibitions mediated by glycine and GABA (though differing in some respects—*see* p. 255), and the synaptic excitations likely to be mediated by glutamate and aspartate, are of relatively rapid time course and consistent on repeated stimuli. Some of these inhibitions and excitations are known to be involved in the control of movement and sensory inputs. It is therefore reasonable to speculate that amino-acid transmitters are predominantly involved in all the central functions which are essential to the immediate survival of the animal, especially as amino-acids affect the large majority of central neurons, and are likely to be transmitters in all nuclei of the C.N.S. These functions include voluntary movement, reflexes, mediation of sensory inputs, and others. Some neurochemical evidence is consistent with this concept. For example, in the spinal cord of a number of different vertebrates, the concentration of glycine is highest in the segments supplying the limbs (Aprison and others, 1969). In the brain, structures belonging to the central motor system (such as the substantia nigra, pallidum, putamen, caudate nucleus, dentate nucleus, oculomotor nucleus, and superior colliculus) are particularly rich in GABA (Okada and others, 1971).

These may not be the only functions of amino-acid transmitters, of course. Studies of the release of amino-acids and ACh from the cerebral cortex following electrical stimulation of the brain-stem have led to the suggestion that glutamate is involved (directly or indirectly) in the chemical mediation of reticulo-cortical activation (Jasper and others, 1965; Jasper and Koyama, 1969), and other types of studies may reveal even more functions for the amino-acids.

In conclusion, it is important to bear in mind that other amino-acids which have hitherto received little or no attention as possible transmitters may well prove to have such a role. In particular, evidence is accumulating rapidly that taurine may be an inhibitory transmitter in the mammalian C.N.S., although this is being recognized slowly by many laboratories. Taurine is one of the most abundant amino-acids in the nervous system, being present in extremely high concentrations at birth (Tallan, 1962; Agrawal, Davis, and Himwich, 1966; Agrawal and others, 1971). The particularly high levels of taurine in the cerebral and cerebellar cortices, the lateral geniculate nuclei, and the retina have been mentioned in preceding chapters. Taurine depresses the firing of neurons in various areas of the C.N.S. (Curtis and Watkins, 1965). *In vitro* studies of the cerebral cortex demonstrating the release and uptake of taurine, as well as the presence of taurine and its synthesizing enzyme in nerve-endings (*see* Chapter 7) are consistent with the hypothesis that taurine is an inhibitory transmitter. Since the depressant action of taurine in the brain is blocked by both strychnine and bicuculline (Curtis, Duggan, Felix and others, 1971; Curtis and Tebēcis, 1972), it is necessary to find synaptic inhibitions sensitive to both these convulsants in order to establish that taurine is the transmitter involved. Other neurochemical studies of taurine are cited in a recent article by Guidotti and others (1972).

Various other amino-acids may prove to be central transmitters, although there is at present little more evidence other than their presence in nervous tissue and their actions on the firing of neurons. These include β-alanine, cysteic acid, cysteine sulphinic acid, cystathionine, imidazole acetic acid, L-α-alanine, L-serine, β-hydroxy-γ-aminobutyric acid, GABA choline, and *N*-acetyl-L-aspartic acid (*see* reviews by Tallan, 1962; Curtis and Johnston, 1970; Cooper and others, 1970; Werman, 1972b; and recent papers by Howells, 1971; Johnston and Curtis, 1972; Nadler and Cooper, 1972).

The identity of the excitatory transmitter released from optic nerve terminals on to lateral geniculate and superior collicular neurons remains a mystery. If the postsynaptic action of this transmitter is antagonized by 5-HT, the transmitter is unlikely to be an amino-acid (*see* Chapter 3), but no firm conclusions may be drawn at this stage.

Acetylcholine

ACh is also a transmitter in the mammalian C.N.S., but differs from the amino-acids in several characteristic ways. First, ACh has actions on only certain types of neurons and therefore cannot be as universal in function as the amino-acids. The proportion of cholinoceptive neurons in any structure is relatively small, although particular types of neurons (identified physiologically) may all respond to ACh consistently. Unlike the amino-acids, each of which has only one type of action, ACh can either excite or depress the firing of a neuron, and sometimes both actions are apparent on the same neuron. Characteristically, both the excitation and depression by ACh are usually of a slower time course than those by the amino-acids, except in the case of Renshaw cells and many lateral geniculate neurons, which are excited by ACh with a time course similar to that of excitant amino-acids. Interpretation of the central actions of ACh are further complicated by the relative proportions of muscarinic and nicotinic properties of the neurons studied; these vary with the types of neurons and the structures in which they are located. In addition, some types of depression by ACh appear to be mediated by neither muscarinic nor nicotinic receptors.

Studies involving intracellular recording have revealed that the depolarization of cortical neurons by ACh is produced by an entirely different mechanism from the depolarization by glutamate. The depolarization by glutamate is accompanied by an increase in membrane *conductance*, but that by ACh (which is mediated predominantly by muscarinic receptors on cortical neurons) is associated with an increase in membrane *resistance*, thought to be the result of a decrease in permeability to potassium ions (Krnjević and others, 1971b). The muscarinic excitation of Renshaw cells appears to involve a different mechanism (Curtis and Felix, 1971c). The ionic mechanism underlying the depressant action of ACh has not been studied. Clearly, further investigations are required of the mechanisms underlying the actions of ACh in different areas of the C.N.S.

Although there is no reason to believe that all cholinoceptive neurons are sites of cholinergic transmission, the marked differential distribution of cholinoceptive cells, together with other data, provide important clues as to the probable sites where ACh may function as a transmitter. This is supported by cases where direct pharmacological evidence for cholinergic transmission has been demonstrated. One attempt to map the cholinergic systems of the brain has been by tracing the paths of AChE-containing fibres, as illustrated in *Figs*. 66, 74, and 80. However, the presence of AChE in non-nervous elements of the C.N.S., the wide variability in its distribution and intensity between different species, as well as other reasons (Silver, 1967; Phillis, 1970), have indicated that the histochemical data for the presence of AChE alone are not reliable criteria for cholinergic pathways, although they are extremely useful in delineating certain pathways, and in some instances may well represent cholinergic neurons. The presence of choline acetyltransferase in nervous tissue is regarded as much stronger evidence for the likelihood of cholinergic synapses, and the development of greater specificity for the histochemical demonstration of this enzyme (Kása, Mann, and Hebb, 1970) should lead to a breakthrough in the mapping of cholinergic pathways.

In conclusion, it seems that ACh is a transmitter on only certain populations of neurons, which may be predominantly involved in slower-conducting pathways and more 'abstract' functions than the amino-acids. The actions of ACh on Renshaw cells (administered electrophoretically and released synaptically) are unique in a number of respects. In the brain, ACh is thought to be important in consciousness, arousal, and memory (amongst other functions). Understandably, such hypotheses are difficult to test, but are consistent with the biochemical, histochemical, and pharmacological data available. For example, stimulation of the mesial thalamic area can result in an increase in the rate of liberation of ACh from the cerebral cortex without increasing that of glutamate in the same preparation, and an evaluation of these findings in relation to the state of the animal has led to the suggestion that reticulo-cortical arousal responses are mediated by different pathways—those utilizing ACh and those utilizing glutamate (Jasper and others, 1965; Jasper and Koyama, 1969). The cholinesterase-containing pathways described (Lewis and Shute, 1967; Shute and Lewis, 1967) originate mainly in subcortical structures (including the brain-stem) and project widely to all higher structures. The few cases where reasonable pharmacological evidence for cholinergic transmission in the brain has been presented have involved studies of long-latency synaptic responses, suggesting either that polysynaptic pathways were investigated, or that slowly conducting pathways were stimulated, or both. Much work remains to be done before any firm conclusions may be made about the possible functions of ACh in the brain. In fact, much work is required to establish clearly whether ACh, and not some other naturally occurring cholinester functions as a transmitter in the different regions of the brain, particularly in view of the recent demonstration of the presence (Hosein, Kato, Vine, and Hill, 1970) and the ACh-like actions (Falchetto, Kato, and Provini, 1971) of acetyl-L-carnitylcholine and other carnitines in the brain.

Monoamines

NA is an inhibitory transmitter released on to cerebellar Purkinje cells from fibres which originate in locus coeruleus of the brain-stem (Chapter 3). This is the only example in the whole mammalian C.N.S. of convincing evidence for a monoamine as a transmitter. However, since NA-containing fibres project from the locus coeruleus and other areas of the brain-stem to many structures of the C.N.S. (Andén and others, 1966; Ungerstedt, 1971), it is probable that the terminations of all these fibres represent noradrenergic synapses.

A comparison of *Fig.* 102, which illustrates the ascending NA-containing pathways of the rat brain, with the DA-containing pathways in *Fig.* 63 clearly indicates differences in the distribution of these catecholamine-containing fibres. Similarly, the areas

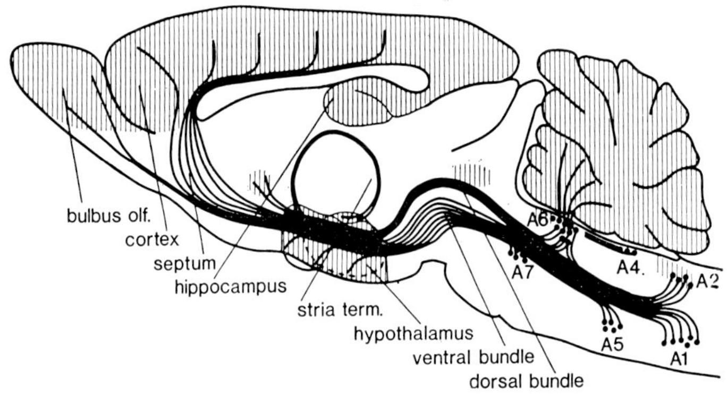

Fig. 102. Sagittal view of the rat brain, showing the ascending NA-containing pathways diagrammatically. Locations of the NA-containing cell bodies are indicated by the black dots; nerve terminals, by the shaded areas. The descending projection is not represented. (Ungerstedt, 1971).

where NA- and DA-containing fibres terminate are the same as those where NA and DA respectively have been postulated to be transmitters from biochemical and pharmacological studies. The origins and distribution of 5-HT-containing nerve-fibres correspond relatively closely to those of NA-containing fibres, but differences nevertheless exist. A general scheme of these three monoamine-containing pathways is shown in *Fig.* 103. In addition, recent evidence suggests the presence of yet another monoamine (possibly a tryptamine)-containing system with a course similar to that of the 5-HT-containing pathways (Björklund and others, 1970, 1971; *see also* Saavedra and Axelrod, 1972).

The actions of the monoamines differ in several characteristic respects from those of amino-acids and ACh. The most obvious difference is that both the depressant and excitant actions of the monoamines are typically of a slower time course than those of amino-acids and, to a lesser extent, of ACh. Studies of the depressant action of NA involving intracellular recording have revealed that NA hyperpolarizes spinal motoneurons and cerebellar Purkinje cells, the hyperpolarization being associated with little change, or an increase in membrane *resistance*. In the

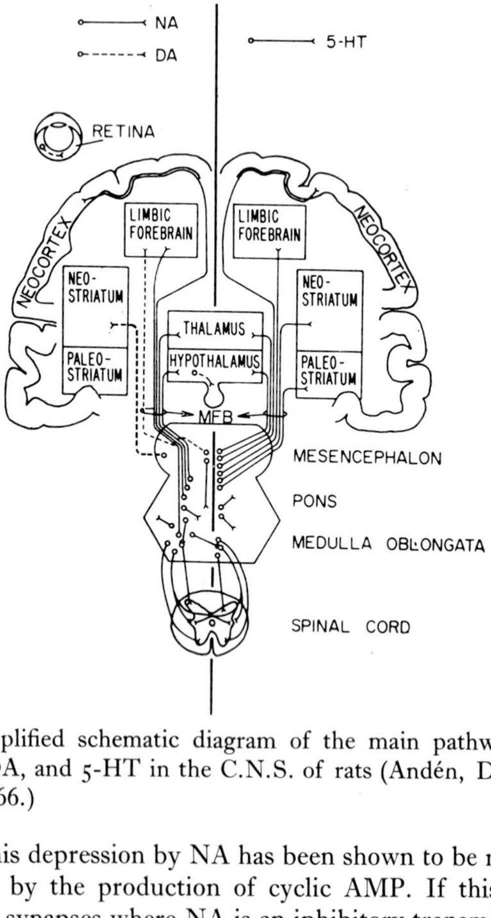

Fig. 103. Simplified schematic diagram of the main pathways containing NA, DA, and 5-HT in the C.N.S. of rats (Andén, Dahlström and others, 1966.)

cerebellum this depression by NA has been shown to be mediated metabolically by the production of cyclic AMP. If this is true for all central synapses where NA is an inhibitory transmitter, the mechanism underlying noradrenergic transmission clearly differs from that underlying the 'classical' synaptic transmission mediated by amino-acids, and possibly also from cholinergic transmission in most cases. It is too premature to speculate on the mechanisms underlying the depressant actions of 5-HT and DA, but these possibly also involve metabolic mechanisms.

The excitant actions of the monoamines are of particularly slow time course, and often decrease progressively on repeated applications. The significance of the monoamine excitations is not at all known, as no definitive pharmacological evidence has been obtained for monoaminergic excitation. Indeed, the central excitant actions of NA and 5-HT may well be the result of indirect effects secondary to localized vascular changes (Stone, 1971, 1972c), or produced by excessive expulsion of H^+ ions (Frederickson and others, 1971; Jordan and others, 1972), or both these effects may be important in some cases, at least.

If it is assumed that the monoamine-containing nerve-fibres represent monoaminergic pathways, some useful generalizations may be made. First, the fibres are of extremely small diameter and are either poorly myelinated or unmyelinated, allowing only very slow conduction velocities. The physiological data on the postulated monoaminergic pathways terminating in the cerebellar cortex (Chapter 3) and neostriatum (Chapter 5) are in agreement with the existence of such slowly conducting fibres. Similarly, as mentioned above, the time course of action of NA, 5-HT, and DA is extremely slow. Since depletion of monoamine stores by pharmacological agents does not result in any drastic changes in the behaviour of the animal, it is likely that monoamine transmitters are involved in central functions not essential for the immediate survival of the animal. Whatever the functions may be, they are widespread and appear to be controlled by the brainstem. For example, histochemical studies suggest that a single NA-containing neuron in the dorsolateral locus coeruleus can monosynaptically innervate the cerebellar and cerebral cortices, as well as subcortical structures, by means of abundant axon collaterals (Olson and Fuxe, 1971; Ungerstedt, 1971). Amongst the functions suggested for the monoamines are fine control of movement, learning, memory, sleep, cortical arousal, conditioned avoidance behaviour, emotion, neuroendocrine activity, thinking, temperature regulation, and sensory perception (*see* reviews by Bloom, 1969; Fuxe and others, 1970).

Concluding Comments

Although it is tempting to speculate on the conceptual framework of transmitter systems, it is too premature to regard these views as anything more than a working hypothesis. Certainly, characteristic differences exist between the amino-acids, monoamines, and ACh, both as regards their actions when applied artificially to neurons, and when released synaptically. The functions of these neurons also appear to be largely different, although in many cases they overlap. Further studies are required to map the transmitter systems of the C.N.S., using biochemical, histochemical, and pharmacological techniques in a number of different mammals. The interspecies differences in the distribution of catecholamine fluorescence in the brain-stem of the rat and cat (Pin and others, 1968; Sladek, 1971) raise the possibility that some of the conclusions based on histochemical studies of rats may not hold generally. Lastly, it is of course necessary to maintain an open mind about the possibility of discovering entirely new transmitters that differ both chemically and physiologically from the transmitters discussed in this book.

REFERENCES

ADES, H. W. (1941), 'Connections of the medial geniculate body in the cat', *Archs Neurol. Psychiat.*, **45,** 138–150.

AGHAJANIAN, G. K. (1972), 'Influence of drugs on the firing of serotonin-containing neurons in brain, *Fedn Proc. Fedn Am. Socs exp. Biol.*, **31,** 91–96.

AGHAJANIAN, G. K., and ASHER, I. M. (1971), 'Histochemical fluorescence of raphe neurons: selective enhancement by tryptophan', *Science, N.Y.*, **172,** 1159–1161.

AGHAJANIAN, G. K., and BLOOM, F. E. (1967), 'Localization of tritiated serotonin in rat brain by electron-microscopic autoradiography', *J. Pharmac. exp. Ther.*, **156,** 23–30.

AGHAJANIAN, G. K., BLOOM, F. E., and SHEARD, M. H. (1969), 'Electron microscopy of degeration within the serotonin pathway of rat brain', *Brain Res.*, **13,** 266–273.

AGHAJANIAN, G. K., FOOTE, W. E., and SHEARD, M. H. (1968), 'Lysergic acid diethylamide: sensitive neuronal units in the midbrain raphé', *Science, N.Y.*, **161,** 706–708.

AGHAJANIAN, G. K., ROSECRANS, J. A., and SHEARD, M. H. (1967), 'Serotonin: release in the forebrain by stimulation of the midbrain raphé', *Science, N.Y.*, **156,** 402–403.

AGRAWAL, H. C., DAVIS, J. M., and HIMWICH, W. A. (1966), 'Postnatal changes in free amino acid pool of rat brain', *J. Neurochem.*, **13,** 607–615.

AGRAWAL, H. C., DAVISON, A. N., and KACZMAREK, L. K. (1971), 'Subcellular distribution of taurine and cysteinesulphinate decarboxylase in developing rat brain', *Biochem. J.*, **122,** 759–763.

AHLMAN, H., GRILLNER, S., and UDO, M. (1971), 'The effect of 5-HTP on the static fusimotor activity and the tonic stretch reflex of an extensor muscle', *Brain Res.*, **27,** 393–396.

ALBERS, R. W., and BRADY, R. O. (1959), 'The distribution of glutamic decarboxylase in the nervous system of the rhesus monkey,' *J. biol. Chem.*, **234,** 926–928.

ALTMAN, J., and CARPENTER, M. B. (1961), 'Fiber projections of the superior colliculus in the cat', *J. comp. Neurol.*, **116,** 157–178.

ALTMANN, H., BRUGGENCATE, G. TEN, and SONNHOF, U. (1972), 'Differential strength of action of glycine and GABA in hypoglossus nucleus', *Pflügers Arch.—Eur. J. Physiol.*, **331,** 90–94.

AMES, A., and POLLEN, D. A. (1969), 'Neurotransmission in central nervous tissue: a study of isolated rabbit retina', *J. Neurophysiol.*, **32,** 424–442.

AMIN, A. H., CRAWFORD, T. B. B., and GADDUM, J. H. (1954), 'The distribution of substance P and 5-hydroxytryptamine in the central nervous system of the dog', *J. Physiol., Lond.*, **126,** 596–618.

ANDÉN, N-E., CARLSSON, A., DAHLSTRÖM, A., FUXE, K., HILLARP, N-Å., and LARSSON, K. (1964), 'Demonstration and mapping out of nigro-neostriatal dopamine neurons', *Life Sci. (Oxford)*, **3,** 523–530.

ANDÉN, N-E., CARLSSON, A., HILLARP, N-Å. and MAGNUSSON, T. (1964). 5-hydroxytryptamine release by nerve stimulation of the spinal cord', *Life Sci. (Oxford)*, **3,** 473–478.

ANDÉN, N-E., CARLSSON, A., HILLARP, N-Å., and MAGNUSSON, T. (1965), 'Noradrenaline release by nerve stimulation of the spinal cord', *Life Sci. (Oxford)*, **4,** 129–132.

ANDÉN, N-E., DAHLSTRÖM, A., FUXE, K., LARSSON, K., OLSON, L., and UNGERSTEDT, U. (1966), 'Ascending monoamine neurons to the telencephalon and diencephalon', *Acta physiol. scand.*, **67,** 313–326.

ANDÉN, N-E., FUXE, K., and UNGERSTEDT, U. (1967), 'Monoamine pathways to the cerebellum and cerebral cortex', *Experientia*, **23,** 838–839.

ANDÉN, N-E., HÄGGENDAL, J., MAGNUSSON, T., and ROSENGREN, E. (1964), 'The time course of the disappearance of noradrenaline and 5-hydroxytryptamine in the spinal cord after transection', *Acta physiol. scand.*, **62,** 115–118.

ANDÉN, N-E., JUKES, M. G. M., and LUNDBERG, A. (1966), 'The effect of DOPA on the spinal cord. 2. A pharmacological analysis', *Acta physiol. scand.*, **67,** 387–397.

ANDÉN, N-E., JUKES, M. G. M., LUNDBERG, A., and VYKLICKÝ, L. (1966a), 'The effect of DOPA on the spinal cord. 1. Influence on transmission from primary afferents', *Acta physiol. scand.*, **67,** 373–386.

ANDÉN, N-E., JUKES, M. G. M., LUNDBERG, A., and VYKLICKÝ, L. (1966b), 'The effect of DOPA on the spinal cord. 3. Depolarization evoked in the central terminals of ipsilateral Ia afferents by volleys in the flexor reflex afferents', *Acta physiol. scand.*, **68,** 322–336.

ANDERSEN, P., BRULAND, H., and KAADA, B. R. (1961), 'Activation of the field CA1 of the hippocampus by septal stimulation', *Acta physiol. scand.*, **51,** 29–40.

ANDERSEN, P., and CURTIS, D. R. (1964a), 'The excitation of thalamic neurones by acetylcholine', *Acta physiol. scand.*, **61,** 85–99.

ANDERSEN, P., and CURTIS, D. R. (1964b), 'The pharmacology of the synaptic and acetylcholine-induced excitation of ventrobasal thalamic neurones', *Acta physiol. scand.*, **61,** 100–120.

ANDERSEN, P., ECCLES, J. C., and LØYNING, Y. (1964), 'Pathway of postsynaptic inhibition in the hippocampus', *J. Neurophysiol.*, **27,** 608–619.

ANDERSEN, P., ECCLES, J. C., LØYNING, Y., and VOORHOEVE, P. E. (1963), 'Strychnine-resistant inhibition in the brain', *Nature, Lond.*, **200,** 843–845.

ANDERSEN, P., ECCLES, J. C., OSHIMA, T., and SCHMIDT, R. F. (1964), 'Mechanisms of synaptic transmission in the cuneate nucleus, *J. Neurophysiol.*, **27,** 1096–1116.

ANDERSEN, P., ECCLES, J. C., SCHMIDT, R. F., and YOKOTA, T. (1964), 'Identification of relay cells and interneurons in the cuneate nucleus', *J. Neurophysiol.*, **27,** 1080–1095.

ANDERSEN, P., ECCLES, J. C., and SEARS, T. A. (1964), 'The ventro-basal complex of the thalamus: types of cells, their responses and their functional organization', *J. Physiol., Lond.*, **174,** 370–399.

ANDERSEN, P., ECCLES, J. C., and VOORHOEVE, P. E. (1964), 'Post-synaptic inhibition of cerebellar Purkinje cells', *J. Neurophysiol.*, **27,** 1138–1153.

ANDERSEN, P., ETHOLM, B., and GORDON, G. (1970), 'Presynaptic and post-synaptic inhibition elicited in the cat's dorsal column nuclei by mechanical stimulation of skin', *J. Physiol., Lond.*, **210,** 433–455.

ANDERSEN, P., HOLMQVIST, B., and VOORHOEVE, P. E. (1966), 'Entorhinal activation of dentate granule cells', *Acta physiol. scand.*, **66,** 448–460.

ANDERSON, E. G. (1972), 'Bulbospinal serotonin-containing neurons and motor control', *Fedn Proc. Fedn Am. Socs exp. Biol.*, **31,** 107–112.

ANDERSON, E. G., BAKER, R. G., and BANNA, N. R. (1967), 'The effects of monoamine oxidase inhibitors on spinal synaptic activity', *J. Pharmac. exp. Ther.*, **158,** 405–415.

ANDERSON, E. G., and HOLGERSON, L. O. (1966), 'The distribution of 5-hydroxytryptamine and norepinephrine in cat spinal cord', *J. Neurochem.*, **13,** 479–485.

ANDERSON, E. G., and SHIBUYA, T. (1966), 'The effects of 5-hydroxytryptophan and L-tryptophan on spinal synaptic activity', *J. Pharmac. exp. Ther.*, **153,** 352–360.

ANDRES, K. H. (1965), Der Feinbau des Bulbus olfactorius der Ratte unter besonderer Berücksichtigung der synaptischen Verbindungen', *Z. Zellforsch., mikrosk. Anat.*, **65,** 530–561.

ANGELUCCI, L. (1956), 'Experiments with perfused frog's spinal cord', *Br. J. Pharmac. chemother.*, **11,** 161–170.

APRISON, M. H. (1970), 'Evidence of the release of ^{14}C-glycine from hemisectioned toad spinal cord with dorsal root stimulation', *Pharmacologist*, **12,** 222P.

APRISON, M. H., DAVIDOFF, R. A., and WERMAN, R. (1970), 'Glycine: its metabolic and possible transmitter roles in nervous tissue', in *Handbook of Neurochemistry, Metabolic Reactions in the Nervous System* (ed. Lajtha), vol. 13, pp. 381–397. New York: Plenum Press.

Aprison, M. H., and Hingtgen, J. N. (1972), 'Serotonin and behavior: a brief summary', *Fedn Proc. Fedn Am. Socs exp. Biol.*, **31**, 121–129.

Aprison, M. H., Shank, R. P., and Davidoff, R. A. (1969), 'A comparison of the concentration of glycine, a transmitter suspect, in different areas of the brain and spinal cord in seven different vertebrates', *Comp. Biochem. Physiol.*, **28**, 1345–1355.

Aprison, M. H., and Werman, R. (1965), 'The distribution of glycine in cat spinal cord roots', *Life Sci. (Oxford)*, **4**, 2075–2083.

Arnfred, T., and Hertz, L. (1971), 'Effects of potassium and glutamate on brain cortex slices: uptake and release of glutamic and other amino acids', *J. Neurochem.*, **18**, 259–265.

Ashkenazi, R., Holman, R. B., and Vogt, M. (1972), 'Release of transmitters on stimulation of the nucleus linearis raphe in the cat', *J. Physiol., Lond.*, **223**, 255–259.

Austin, L., and Phillis, J. W. (1965), 'The distribution of cerebellar cholinesterases in several species', *J. Neurochem.*, **12**, 709–727.

Avanzino, G. L., Bradley, P. B., Comis, S. D., and Wolstencroft, J. H. (1966), 'A comparison of the actions of ergothioneine and chlorpromazine applied to single neurons by two different methods', *Int. J. Neuropharmac.*, **5**, 331–333.

Avanzino, G. L., Bradley, P. B., and Wolstencroft, J. H. (1966), 'Pharmacological properties of neurons of the paramedian reticular nucleas', *Experientia*, **22**, 410.

Awapara, J., Landua, A. J., Fuerst, R., and Seale, B. (1950), 'Free γ-aminobutyric acid in brain', *J. biol. Chem.*, **187**, 35–39.

Axelrod, J., Albers, W., and Clemente, C. D. (1959), 'Distribution of catechol-*O*-methyl transferase in the nervous system and other tissues', *J. Neurochem.*, **5**, 68–72.

Bak, I. J., Hassler, R., Parizek, J., and Wagner, A. (1970), 'Electron microscopy of synaptic organization in striatum and substantia nigra of rat', *Proc. VII Cong. int. Microsc. elect.*, Grenoble, vol. **3**, 697–698.

Baker, R. G., and Anderson, E. G. (1970a), 'The effects of L-3, 4-dihydroxyphenylalanine on spinal reflex activity', *J. Pharmac. exp. Ther.*, **173**, 212–223.

Baker, R. G., and Anderson, E. G. (1970b), 'The antagonism of the effects of L-3,4-dihydroxyphenylalanine on spinal reflexes by adrenergic blocking agents', *J. Pharmac. exp. Ther.*, **173**, 224–231.

Banna, N. R., and Anderson, E. G. (1968), 'The effects of 5-hydroxytryptamine antagonists on spinal neuronal activity', *J. Pharmac. exp. Ther.*, **162**, 319–325.

Banna, N. R., and Jabbur, S. J. (1969), 'Pharmacological studies on inhibition in the cuneate nucleus of the cat', *Int. J. Neuropharmac.*, **8**, 299–307.

BANNA, N. R., NACCACHE, A., and JABBUR, S. J. (1972), 'Picrotoxin-like action of bicuculline', *Eur. J. Pharmac.*, **17,** 301–302.

BARBEAU, A., and McDOWELL, F. H. (1970) (ed.), *L-DOPA and Parkinsonism*, pp. 1–433. Philadelphia: F. A. Davis Co.

BARGMANN, W., and SCHARRER, B. (1970), 'Aspects of Neuroendocrinology', in V. *Int. Symp. on Neurosecretion*. Berlin: Springer.

BARKER, J. L., CRAYTON, J. W., and NICOLL, R. A. (1971a), 'Supraoptic neurosecretory cells: autonomic modulation', *Science, N.Y.*, **171,** 206–207.

BARKER, J. L., CRAYTON, J. W., and NICOLL, R. A. (1971b), 'Supraoptic neurosecretory cells: adrenergic and cholinergic sensitivity', *Science, N.Y.*, **171,** 208–210.

BARKER, J. L., CRAYTON, J. W., and NICOLL, R. A. (1971c), 'Noradrenaline and acetylcholine responses of supraoptic neurosecretory cells', *J. Physiol. Lond.*, **218,** 19–32.

BARKER, J. L., and NICOLL, R. A. (1972), 'Gamma-aminobutyric acid: role in primary afferent depolarization', *Science, N.Y.*, **176,** 1043–1045.

BARNARD, J. W. (1940), 'The hypoglossal complex of vertebrates', *J. comp. Neurol.*, **72,** 489–524.

BARNES, C. D., and POMPEIANO, O. (1971a), 'Vestibular nerve activation of a brain stem cholinergic system influencing the spinal cord', *Neuropharmacology*, **10,** 425–436.

BARNES, C. D., and POMPEIANO, O. (1971b), 'The interaction of brain stem adrenergic systems and VIIIth nerve stimulation on the spinal cord', *Neuropharmacology*, **10,** 437–446.

BARTOLINI, A., and PEPEU, G. (1967), 'Investigations into the acetylcholine output from the cerebral cortex of the cat in the presence of hyoscine', *Br. J. Pharmac. Chemother.*, **31,** 66–73.

BARTOLINI, A., WEISENTHAL, L. M., and DOMINO, E. F. (1972), 'Effect of photic stimulation on acetycholine release from cat cerebral cortex', *Neuropharmacology*, **11,** 113–122.

BATTISTIN, L., GRYNBAUM, A., and LAJTHA, A. (1969), 'Distribution and uptake of amino acids in various regions of the cat brain *in vitro*', *J. Neurochem.*, **16,** 1459–1468.

BAXTER, C. F., and ROBERTS, E. (1958), 'The γ-aminobutyric acid α-ketoglutaric acid transminase of beef brain', *J. biol. Chem.*, **233,** 1135–1139.

BAXTER, C. F., and ROBERTS, E. (1961), 'Elevation of γ-aminobutyric acid in brain: selective inhibition of γ-aminobutyric-α-ketoglutaric acid transaminase', *J. biol. Chem.*, **236,** 3287–3294.

BAZEMORE, A. W., ELLIOTT, K. A. C., and FLOREY, E. (1957), 'Isolation of factor I', *J. Neurochem.*, **1,** 334–339.

BEANI, L., BIANCHI, C., SANTINOCETO, L., and MARCHETTI, P. (1968), 'The cerebral acetylcholine release in conscious rabbits with semi-permanently implanted epidural cups', *Int. J. Neuropharmac.*, **7,** 469–481.

BECKMAN, A. L., and EISENMAN, J. S. (1970), 'Microelectrophoresis of biogenic amines on hypothalamic thermosensitive cells', *Science, N.Y.*, **170,** 334–336.

BELESLIN, D. B., and MYERS, R. D. (1970), 'The release of acetylcholine and 5-hydroxytryptamine from the mesencephalon of the unanaesthetized rhesus monkey', *Brain Res.*, **23,** 437–442.

BELESLIN, D., POLAK, R. L., and SPROULL, D. H. (1965), 'The release of acetylcholine into the cerebral subarachnoid space of anaesthetized cats', *J. Physiol, Lond.*, **177,** 420–428.

BERNARDI, G., ZIEGLGÄNSBERGER, W., HERZ, A., and PUIL, E. A. (1972), 'Intracellular studies on the action of L-glutamic acid on spinal neurones of the cat', *Brain Res.*, **39,** 523–525.

BERTLER, Å. (1961), 'Occurrence and localization of catechol amines in the human brain', *Acta physiol. scand.*, **51,** 97–107.

BERTLER, Å., FALCK, B., GOTTFRIES, C. G., LJUNGGREN, L., and ROSENGREN, E. (1964), 'Some observations on adrenergic connections between mesencephalon and cerebral hemispheres', *Acta pharmac. toxic.*, **21,** 283–289.

BERTLER, Å., and ROSENGREN, E. (1959), 'Occurrence and distribution of catechol amines in brain', *Acta physiol. scand.*, **47,** 350–361.

BESSON, M-J., CHERAMY, A., FELTZ, P., and GLOWINSKI, J. (1971), 'Dopamine: spontaneous and drug-induced release from the caudate nucleus in the cat', *Brain Res.*, **32,** 407–424.

BISCOE, T. J., and CURTIS, D. R. (1966), 'Noradrenaline and inhibition of Renshaw cells', *Science, N.Y.*, **151,** 1230–1231.

BISCOE, T. J., and CURTIS, D. R. (1967), 'Strychnine and cortical inhibition', *Nature, Lond.*, **214,** 914–915.

BISCOE, T. J., CURTIS, D. R., and RYALL, R. W. (1966), 'An investigation of catecholamine receptors of spinal interneurones', *Int. J. Neuropharmac.*, **5,** 429–434.

BISCOE, T. J., and KRNJEVIĆ, K. (1963), 'Chloralose and the activity of Renshaw cells', *Expl Neurol.*, **8,** 395–405.

BISCOE, T. J., and STRAUGHAN, D. W. (1966), 'Micro-electrophoretic studies of neurones in the cat hippocampus', *J. Physiol., Lond.*, **183,** 341–359.

BISHOP, G. H., CLARE, M. H., and LANDAU, W. M. (1969), 'Further analysis of fiber groups in the optic tract of the cat', *Expl Neurol.*, **24,** 386–399.

BISHOP, P. O. (1953), 'Synaptic transmission. An analysis of the electrical activity of the lateral geniculate nucleus in the cat after optic nerve stimulation', *Proc. R. Soc. B.*, **141,** 362–392.

BISHOP, P. O., BURKE, W., and DAVIS, R. (1962), 'Single-unit recording from antidromically activated optic radiation neurones', *J. Physiol., Lond.*, **162,** 432–450.

Bishop, P. O., Burke, W., and Hayhow, W. R. (1959), 'Lysergic acid diethylamide block of lateral geniculate synapses and relief by repetitive stimulation', *Expl Neurol.*, **1,** 556–568.

Bishop, P. O., and Davis, R. (1960), 'Synaptic potentials, after-potentials and slow rhythms of lateral geniculate neurones', *J. Physiol., Lond.*, **154,** 514–546.

Bishop, P. O., Field, G., Hennessy, B. L., and Smith, J. R. (1958), 'Action of D-lysergic acid diethylamide on lateral geniculate synapses', *J. Neurophysiol.*, **21,** 529–549.

Bishop, P. O., Jeremy, D., and McLeod, J. G. (1953), 'Phenomenon of repetitive firing in lateral geniculate of cat', *J. Neurophysiol.*, **16,** 437–447.

Bishop, P. O., and McLeod, J. G. (1954), 'Nature of potentials associated with synaptic transmission in lateral geniculate of cat', *J. Neurophysiol.*, **17,** 387–414.

Bisti, S., Iosif, G., Marchesi, G. F., and Strata, P. (1971), 'Pharmacological properties of inhibitions in the cerebellar cortex', *Expl Brain Res.*, **14,** 24–37.

Bisti, S., Iosif, G., and Strata, P. (1971), 'Suppression of inhibition in the cerebellar cortex by picrotoxin and bicuculline', *Brain Res.*, **28,** 591–593.

Björklund, A., Falck, B., and Stenevi, U. (1970), 'On the possible existence of a new intraneuronal monoamine in the spinal cord of the rat', *J. Pharmac. exp. Ther.*, **175,** 525–532.

Björklund, A., Falck, B., and Stenevi, U. (1971), 'Classification of monoamine neurones in the rat mesencephalon: distribution of a new monoamine neurone system', *Brain Res.*, **32,** 269–285.

Blackstad, T. W., Fuxe, K., and Hökfelt, T. (1967), 'Noradrenaline nerve terminals in the hippocampal region of the rat and the guinea pig', *Z. Zellforsch. mikrosk. Anat.*, **78,** 463–473.

Bloom, F. E. (1969), 'Serotonin neurons: localization and possible physiological role', in *Advances in Biochemical Psychopharmacology*, (ed. Costa and Greengard), vol. I, pp. 27–47. New York: Raven Press.

Bloom, F. E., Algeri, S., Groppetti, A., Revuelta, A., and Costa, E. (1969), 'Lesions of central norepinephrine terminals with 6-OH-dopamine: biochemistry and fine structure', *Science, N.Y.*, **166,** 1284–1286.

Bloom, F. E., Costa, E., and Salmoiraghi, G. C. (1964), 'Analysis of individual rabbit olfactory bulb neuron responses to the micro-electrophoresis of acetylcholine, norepinephrine and serotonin synergists and antagonists', *J. Pharmac. exp. Ther.*, **146,** 16–23.

Bloom, F. E., Costa, E., and Salmoiraghi, G. C. (1965), 'Anesthesia and the responsiveness of individual neurons of the caudate nucleus of the cat to acetylcholine, norepinephrine and dopamine administered by microelectrophoresis', *J. Pharmac. exp. Ther.*, **150,** 244–252.

Bloom, F. E., Hoffer, B. J., Battenberg, E. R., Siggins, G. R., Steiner, A. L., Parker, C. W., and Wedner, H. J. (1972), 'Adenosine 3′, 5′-monophosphate is localized in cerebellar neurons: immunofluorescence evidence', *Science, N.Y.*, **177**, 436–438.

Bloom, F. E., Hoffer, B. J., and Siggins, G. R. (1971), 'Studies on norepinephrine-containing afferents to Purkinje cells of rat cerebellum. 1. Localization of the fibers and their synapses', *Brain Res.*, **25**, 501–521.

Bloom, F. E., Hoffer, B. J., and Siggins, G. R. (1972), 'Norepinephrine mediated cerebellar synapses: a model system for neuropsychopharmacology', *Biol. Psychiat.*, **4**, 157–177.

Bloom, F. E., Hoffer, B. J., Siggins, G. R., Barker, J. L., and Nicoll, R. A. (1972), 'Effects of serotonin on central neurons: microiontophoretic administration', *Fedn Proc. Fedn Am. Socs exp. Biol.*, **31**, 97–106.

Bloom, F. E., and Iversen, L. L. (1971), 'Localizing ^{3}H-GABA in nerve terminals of rat cerebral cortex by electron microscopic autoradiography', *Nature, Lond.*, **229**, 628–630.

Bloom, F. E., Oliver, A. P., and Salmoiraghi, G. C. (1963), 'The responsiveness of individual hypothalamic neurons to microelectrophoretically administered endogenous amines', *Int. J. Neuropharmac.*, **2**, 181–193.

Boakes, R. J., Bradley, P. B., Briggs, I., and Dray, A. (1970), 'Antagonism of 5-hydroxytryptamine by LSD 25 in the central nervous system: a possible neuronal basis for the actions of LSD 25', *Br. J. Pharmac.*, **40**, 202–218.

Boakes, R. J., Bradley, P. B., Brookes, N., Candy, J. M., and Wolstencroft, J. H. (1971), 'Actions of noradrenaline, other sympathomimetic amines and antagonists on neurons in the brain stem of the cat', *Br. J. Pharmac.*, **41**, 462–479.

Bobbin, R. P., and Guth, P. S. (1970), 'Evidence that gamma-aminobutyric acid is not the inhibitory transmitter at the crossed olivocochlear nerve-hair cell junction', *Neuropharmacology*, **9**, 567–574.

Bodian, D. (1967), 'Synaptic types on spinal motoneurons: an electron microscopic study', *Johns Hopk. Hosp. Bull.*, **119**, 16–45.

Bogdanski, D. F., Weissbach, H. and Udenfriend, S. (1957), 'The distribution of serotonin, 5-hydroxytryptophan decarboxylase, and monoamine oxidase in brain', *J. Neurochem.*, **1**, 272–278.

Bornschein, H., and Heiss, W-D. (1966), 'Strychnine-resistant inhibition in the retina', *Experientia*, **22**, 49.

Boyd, E. S., Meritt, D. A., and Gardner, L. C. (1966), 'The effect of convulsant drugs on transmission through the cuneate nucleus', *J. Pharmac. exp. Ther.*, **154**, 398–409.

BRADLEY, P. B. (1968), 'Synaptic transmission in the central nervous system and its relevance for drug action', *Int. Rev. Neurobiol.*, **11**, 1–56.

BRADLEY, P. B., DHAWAN, B. N., and WOLSTENCROFT, J. H. (1966), 'Pharmacological properties of cholinoceptive neurones in the medulla and pons of the cat', *J. Physiol. Lond.*, **183**, 658–674.

BRADLEY, P. B., and WOLSTENCROFT, J. H. (1965), 'Actions of drugs on single neurones in the brain-stem', *Br. med. Bull.*, **21**, 15–18.

BRADLEY, P. B., WOLSTENCROFT, J. H., HÖSLI, L., and AVANZINO, G. L. (1966), 'Neuronal basis for the central action of chlorpromazine', *Nature, Lond.*, **212**, 1425–1427.

BRADSHAW, C. M., ROBERTS, M. H. T., and SZABADI, E. (1971), 'Effect of tricyclic antidepressants on monoamine responses of single cortical neurones', *Br. J. Pharmac.*, **41**, 394–395P.

BREMER, F. (1953), 'Strychnine tetanus of the spinal cord', in *The Spinal Cord* (Ciba Found. Symp.), pp. 78–83. London: Churchill.

BRIGGS, I. (1972), 'Ergothioneine in the central nervous system', *J. Neurochem.*, **19**, 27–35.

BRODAL, A. (1953), 'Reticulo-cerebellar connections in the cat. An experimental study', *J. comp. Neurol.*, **98**, 113–153.

BRODAL, A., and GOGSTAD, A. C. (1954), 'Rubro-cerebellar connections. An experimental study in the cat', *Anat. Rec.*, **118**, 455–486.

BRODAL, A., POMPEIANO, O., and WALBERG, F. (1962), *The Vestibular Nuclei and Their Connections*. Springfield, Ill.: Charles C. Thomas.

BRODAL, A., and ROSSI, G. F. (1955), 'Ascending fibers in brain stem reticular formation of cat', *Archs Neurol. Psychiat.*, **74**, 68–87.

BRODAL, A., TABER, E., and WALBERG, F. (1960), 'The raphe nuclei of the brain stem in the cat. II. Efferent connections', *J. comp. Neurol.*, **114**, 239 260.

BRODAL, A., and TORVIK, A. (1954), 'Cerebellar projection of paramedian reticular nucleus of medulla oblongata in cat', *J. Neurophysiol.*, **17**, 484–495.

BRODAL, A., WALBERG, F., and TABER, E. (1960), 'The raphe nuclei of the brain stem in the cat. III. Afferent connections', *J. comp. Neurol.*, **114**, 261–282.

BROOKS, C.McC., ISHIKAWA, T., KOIZUMI, K., and LU, H–H. (1966), 'Activity of neurones in the paraventricular nucleus of the hypothalamus and its control', *J. Physiol., Lond.*, **182**, 217–231.

BROOKS, C. McC., USHIYAMA, J., and LANGE, G. (1962), 'Reactions of neurons in or near the supraoptic nuclei', *Am. J. Physiol.*, **202**, 487–490.

BROOKS, V. B., CURTIS, D. R., and ECCLES, J. C. (1957), 'The action of tetanus toxin on the inhibition of motoneurones', *J. Physiol., Lond.*, **135**, 655–672.

Brooks, V. B., and Wilson, V. J. (1959), 'Recurrent inhibition in the cat's spinal cord', *J. Physiol., Lond.*, **146,** 380–391.

Brown, R. D., Daigneault, E. A., and Pruett, J. R. (1969), 'The effects of selected cholinergic drugs and strychnine on cochlear responses and olivo-cochlear inhibition', *J. Pharmac. exp. Ther.*, **165,** 300–309.

Brücke, F., Gogolák, G., and Stumpf, Ch. (1963), 'Mikroelektrodenuntersuchung der Reizantwort und der Zelltätigkeit im Hippocampus bei Septumreizung', *Pflügers Arch.*, **276,** 456–470.

Bruggencate, G. ten, and Engberg, I. (1968), 'Analysis of glycine actions on spinal interneurones by intracellular recording', *Brain Res.*, **11,** 446–450.

Bruggencate, G. ten, and Engberg, I. (1971), 'Iontophoretic studies in Deiters' nucleus of the inhibitory actions of GABA and related amino acids and the interactions of strychnine and picrotoxin', *Brain Res.*, **25,** 431–448.

Bruggencate, G. ten, and Sonnhof, U. (1971), 'Glycine and GABA actions in hypoglossus nucleus and blocking effects of strychnine and picrotoxin', *Experientia*, **27,** 1109.

Bruggencate, G. ten, and Sonnhof, U. (1972), 'Effects of glycine and GABA, and blocking actions of strychnine and picrotoxin in the hypoglossus nucleus', *Pflügers Arch.—Eur. J. Physiol.*, **334,** 240–252.

Bruun, A., and Ehinger, B. (1972), 'Uptake of the putative neurotransmitter, glycine, into the rabbit retina', *Invest. Ophthalmol.*, **11,** 191–198.

Bülbring, E., and Burn, J. H. (1941), 'Observations bearing on synaptic transmission by acetylcholine in the spinal cord', *J. Physiol., Lond.*, **100,** 337–368.

Burgen, A. S. V., and Chipman, L. M. (1951), 'Cholinesterase and succinic dehydrogenase in the central nervous system of the dog', *J. Physiol., Lond.*, **114,** 296–305.

Burke, W., and Sefton, A. J. (1966a), 'Recovery of responsiveness of cells of lateral geniculate nucleus of rat', *J. Physiol., Lond.*, **187,** 213–229.

Burke, W., and Sefton, A. J. (1966b), 'Inhibitory mechanisms in lateral geniculate nucleus of rat', *J. Physiol., Lond.*, **187,** 231–246.

Burnstock, G. (1972), 'Purinergic nerves', *Pharmac. Rev.*, **24,** 509–581.

Busch, H. F. M. (1964), 'Anatomical aspects of the anterior and lateral funiculi at the spinobulbar junction', in: *Prog. Brain Res., Organization of the Spinal Cord* (ed. Eccles and Schadé), vol. 11, pp. 223–237. Amsterdam: Elsevier.

Carlsson, A., Falck, B., Fuxe, K., and Hillarp, N-Å. (1964), 'Cellular localization of monoamines in the spinal cord', *Acta physiol. scand.*, **60,** 112–119.

CARLSSON, A., FALCK, B., and HILLARP, N-Å. (1962), 'Cellular localization of brain monoamines', *Acta physiol. scand.*, **56,** suppl. 196, 1–27.

CELESIA, G. G., and JASPER, H. H. (1966), 'Acetylcholine released from cerebral cortex in relation to state of activation', *Neurology, Minneap.*, **16,** 1053–1064.

CHALMERS, A., MCGEER, E. G., WICKSON, V., and MCGEER, P. L. (1970), 'Distribution of glutamic acid decarboxylase in the brain of various mammalian species', *Comp. gen. Pharmac.*, **1,** 385–390.

CHAPMAN, J. B., and MCCANCE, I. (1967), 'Acetylcholine sensitive cells in the intracerebellar nuclei of the cat', *Brain Res.*, **5,** 535–538.

CHATFIELD, P. O., and DEMPSEY, E. W. (1942), 'Some effects of prostigmine and acetylcholine on cortical potentials', *Am. J. Physiol.*, **135,** 633–640.

CHATFIELD, P. O., and PURPURA, D. P. (1954), 'Augmentation of evoked cortical potentials by topical application of prostigmine and acetylcholine after atropinization of cortex', *Electroenceph. clin. Neurophysiol.*, **6,** 287–298.

CHRIST, J. F., and BAK, I. J. (1970), 'Some fine structural observations on the small vesicular components in the posterior pituitary nerve fibres of rabbit', *Z.mikr.-anat. Forsch.*, **81,** 329–344.

CHU, S. (1968), 'Strychnine-sensitive and -insensitive inhibitions in cat's retina', *Tohoku J. exp. Med.*, **96,** 37–43.

CLARE, M. H., LANDAU, W. M., and BISHOP, G. H. (1969), 'The relationship of optic nerve fibre groups activated by electrical stimulation to the consequent central postsynaptic events', *Expl Neurol.*, **24,** 400–420.

CLINESCHMIDT, B. V., and ANDERSON, E. G. (1970), 'The blockade of bulbospinal inhibition by 5-hydroxytryptamine antagonists', *Expl Brain Res.*, **11,** 175–186.

CLINESCHMIDT, B. V., PIERCE, J. E., and LOVENBERG, W. (1971), 'Tryptophan hydroxylase and serotonin in spinal cord and brain stem before and after chronic transection', *J. Neurochem.*, **18,** 1593–1596.

COBBIN, L. B., LEEDER, S., and POLLARD, J. (1965), 'Smooth muscle stimulants in extracts of optic nerves, optic tracts and lateral geniculate bodies of sheep', *Br. J. Pharmac. Chemother.*, **25,** 295–306.

COLLIER, B., and MITCHELL, J. F. (1966), 'The central release of acetylcholine during stimulation of the visual pathway', *J. Physiol., Lond.*, **184,** 239–254.

COLLIER, B., and MITCHELL, J. F. (1967), 'The central release of acetylcholine during consciousness and after brain lesions', *J. Physiol., Lond.*, **188,** 83–98.

COMIS, S. D. (1970), 'Centrifugal inhibitory process affecting neurones in the cat cochlear nucleus', *J. Physiol., Lond.*, **210,** 751–760.

COMIS, S. D., and DAVIES, W. E. (1969), 'Acetylcholine as a transmitter in the cat auditory system', *J. Neurochem.*, **16,** 423–429.

COMIS, S. D., EVANS, E. F., and WHITFIELD, I. C. (1964), 'A micro-tap for controlling the application of drugs to single neurones', *J. Physiol., Lond.*, **173,** 4–6P.

COMIS, S. D., and WHITFIELD, I. C. (1966), 'The effect of acetyl choline on neurones of the cochlear nucleus', *J. Physiol., Lond.*, **183,** 22–23P.

COMIS, S. D., and WHITFIELD, I. C. (1967), 'Centrifugal excitation and inhibition in the cochlear nucleus', *J. Physiol., Lond.*, **188,** 34–35P.

COMIS, S. D., and WHITFIELD, I. C. (1968), 'Influence of centrifugal pathways on unit activity in the cochlear nucleus', *J. Neurophysiol.* **31,** 62–68.

CONNOR, J. D. (1968), 'Caudate unit responses to nigral stimuli: evidence for a possible nigro-neostriatal pathway', *Science, N.Y.*, **160,** 899–900.

CONNOR, J. D. (1970), 'Caudate nucleus neurones: correlation of the effects of substantia nigra stimulation with iontophoretic dopamine', *J. Physiol., Lond.*, **208,** 691–703.

CONRADI, S. (1969), 'On motoneuron synaptology in adult cats. An electron microscopic study of the structure and location of neuronal and glial elements on cat lumbosacral motoneurons in the normal state and after dorsal root section', *Acta physiol. scand.*, suppl. 332, 1–115.

COOMBS, J. S., ECCLES, J. C., and FATT, P. (1955). 'Excitatory synaptic action in motoneurones', *J. Physiol., Lond.*, **130,** 374–395.

COOPER, J. R., BLOOM, F. E., and ROTH, R. H. (1970), *The Biochemical Basis of Neuropharmacology*, pp. 1–220. London: Oxford University Press.

COUCH, J. R., jun. (1970), 'Responses of neurons in the raphe muclei to serotonin, norepinephrine and acetylcholine and their correlation with an excitatory synaptic input', *Brain Res.*, **19,** 137–150.

COURVILLE, J. (1966), 'Somatotopical organization of the projection from the nucleus interpositus anterior of the cerebellum to the red nucleus. An experimental study in the cat with silver impregnation methods', *Expl Brain Res.*, **2,** 191–215.

COYLE, J. T., and SNYDER, S. H. (1969), 'Catecholamine uptake by synaptosomes in homogenates of rat brain: stereospecificity in different areas', *J. Pharmac. exp. Ther.*, **170,** 221–231.

CRANMER, J. I., BRANN, A. W., and BACH, L. M. N. (1959), 'An adrenergic basis for bulbar inhibition', *Am. J. Physiol.*, **197,** 835–838.

CRAWFORD, J. M. (1970), 'The sensitivity of cortical neurones to acidic amino acids and acetylcholine', *Brain Res.*, **17,** 287–296.

Crawford, J. M., and Curtis, D. R. (1964), 'The excitation and depression of mammalian cortical neurones by amino acids', *Br. J. Pharmac. Chemother.*, **23,** 313–329.

Crawford, J. M., and Curtis, D. R. (1966), 'Pharmacological studies on feline Betz cells', *J. Physiol., Lond.*, **186,** 121–138.

Crawford, J. M., Curtis, D. R., Voorhoeve, P. E., and Wilson, V. J. (1963a), 'Strychnine and cortical inhibition', *Nature, Lond.*, **200,** 845–846.

Crawford, J. M., Curtis, D. R., Voorhoeve, P. E., and Wilson, V. J. (1963b), 'Excitation of cerebellar neurones by acetylcholine', *Nature, Lond.*, **200,** 579–580.

Crawford, J. M., Curtis, D. R., Voorhoeve, P. E., and Wilson, V. J. (1966), 'Acetylcholine sensitivity of cerebellar neurones in the cat', *J. Physiol., Lond.*, **186,** 139–165.

Cross, B. A., and Green, J. D. (1959), 'Activity of single neurones in the hypothalamus: effect of osmotic and other stimuli', *J. Physiol., Lond.*, **148,** 554–569.

Cross, B. A., Moss, R. L., and Urban, I. (1971), 'Effect of iontophoretic application of acetylcholine and noradrenaline to antidromically identified paraventricular neurones', *J. Physiol., Lond.*, **214,** 28–30P.

Cross, B. A., and Silver, I. A. (1966), 'Electrophysiological studies on the hypothalamus', *Br. med. Bull.*, **22,** 254–260.

Crossland, J., and Mitchell, J. F. (1956), 'The effect on the electrical activity of the cerebellum of a substance present in cerebellar extracts', *J. Physiol., Lond.*, **132,** 391–405.

Crossland, J., Mitchell, J. F., and Woodruff, G. N. (1966), 'The presence of ergothioneine in the central nervous system and its probable identity with the cerebellar factor', *J. Physiol., Lond.*, **182,** 427–438.

Crossland, J., Woodruff, G. N., and Mitchell, J. F. (1964), 'Identity of the cerebellar factor', *Nature, Lond.*, **203,** 1388–1389.

Crowshaw, K., Jessup, S. J., and Ramwell, P. W. (1967), 'Thin-layer chromatography of l-dimethylamino-naphthalene-5-sulphonyl derivatives of amino acids present in superfusates of cat cerebral cortex', *Biochem. J.*, **103,** 79–85.

Curtis, D. R. (1961), 'The identification of mammalian inhibitory transmitters', in *Nervous Inhibition* (ed. Florey), Proc. 2nd Friday Harbor Symp., pp. 342–349. Oxford: Pergamon.

Curtis, D. R. (1963), 'The pharmacology of central and peripheral inhibition', *Pharmac. Rev.*, **15,** 333–364.

Curtis, D. R. (1964), 'Microelectrophoresis', in *Physical Techniques in Biological Research* (Electrophysiological Methods, Pt. A, ed. Nastuk), vol. 5, pp. 144–190. New York: Academic Press.

CURTIS, D. R. (1965), 'The actions of amino acids upon mammalian neurones', in *Studies in Physiology* (ed. Curtis and McIntyre), pp. 34–42. Berlin: Springer-Verlag.

CURTIS, D. R. (1968), 'Pharmacology and neurochemistry of mammalian central inhibitory processes', in *Structure and Functions of Inhibitory Neuronal Mechanisms* (ed. von Euler), pp. 429–456. Oxford: Pergamon.

CURTIS, D. R. (1969), 'The pharmacology of spinal postsynaptic inhibition', in *Prog. Brain Res., Mechanisms of Synaptic Transmission* (ed. Akert and Waser) vol. 31, pp. 171–189. Amsterdam: Elsevier.

CURTIS, D. R., and CRAWFORD, J. M. (1969), 'Central synaptic transmission—microelectrophoretic studies', *A. Rev. Pharmac.*, **9**, 209–240.

CURTIS, D. R., and DAVIS, R. (1962), 'Pharmacological studies upon neurones of the lateral geniculate nucleus of the cat', *Br. J. Pharmac. Chemother.*, **18**, 217–246.

CURTIS, D. R., and DAVIS, R. (1963), 'Excitation of lateral geniculate neurones by quaternary ammonium derivatives', *J. Physiol., Lond.*, **165**, 62–82.

CURTIS, D. R., and DE GROAT, W. C. (1968), 'Tetanus toxin and spinal inhibition', *Brain Res.*, **10**, 208–212.

CURTIS, D. R., DUGGAN, A. W., and FELIX, D. (1970), 'GABA and inhibition of Deiters' neurones', *Brain Res.*, **23**, 117–120.

CURTIS, D. R., DUGGAN, A. W., FELIX, D., and JOHNSTON, G. A. R. (1970), 'GABA, bicuculline and central inhibition', *Nature, Lond.*, **226**, 1222–1224.

CURTIS, D. R., DUGGAN, A. W., FELIX, D., and JOHNSTON, G. A. R. (1971), 'Bicuculline, an antagonist of GABA and synaptic inhibition in the spinal cord of the cat', *Brain Res.*, **32**, 69–96.

CURTIS, D. R., DUGGAN, A. W., FELIX, D., JOHNSTON, G. A. R., and MCLENNAN, H. (1971), 'Antagonism between bicuculline and GABA in the cat brain', *Brain Res.*, **33**, 57–73.

CURTIS, D. R., DUGGAN, A. W., FELIX, D., JOHNSTON, G. A. R., TEBĒCIS, A. K., and WATKINS, J. C. (1972), 'Excitation of mammalian central neurones by acidic amino acids', *Brain Res.*, **41**, 283–301.

CURTIS, D. R., DUGGAN, A. W., and JOHNSTON, G. A. R. (1970), 'The inactivation of extracellularly administered amino acids in the feline spinal cord', *Expl Brain Res.*, **10**, 447–462.

CURTIS, D. R., and ECCLES, R. M. (1958a), 'The excitation of Renshaw cells by pharmacological agents applied electrophoretically', *J. Physiol., Lond.*, **141**, 435–445.

CURTIS, D. R., and ECCLES, R. M. (1958b), 'The effect of diffusional barriers upon the pharmacology of cells within the central nervous system', *J. Physiol., Lond.*, **141**, 446–463.

CURTIS, D. R., and FELIX, D. (1971a), 'GABA and prolonged spinal inhibition', *Nature, Lond.* (*New Biology*), **231**, 187–188.

CURTIS, D. R., and FELIX, D. (1971b), 'The effect of bicuculline upon synaptic inhibition in the cerebral and cerebellar cortices of the cat', *Brain Res.*, **34**, 301–321.

CURTIS, D. R., and FELIX, D. (1971c), 'Muscarinic action of acetylcholine', *Nature, Lond.*, **234**, 355.

CURTIS, D. R., FELIX, D., and MCLENNAN, H. (1970), 'GABA and hippocampal inhibition', *Br. J. Pharmac.*, **40**, 881–883.

CURTIS, D. R., HÖSLI, L., and JOHNSTON, G. A. R. (1968), 'A pharmacological study of the depression of spinal neurones by glycine and related amino acids', *Expl Brain Res.*, **6**, 1–18.

CURTIS, D. R., HÖSLI, L., JOHNSTON, G. A. R., and JOHNSTON, I. H. (1968), 'The hyperpolarization of spinal motoneurones by glycine and related amino acids', *Expl Brain Res.*, **5**, 235–258.

CURTIS, D. R., and JOHNSTON, G. A. R. (1970), 'Amino acid transmitters', in *Handbook of Neurochemistry* (ed. Lajtha), vol. 4, Central Mechanisms in the Nervous System, pp. 115–134. New York: Plenum Press.

CURTIS, D. R., and KOIZUMI, K. (1961), 'Chemical transmitter substances in brain stem of cat', *J. Neurophysiol.*, **24**, 80–90.

CURTIS, D. R., PHILLIS, J. W., and WATKINS, J. C. (1959), 'The depression of spinal neurones by γ-amino-*N*-butyric acid and β-alanine', *J. Physiol., Lond.*, **146**, 185–203.

CURTIS, D. R., PHILLIS, J. W., and WATKINS, J. C. (1960), 'The chemical excitation of spinal neurones by certain acidic amino acids', *J. Physiol., Lond.*, **150**, 656–682.

CURTIS, D. R., PHILLIS, J. W., and WATKINS, J. C. (1961), 'Cholinergic and non-cholinergic transmission in the mammalian spinal cord', *J. Physiol., Lond.*, **158**, 296–323.

CURTIS, D. R., and RYALL, R. W. (1966a), 'The excitation of Renshaw cells by cholinomimetics', *Expl Brain Res.*, **2**, 49–65.

CURTIS, D. R., and RYALL, R. W. (1966b), 'The acetylcholine receptors of Renshaw cells', *Expl Brain Res.*, **2**, 66–80.

CURTIS, D. R., and RYALL, R. W. (1966c), 'The synaptic excitation of Renshaw cells', *Expl Brain Res.*, **2**, 81–96.

CURTIS, D. R., and RYALL, R. W. (1966d), 'Pharmacological studies upon spinal presynaptic fibres', *Expl Brain Res.*, **1**, 195–204.

CURTIS, D. R., RYALL, R. W., and WATKINS, J. C. (1966), 'The action of cholinomimetics on spinal interneurones', *Expl Brain Res.*, **2**, 97–106.

CURTIS, D. R., and TEBĒCIS, A. K. (1972), 'Bicuculline and thalamic inhibition', *Expl Brain Res.*, **16**, 210–218.

CURTIS, D. R., and WATKINS, J. C. (1960), 'The excitation and depression of spinal neurones by structurally related amino acids', *J. Neurochem.*, **6**, 117–141.

CURTIS, D. R., and WATKINS, J. C. (1963), 'Acidic amino acids with strong excitatory actions on mammalian neurones', *J. Physiol., Lond.*, **166,** 1–14.

CURTIS, D. R., and WATKINS, J. C. (1965), 'The pharmacology of amino acids related to gamma-aminobutyric acid', *Pharmac. Rev.*, **17,** 347–391.

DAHLSTRÖM, A., and FUXE, K. (1965a), 'Evidence for the existence of monoamine-containing neurons in the central nervous system. I. Demonstration of monoamines in the cell bodies of brain stem neurones', *Acta physiol. scand.*, **62,** supp. 232, 1–55.

DAHLSTRÖM, A., and FUXE, K. (1965b), 'Evidence for the existence of monoamine neurons in the central nervous system. II. Experimentally induced changes in the intra-neuronal amine levels of bulbospinal neuron systems', *Acta physiol. Scand.*, **64,** supp. 247, 1–36.

DAHLSTRÖM, A., FUXE, K., KERNELL, D., and SEDVALL, G. (1965), 'Reduction of the monoamine stores in the terminals of bulbospinal neurones following stimulation in the medulla oblongata', *Life Sci. (Oxford)*, **4,** 1207–1212.

DAHLSTRÖM, A., FUXE, K., OLSON, L., and UNGERSTEDT, U. (1965), 'On the distribution and possible function of monoamine nerve terminals in the olfactory bulb of the rabbit', *Life Sci. (Oxford)*, **4,** 2071–2074.

DALE, H. H. (1934), 'Pharmacology and nerve endings', *Proc. R. Soc. Med.*, **28,** 319–332.

DAVID, J. P., MURAYAMA, S., MACHNE, X., and UNNA, K. R. (1963), 'Evidence supporting cholinergic transmission at the lateral geniculate body of the cat', *Int. J. Neuropharmac.*, **2,** 113–125.

DAVIDOFF, R. A. (1972a), 'Gamma-aminobutyric acid antagonism and presynaptic inhibition in the frog spinal cord', *Science, N.Y.*, **175,** 331–333.

DAVIDOFF, R. A. (1972b), 'The effects of bicuculline on the isolated spinal cord of the frog', *Expl Neurol.*, **35,** 179–193.

DAVIDOFF, R. A., GRAHAM, L. T., jun., SHANK, R. P., WERMAN, R., and APRISON, M. H. (1967), 'Changes in amino acid concentrations associated with loss of spinal interneurons', *J. Neurochem.*, **14,** 1025–1031.

DAVIDSON, N. and REISINE, H. (1971), 'Presynaptic inhibition in cuneate blocked by GABA antagonists', *Nature, Lond. (New Biology)*, **234,** 223–224.

DAVIDSON, N., and SOUTHWICK, C. A. P. (1971), 'Amino acids and presynaptic inhibition in the rat cuneate nucleus', *J. Physiol., Lond.*, **219,** 689–708.

DAVIES, J., and WATKINS, J. C. (1972), 'Is 1-hydroxy-3-aminopyrrolidone-2 (HA-966) a selective excitatory amino-acid antagonist?', *Nature, Lond. (New Biology)*, **238,** 61–63.

Davis, R. (1966), 'Acetylcholine-sensitive neurons in ventro-lateral thalamic nucleus', in *The Thalamus* (ed. Purpura and Yahr), pp. 193–195. New York: Columbia University Press.

Davis, R. (1969), 'Brachio-rubral and rubro-brachial activity in the cat', *Brain Res.*, **15**, 157–173.

Davis, R., and Huffman, R. D. (1969), 'Pharmacology of the brachium conjunctivum-red nucleus synaptic system in the baboon', *Fedn Proc. Fedn Am. Socs exp. Biol.*, **28**, 775.

Davis, R., and Vaughan, P. C. (1969), 'Pharmacological properties of feline red nucleus', *Int. J. Neuropharmac.*, **8**, 475–488.

Davison, A. N., and Kaczmarek, L. K. (1971), 'Taurine—a possible neurotransmitter?', *Nature, Lond.*, **234**, 107–108.

De Feudis, F. V., Delgado, J. M. R., and Roth, R. H. (1970), 'Content, synthesis and collectability of amino acids in various structures of the brains of rhesus monkeys', *Brain Res.*, **18**, 15–23.

Deffenu, G., Bertaccini, G., and Pepeu, G. (1967), 'Acetylcholine and 5-hydroxytryptamine levels of the lateral geniculate bodies and superior colliculus of cats after visual deafferentation', *Expl Neurol.*, **17**, 203–209.

De Groat, W. C. (1970), 'The effects of glycine, GABA and strychnine on sacral parasympathetic preganglionic neurones', *Brain Res.*, **18**, 542–544.

De Groat, W. C., and Ryall, R. W. (1967), 'An excitatory action of 5-hydroxytryptamine on sympathetic preganglionic neurones', *Expl Brain Res.*, **3**, 299–305.

De Groat, W. C., and Ryall, R. W. (1968a), 'The identification and characteristics of sacral parasympathetic preganglionic neurones', *J. Physiol., Lond.*, **196**, 563–577.

De Groat, W. C., and Ryall, R. W. (1968b), 'Recurrent inhibition in sacral parasympathetic pathways to the bladder', *J. Physiol., Lond.*, **196**, 579–591.

De Groot, J. (1959), 'The fat forebrain in stereotaxic coordinates', *Trans. R. Neth. Acad. Sci.*, **52**, 1–40.

De Roetth, A., jun. (1951), 'Role of acetylcholine in nerve activity', *J. Neurophysiol.*, **14**, 55–57.

Desmedt, J. E., and La Grutta, V. (1963), 'Function of the uncrossed efferent olivo-cochlear fibres in the cat', *Nature, Lond.*, **200**, 472–474.

Desmedt, J. E., and Monaco, P. (1961), 'Mode of action of the efferent olivo-cochlear bundle on the inner ear', *Nature, Lond.*, **192**, 1263–1265.

Dickson, D. H., Flumerfelt, B. A., Hollenberg, M. J., and Gwyn, D. G. (1971), 'Ultrastructural localization of cholinesterase activity in the outer plexiform layer of the newt retina', *Brain Res.*, **35**, 299–303.

DONNER, K. O., and REUTER, T. (1965), 'The dark-adaptation of single units in the frog's retina and its relation to the regeneration of rhodopsin', *Vision Res.*, **5,** 615–632.

DOWLING, J. E. (1967), 'The site of visual adaptation', *Science, N.Y.*, **155,** 273–279.

DREIFUSS, J. J., KALNINS, I., KELLY, J. S., and RUF, K. B. (1971), 'Action potentials and release of neurohypophysial hormones *in vitro*', *J. Physiol., Lond.*, **215,** 805–817.

DREIFUSS, J. J., and KELLY, J. S. (1970), 'Excitation of identified supraoptic neurones by the iontophoretic application of acetylcholine', *J. Physiol., Lond.*, **210,** 170–172P.

DREIFUSS, J. J., and KELLY, J. S. (1972a), 'Recurrent inhibition of antidromically identified rat supraoptic neurones', *J. Physiol., Lond.*, **220,** 87–103.

DREIFUSS, J. J., and KELLY, J. S. (1972b), 'The activity of identified supraoptic neurones and their response to acetylcholine applied by iontophoresis', *J. Physiol., Lond.*, **220,** 105–118.

DREIFUSS, J. J., KELLY, J. S., and KRNJEVIĆ, K. (1969), 'Cortical inhibition and γ-aminobutyric acid', *Expl Brain Res.*, **9,** 137–154.

DRUJAN, B. D., DÍAZ BORGES, J. M., and ALVAREZ, N. (1965), 'Relationship between the contents of adrenaline, noradrenaline and dopamine in the retina and its adaptional state', *Life Sci.*, **4,** 473–477.

DUDAR, J. D., and SZERB, J. C. (1969), 'The effect of topically applied atropine on resting and evoked cortical acetylcholine release', *J. Physiol., Lond.*, **203,** 741–762.

DUGGAN, A. W., and JOHNSTON, G. A. R. (1970), 'Glutamate and related amino acids in cat spinal roots, dorsal root ganglia and peripheral nerves', *J. Neurochem.*, **17,** 1205–1208.

DUGGAN, A. W., and MCLENNAN, H. (1971), 'Bicuculline and inhibition in the thalamus', *Brain Res.*, **25,** 188–191.

DYBALL, R. E. J. (1971), 'Oxytocin and ADH secretion in relation to electrical activity in antidromically identified supraoptic and paraventricular units', *J. Physiol., Lond.*, **214,** 245–246.

DYBALL, R. E. J., and KOIZUMI, K. (1969), 'Electrical activity in the supraoptic and paraventricular nuclei associated with neurohypophysial hormone release', *J. Physiol., Lond.*, **201,** 711–722.

ECCLES, J. C. (1964), *The Physiology of Synapses.* Berlin: Springer.

ECCLES, J. C. (1969), *The Inhibitory Pathways of the Central Nervous System*, The Sherrington Lectures IX, pp. 1–135. Springfield, Ill.: Charles C. Thomas.

ECCLES, J. C., ECCLES, R. M., and FATT, P. (1956), 'Pharmacological investigations on a central synapse operated by acetylcholine', *J. Physiol., Lond.*, **131,** 154–169.

ECCLES, J. C., FATT, P., and KOKETSU, K. (1954), 'Cholinergic and inhibitory synapses in a pathway from motor axon collaterals to motoneurones', *J. Physiol., Lond.*, **126,** 524–562.

ECCLES, J. C., ITO, M., and SZENTÁGOTHAI, J. (1967), *The Cerebellum as a Neuronal Machine.* Berlin: Springer.

ECCLES, J. C., LLINÁS, R., and SASAKI, K. (1966a), 'The inhibitory interneurones within the cerebellar cortex', *Expl Brain Res.*, **1,** 1–16.

ECCLES, J. C., LLINÁS, R., and SASAKI, K. (1966b), 'Parallel fibre stimulation and the responses induced thereby in the Purkinje cells of the cerebellum', *Expl Brain Res.*, **1,** 17–39.

ECCLES, J. C., LLINÁS, R., and SASAKI, K. (1966c), 'The mossy fibre-granule cell relay of the cerebellum and its inhibitory control by Golgi cells', *Expl Brain Res.*, **1,** 82–101.

ECCLES, J. C., LLINÁS, R., and SASAKI, K. (1966d), 'The excitatory synaptic action of climbing fibres on the Purkinje cells of the cerebellum', *J. Physiol., Lond.*, **182,** 268–296.

ECCLES, J. C., LLINÁS, R., and SASAKI, K. (1966e), 'The action of antidromic impulses on the cerebellar Purkinje cells', *J. Physiol., Lond.*, **182,** 316–345.

ECCLES, J. C., LLINÁS, R., and SASAKI, K. (1966f), 'Intracellularly recorded responses of the cerebellar Purkinje cells', *Expl Brain Res.*, **1,** 161–183.

ECCLES, J. C., SCHMIDT, R. F., and WILLIS, W. D. (1963), 'Pharmacological studies on presynaptic inhibition', *J. Physiol., Lond.*, **168,** 500–530.

EDERY, H., and LEVINGER, I. M. (1971), 'Acetylcholine release into the perfused intermeningeal spaces of the cat spinal cord', *Neuropharmacology*, **10,** 239–246.

EHINGER, B. (1966a), 'Distribution of adrenergic nerves in the eye and some related structures in the cat', *Acta physiol. scand.*, **66,** 123–128.

EHINGER, B. (1966b), 'Adrenergic retinal neurons', *Z. Zellforsch. mikrosk. Anat.*, **71,** 146–152.

EHINGER, B. (1970), 'Autoradiographic identification of rabbit retinal neurons that take up GABA', *Experientia*, **26,** 1063–1064.

EHINGER, B., and FALCK, B. (1971), 'Autoradiography of some suspected neurotransmitter substances: GABA, glycine, glutamic acid, histamine, dopamine, and L-DOPA', *Brain Res.*, **33,** 157–172.

EIDELBERG, E., GOLDSTEIN, G. P., and DEZA, L. (1967), 'Evidence for serotonin as a possible inhibitory transmitter in some limbic structures', *Expl Brain Res.*, **4,** 73–80.

ELLIOTT, K. A. C., and FLOREY, E. (1956), 'Factor 1—inhibitory factor from brain. Assay. Conditions in brain. Simulating and antagonizing substances', *J. Neurochem.*, **1,** 181–191.

ENGBERG, I., LUNDBERG, A., and RYALL, R. W. (1968a), 'The effect of reserpine on transmission in the spinal cord', *Acta physiol. scand.*, **72,** 115–122.

ENGBERG, I., LUNDBERG, A., and RYALL, R. W. (1968b), 'Is the tonic decerebrate inhibition of reflex paths mediated by monoaminergic pathways?', *Acta physiol. scand.*, **72,** 123–133.

ENGBERG, I., and MARSHALL, K. C. (1971), 'Mechanism of noradrenaline hyperpolarization in spinal cord motoneurones of the cat', *Acta physiol. scand.*, **83,** 142–144.

ENGBERG, I., and RYALL, R. W. (1966), 'The inhibitory action of noradrenaline and other monoamines on spinal neurones', *J. Physiol., Lond.*, **185,** 298–322.

ENGBERG, I., and THALLER, A. (1970a), 'On the interaction of picrotoxin with GABA and glycine in the spinal cord', *Brain Res.*, **19,** 151–154.

ENGBERG, I., and THALLER, A. (1970b), 'Hyperpolarizing actions of noradrenaline in spinal motoneurones', *Acta physiol. scand.*, **80,** 34A–35A.

ERULKAR, S. D., NICHOLS, C. W., POPP, M. B., and KOELLE, G. B. (1968), 'Renshaw elements: localization and acetylcholinesterase content', *J. Histochem. Cytochem.*, **16,** 128–135.

ESPLIN, D. W., ČAPEK, R., and ESPLIN, B. A. (1972), 'Pharmacological studies on dorsal root responses produced by ventral root stimulation in the cat', *Can. J. Physiol., Pharmac.*, **50,** 119–122.

EVARTS, E. V., LANDAU, W., FREYGANG, W., jun., and MARSHALL, W. H. (1955), 'Some effects of lysergic acid diethylamide and bufotenine on electrical activity in the cat's visual system', *Am. J. Physiol.*, **182,** 594–598.

FAHN, S., and CÔTÉ, L. J. (1968a), 'Regional distribution of γ-aminobutyric acid (GABA) in brain of the Rhesus monkey', *J. Neurochem.*, **15,** 209–213.

FAHN, S., and CÔTÉ, L. J. (1968b), 'Regional distribution of choline acetylase in the brain of the rhesus monkey', *Brain Res.*, **7,** 323–325.

FALCHETTO, S., KATO, G., and PROVINI, L. (1971), 'The action of carnitines on cortical neurons', *Can. J. Physiol. Pharmac.*, **49,** 1–7.

FALCK, B., (1964), 'Cellular localization of monoamines', *Prog. Brain Res.*, **8,** 28–44.

FEDINA, L., LUNDBERG, A., and VYKLICKÝ, L. (1971), 'The effect of a noradrenaline liberator (4, alpha-dimethyl-meta-tyramine) on reflex transmission in spinal cats', *Acta physiol. scand.*, **83,** 495–504.

FEDINEC, A. A., and SHANK, R. P. (1971), 'Effect of tetanus toxin on the content of glycine, gamma-aminobutyric acid, glutamate, glutamine and aspartate in the rat spinal cord', *J. Neurochem.*, **18,** 2229–2234.

FELDBERG, W. (1945), 'Present views on the mode of action of acetylcholine in the central nervous system', *Physiol. Rev.*, **25,** 596–642.

FELDBERG, W., and MYERS, R. D. (1966), 'Appearance of 5-hydroxytryptamine and an unidentified pharmacologically active lipid acid in effluent from perfused cerebral ventricles', *J. Physiol., Lond.*, **184,** 837–855.

FELDBERG, W., and VOGT, M. (1948), 'Acetylcholine synthesis in different regions of the central nervous system', *J. Physiol., Lond.*, **107,** 372–381.

FELIX, D., and MCLENNAN, H. (1971), 'The effect of bicuculline on the inhibition of mitral cells of the olfactory bulb', *Brain Res.*, **25,** 661–664.

FELPEL, L. P. (1972), 'Effects of strychnine, bicuculline and picrotoxin on labyrinthine-evoked inhibition in neck motoneurons of the cat', *Expl Brain Res.*, **14,** 494–502.

FELTZ, A., KRNJEVIĆ, K., and LISIEWICZ, A. (1972), 'Intracellular free Ca^{2+} and membrane properties of motoneurones', *Nature, Lond. (New Biology)*, **237,** 179–181.

FELTZ, P. (1971a), 'Monoamines and the excitatory nigro-striatal synaptic linkage', *Experientia*, **27,** 1111–1112.

FELTZ, P. (1971b), 'γ-Aminobutyric acid and a caudato-nigral inhibition', *Can. J. Physiol. Pharmac.*, **49,** 1113–1115.

FELTZ, P., and DE CHAMPLAIN, J. (1972), 'Electrophysiology and pharmacology of nigro-caudate excitation in 6-hydroxydopamine (6–OH–DA) treated cats', *Fedn Proc. Fedn Am. Socs exp. Biol.*, **31,** 603.

FELTZ, P., and MACKENZIE, J. S. (1969), 'Properties of caudate unitary responses to repetitive nigral stimulation', *Brain Res.*, **13,** 612–616.

FEX, J. (1967), 'Efferent inhibition in the cochlea related to hair-cell dc-activity: study of postsynaptic activity of the crossed olivo-cochlear fibres in the cat', *J. acoust. Soc. Am.*, **41,** 666–675.

FLOREY, E. (1954), 'An inhibitory and an excitatory factor of mammalian central nervous system, and their action on a single sensory neuron', *Archs int. Physiol.*, **62,** 33–53.

FLOREY, E. (1960), 'Physiological evidence for naturally occurring inhibitory substances', in *Inhibition in the Nervous System and Gamma-aminobutyric Acid*, (Proc. Int. Symp. held at Duarte, California, May 22–24, 1959), pp. 72–84. Oxford: Pergamon.

FONNUM, F. (1970), 'Topographical and subcellular localization of choline acetyltransferase in rat hippocampal region', *J. Neurochem.*, **17,** 1029–1037.

FONNUM, F., STORM-MATHISEN, J., and WALBERG, F. (1970), 'Glutamate decarboxylase in inhibitory neurons. A study of the enzyme in Purkinje cell axons and boutons in the cat', *Brain Res.*, **20,** 259–275.

FOX, C. A., and BARNARD, J. W. (1957), 'A quantitative study of the Purkinje cell dendritic branchlets and their relationship to afferent fibres', *J. Anat., Lond.*, **91,** 299–313.

Fox, C. A., Siegesmund, K. A., and Dutta, C. R. (1964), 'The Purkinje cell dendritic branchlets and their relation with the parallel fibres: light and electron microscopic observations', in *Morphological and Biochemical Correlates of Neural Activity* (ed. Cohen and Snider), pp. 112–141. New York: Harper and Row.

Francis, C. M. (1953), 'Cholinesterase in the retina', *J. Physiol., Lond.*, **120,** 435–439.

Frank, K. (1959), 'Basic mechanisms of synaptic transmission in the central nervous system', *I.R.E. Trans. med. Electronics*, ME–**6,** 85–88.

Frank, K., and Fuortes, M. G. F. (1956), 'Unitary activity of spinal interneurones of cats', *J. Physiol., Lond.*, **131,** 424–435.

Frederickson, R. C. A., Jordan, L. A., and Phillis, J. W. (1971), 'The action of noradrenaline on cortical neurons: effects of pH', *Brain Res.*, **35,** 556–560.

Friede, R. L., and Fleming, L. M. (1964), 'A comparison of cholinesterase distribution in the cerebellum of several species', *J. Neurochem.*, **11,** 1–7.

Frigyesi, T. L., and Purpura, D. P. (1966), 'Acetylcholine sensitivity of thalamic synaptic organizations activated by brachium conjunctivum stimulation', *Archs int. Pharmacodyn. Ther.*, **163,** 110–132.

Frigyesi, T. L., and Purpura, D. P. (1967), 'Electrophysiological analysis of reciprocal caudato-nigral relations', *Brain Res.*, **6,** 440–456.

Fukuda, J., Highstein, S. M., and Ito, M. (1972), 'Cerebellar inhibitory control of the vestibulo-ocular reflex investigated in rabbit IIIrd nucleus', *Expl Brain Res.*, **14,** 511–526.

Funderburk, W. H., and Case, T. J. (1951), 'The effect of atropine on cortical potentials', *Electroenceph. clin. Neurophysiol.*, **3,** 213–223.

Fuster, J. M., Creutzfeldt, O. D., and Straschill, M. (1965), 'Intracellular recording of neuronal activity in the visual system', *Z. vergl. Physiol.*, **49,** 604–622.

Fuxe, K. (1964), 'Cellular localization of monoamines in the median eminence and the infundibular stem of some mammals', *Z. Zellforsch. mikrosk. Anat.*, **61,** 710–724.

Fuxe, K. (1965), 'Evidence for the existence of monoamine neurons in the central nervous system. IV. Distribution of monoamine nerve terminals in the central nervous system', *Acta physiol. scand.*, **64,** suppl. 247, 37–85.

Fuxe, K., Grobecker, H., Hökfelt, T., and Jonsson, G. (1967), 'Identification of dopamine, noradrenaline and 5-hydroxtryptamine varicosities in a fraction containing nerve ending particles', *Brain Res.*, **6,** 475–480.

FUXE, K., HAMBERGER, B., and HÖKFELT, T. (1968), 'Distribution of noradrenaline nerve terminals in cortical areas of the rat', *Brain Res.*, **8,** 125–131.

FUXE, K., HÖKFELT, T., and UNGERSTEDT, U. (1970), 'Morphological and functional aspects of central monoamine neurons', *Int. Rev. Neurobiol.*, **13,** 93–126.

GALINDO, A. (1969), 'GABA-picrotoxin interaction in the mammalian central nervous system', *Brain Res.*, **14,** 763–767.

GALINDO, A., KRNJEVIĆ. K., and SCHWARTZ, S. (1967), 'Micro-iontophoretic studies on neurones in the cuneate nucleus', *J. Physiol., Lond.*, **192,** 359–377.

GALINDO, A., KRNJEVIĆ, K., and SCHWARTZ, S. (1968), 'Patterns of firing in cuneate neurones and some effects of Flaxedil', *Expl Brain Res.*, **5,** 87–101.

GENESER-JENSEN, F. A. (1972), 'Distribution of acetyl cholinesterase in the hippocampal region of the guinea pig. II. Subiculum and hippocampus', *Z. Zellforsch. mikrosk. Anat.*, **124,** 546–560.

GEREBTZOFF, M. A. (1959), *Cholinesterases, A Histochemical Contribution to the Solution of Some Functional Problems* (Int. Ser. Monog. on Pure and App. Biol.), vol. 3. Oxford: Pergamon.

GERSCHENFELD, H. M., TRAMEZZANI, J. H., and DE ROBERTIS, E. (1960), 'Ultrastructure and function in neurohypophysis of the toad', *Endocrinology*, **66,** 741–762.

GILLINGHAM, F. J., and DONALDSON, I. M. L. (1969) (ed.), *Third Symposium on Parkinson's Disease*, held at The Royal College of Surgeons of Edinburgh, May, 1968. Edinburgh: Livingstone.

GODFRAIND, J. M., KAWAMURA, H., KRNJEVIĆ, K., and PUMAIN, R. (1971), 'Actions of dinitrophenol and some other metabolic inhibitors on cortical neurones', *J. Physiol., Lond.*, **215,** 199–222.

GODFRAIND, J. M., KRNJEVIĆ, K., and PUMAIN, R. (1970), 'Unexpected features of the action of dinitrophenol on cortical neurones', *Nature, Lond.*, **228,** 562–564.

GOLDBERG, A. M., and MCCAMAN, R. E. (1967), 'A quantitative microchemical study of choline acetyltransferase and acetylcholinesterase in the cerebellum of several species', *Life Sci.*, **6,** 1493–1500.

GOLDBERG, J. M., and MOORE, R. Y. (1967), 'Ascending projections of the lateral lemniscus in the cat and monkey', *J. comp. Neurol.* **129,** 143–156.

GOLDBERG, L. J. (1972), 'Excitatory and inhibitory effects of lingual nerve stimulation on reflexes controlling the activity of masseteric motoneurons', *Brain Res.*, **39,** 95–108.

GOODCHILD, M., and NEAL, M. J. (1970), 'Uptake of ^{3}H-gamma-aminobutyric acid (GABA) by rat retina', *J. Physiol., Lond.*, **210,** 182–183P.

GOTTESFELD, Z., KELLY, J. S., and RENAUD, L. P. (1972), 'The *in vivo* neuropharmacology of amino-oxyacetic acid in the cerebral cortex of the cat', *Brain Res.*, **42,** 319–335.

GRAHAM, L. T., jun. (1972), 'Intraretinal distribution of GABA content and GAD activity', *Brain Res.*, **36,** 476–479.

GRAHAM, L. T., jun., and APRISON, M. H. (1969), 'Distribution of some enzymes associated with the metabolism of glutamate, aspartate, γ-aminobutyrate and glutamine in cat spinal cord', *J. Neurochem.*, **16,** 559–566.

GRAHAM, L. T., jun., BAXTER, C. F., and LOLLEY, R. N. (1970), '*In vivo* influence of light or darkness on the GABA system in the retina of the frog (*Rana pipiens*)', *Brain Res.*, **20,** 379–388.

GRAHAM, L. T., jun., SHANK, R. P., WERMAN, R., and APRISON, M. H. (1967), 'Distribution of some synaptic transmitter suspects in cat spinal cord: glutamic acid, aspartic acid, γ-aminobutyric acid, glycine, and glutamine', *J. Neurochem.*, **14,** 465–472.

GRANIT, R., and PHILLIPS, C. G. (1956), 'Excitatory and inhibitory processes acting upon individual Purkinje cells of the cerebellum in cats', *J. Physiol., Lond.*, **133,** 520–547.

GRANIT, R., and PHILLIPS, C. G. (1957), 'Effects on Purkinje cells of surface stimulation of the cerebellum', *J. Physiol., Lond.*, **135,** 73–92.

GREEN, J. D. (1964), 'The hippocampus', *Physiol. Rev.*, **44,** 561–608.

GREEN, J. D., MANCIA, M., and VON BAUMGARTEN, R. (1962), 'Recurrent inhibition in the olfactory bulb. I. Effects of antidromic stimulation of the lateral olfactory tract', *J. Neurophysiol.*, **25,** 467–488.

GREEN, J. R., HALPERN, L. M., and VAN NIEL, S. (1970), 'Alterations in the activity of selected enzymes in the chronic isolated cerebral cortex of cat', *Brain*, **93,** 57–64.

GRILLNER, S., HONGO, T., and LUNDBERG, A. (1967), 'The effect of DOPA on the spinal cord. 7. Reflex activation of static γ-motoneurones from the flexor reflex afferents', *Acta physiol. scand.*, **70,** 403–411.

GROFOVÁ, I., and RINVIK, E. (1970), 'An experimental electron microscopic study on the striatonigral projection in the cat', *Expl Brain Res.*, **11,** 249–262.

GUIDOTTI, A., BADIANI, G., and PEPEU, G. (1972), 'Taurine distribution in cat brain', *J. Neurochem.*, **19,** 431–435.

GUILLERY, R. W. (1966), 'A study of Golgi preparations from the dorsal lateral geniculate nucleus of the adult cat', *J. comp. Neurol.*, **128,** 21–50.

GUILLERY, R. W. (1969), 'The organization of synaptic interconnections in the laminae of the dorsal lateral geniculate nucleus of the cat', *Z. Zellforsch. mikrosk. Anat.*, **96,** 1–38.

GUTH, P. S., and AMARO, J. (1969), 'A possible cholinergic link in olivo-cochlear inhibition', *Int. J. Neuropharmac.*, **8,** 49–53.

GUTH, P. S., and BOBBIN, R. P. (1971), 'The pharmacology of peripheral auditory processes; cochlear pharmacology', in *Advances in Pharmacology and Chemotherapy* (ed. Garattini, Goldin, Hawking, and Kopin), vol. 9, pp. 93–130. New York: Academic Press.

GWYN, D. G. (1971), 'Acetylcholinesterase activity in the red nucleus of the rat. Effects of rubrospinal tractotomy', *Brain Res.*, **35,** 447–461.

HÄGGENDAL, J., and MALMFORS, T. (1965), 'Identification and cellular localization of the catecholeamines in the retina and choroid of the rabbit', *Acta physiol. scand.*, **64,** 58–66.

HALDEMAN, S., HUFFMAN, R. D., MARSHALL, K. C., and MCLENNAN, H. (1972), 'The antagonism of the glutamate-induced and synaptic excitations of thalamic neurones', *Brain Res.*, **39,** 419–425.

HAYASHI, T. (1954), 'Effects of sodium glutamate on the nervous system', *Keio J. Med.*, **3,** 183–192.

HAYASHI, T. (1956), *Chemical Physiology of Excitation in Muscle and Nerve*. Tokyo: Nakayama-Shoten.

HEBB, C. O. (1957), 'Biochemical evidence for the neural function of acetylcholine', *Physiol. Rev.*, **37,** 196–220.

HEBB, C. O. (1970), 'CNS at the cellular level: identity of transmitter agents', *A. Rev. Physiol.*, **32,** 165–192.

HEBB, C. O., KRNJEVIĆ, K., and SILVER, A. (1963), 'Effect of undercutting on the acetylcholinesterase and choline acetyltransferase activity in the cat's cerebral cortex', *Nature, Lond.*, **198,** 692.

HEBB, C. O., and SILVER, A. (1956), 'Choline acetylase in the central nervous system of man and some other mammals', *J. Physiol., Lond.*, **134,** 718–728.

HEISS, W-D. (1967), 'Daueraktivität retinaler Neurone unter Einwirkung von Strychnin und Pikrotoxin', *Vision Res.*, **7,** 583–598.

HEISS, W-D., and BORNSCHEIN, H. (1965), 'Die Impulsverteilung der Daueraktivität von Einzelfasern des N. opticus. Einflüsse von Licht, Ischämie, Strychnin und Barbiturat', *Pflügers Arch.*, **286,** 1–18.

HELLER, A., SEIDEN, L. S., and MOORE, R. Y. (1966), 'Regional effects of lateral hypothalamic lesions on brain norepinephrine in the cat', *Int. J. Neuropharmac.*, **5,** 91–101.

HERZ, A., and VON FREYTAG-LORINGHOVEN, Hj. (1968), 'Über die synaptische Erregung im Corpus striatum und deren antagonistische Beeinflussung durch mikroelektophoretisch verabfolgte Glutaminsäure und Gamma-aminobuttersäure', *Pflügers Arch.*, **299,** 167–184.

HERZ, A., and GOGOLÁK, G. (1965), 'Mikroelektrophoretische Untersuchungen am Septum des Kaninchens', *Pflügers Arch.*, **285,** 317–330.

HERZ, A., and NACIMIENTO, A. C. (1965), 'Über die Wirkung von Pharmaka auf Neurone des Hippocampus nach mikroelektrophoretischer Verabfolgung', *Naunyn-Schmiedeberg's Arch. Pharmak.*, **251,** 295–314.

HERZ, A., and ZIEGLGÄNSBERGER, W. (1966), 'Synaptic excitation in the corpus striatum inhibited by microelectrophoretically administered dopamine', *Experientia*, **22,** 839–842.

HERZ, A., and ZIEGLGÄNSBERGER, W. (1968), 'The influence of microelectrophoretically applied biogenic amines, cholinomimetics and procaine on synaptic excitation in the corpus striatum', *Int. J. Neuropharmac.*, **7,** 221–230.

HIGHSTEIN, S. M., ITO, M., and TSUCHIYA, T. (1971), 'Synaptic linkage in the vestibulo-ocular reflex pathway of rabbit', *Expl Brain Res.*, **13,** 306–326.

HILL, R. G., SIMMONDS, M. A., and STRAUGHAN, D. W. (1972), 'Antagonism of GABA by picrotoxin in the feline cerebral cortex', *Br. J. Pharmac.*, **44,** 807–809.

HIRSCH, H. E., and ROBINS, E. (1962), 'Distribution of γ-aminobutyric acid in the layers of the cerebral and cerebellar cortex. Implications for its physiological role', *J. Neurochem.*, **9,** 63–70.

HODGKIN, A. L. (1964), *The Conduction of the Nervous Impulse*, the Sherrington Lectures VII, pp. 1–108. Liverpool University Press.

HOFFER, B. J., SIGGINS, G. R., and BLOOM, F. E. (1969), 'Response of single cerebellar Purkinje (P) cells to microelectrophoresis of norepinephrine (NE) and serotonin (5-HT)', *Fedn Proc. Fedn Am. Socs exp. Biol.*, **28,** 443.

HOFFER, B. J., SIGGINS, G. R., and BLOOM, F. E. (1971), 'Studies on norepinephrine-containing afferents to Purkinje cells of rat cerebellum. II. Sensitivity of Purkinje cells to norepinephrine and related substances administered by microiontophoresis', *Brain Res.*, **25,** 523–534.

HOFFER, B. J., SIGGINS, G. R., OLIVER, A. P., and BLOOM, F. E. (1971), 'Cyclic AMP mediation of norepinephrine inhibition in rat cerebellar cortex: a unique class of synaptic responses', *Ann. N.Y. Acad. Sci.*, **185,** 531–549.

HOFFER, B. J., SIGGINS, G. R., OLIVER, A. P., and BLOOM, F. E. (1972), 'Cyclic AMP mediated adrenergic synapses to cerebellar Purkinje cells', in *The Physiology and Pharmacology of Cyclic AMP* (ed. Greengard, Robison, and Paoletti), pp. 411–423. New York: Raven Press.

HOFFER, B. J., SIGGINS, G. R., WOODWARD, D. J., and BLOOM, F. E. (1971), 'Spontaneous discharge of Purkinje neurons after destruction of catecholamine-containing afferents by 6-hydroxydopamine', *Brain Res.*, **30,** 425–430.

HOFFMANN, K-P. (1972), 'The retinal input to the superior colliculus in the cat', *Invest. Ophthal.*, **11,** 467–473.

HÖKFELT, T., and FUXE, K. (1969), 'Cerebellar monoamine nerve terminals, a new type of afferent fibers to the cortex cerebelli', *Expl Brain Res.*, **9,** 63–72.

HÖKFELT, T., JONSSON, G., and LJUNGDAHL, Å. (1970), 'Regional uptake and subcellular localization of (^{3}H)-gamma-aminobutyric acid (GABA) in rat brain slices', *Life Sci.*, **9,** 203–212.

HÖKFELT, T., and LJUNGDAHL, Å. (1970), 'Cellular localization of labeled gamma-aminobutyric acid (^{3}H-GABA) in rat cerebellar cortex: an autoradiographic study', *Brain Res.*, **22,** 391–396.

HÖKFELT, T., and LJUNGDAHL, Å. (1971a), 'Light and electron microscopic autoradiography on spinal cord slices after incubation with labeled glycine', *Brain Res.*, **32,** 189–194.

HÖKFELT, T., and LJUNGDAHL, Å. (1971b), 'Uptake of (^{3}H)noradrenaline and γ-(^{3}H)aminobutyric acid in isolated tissues of rat: an autoradiographic and fluorescence microscopic study', *Prog. Brain Res.*, **34,** 87–102. *Histochemistry of Nervous Transmission* (ed. Eränko), Amsterdam: Elsevier.

HÖKFELT, T., and LJUNGDAHL, Å. (1972), 'Autoradiographic identification of cerebral and cerebellar cortical neurons accumulating labeled gamma-aminobutyric acid (^{3}H-GABA)', *Expl Brain Res.*, **14,** 354–362.

HÖKFELT, T., and UNGERSTEDT, U. (1969), 'Electron and fluorescence microscopical studies on the nucleus caudatus putamen of the rat after unilateral lesions of ascending nigro-neostriatal dopamine neurons', *Acta physiol. scand.*, **76,** 415–426.

HOLMAN, R. B., and VOGT, M. (1972), 'Release of 5-hydroxytryptamine from caudate nucleus and septum', *J. Physiol., Lond.*, **223,** 243–254.

HONGO, T., and RYALL, R. W. (1966), 'Electrophysiological and micro-electrophoretic studies on sympathetic preganglionic neurones in the spinal cord', *Acta physiol. scand.*, **68,** 96–104.

HOPKIN, J., and NEAL, M. J. (1971), 'Effect of electrical stimulation and high potassium concentrations on the efflux of (^{14}C) glycine from slices of spinal cord', *Br. J. Pharmac.*, **42,** 215–223.

HORNYKIEWICZ, O. (1966), 'Dopamine (3-hydroxytyramine) and brain function', *Pharmac. Rev.*, **18,** 925–964.

HORTON, E. W. (1969), 'Hypotheses on physiological roles of prostaglandins', *Physiol. Rev.*, **49,** 122–161.

HOSEIN, E. A., KATO, A., VINE, E., and HILL, A. M. (1970), 'The identification of acetyl-*l*-carnitylcholine in rat brain extracts and the comparison of its cholinomimetic properties with acetylcholine', *Can. J. Physiol. Pharmac.*, **48,** 709–722.

HÖSLI, L., and HAAS, H. L. (1971), 'Effects of histamine, histidine and imidazole acetic acid on neurones of the medulla oblongata of the cat', *Experientia*, **27**, 1311–1312.

HÖSLI, L., and TEBĒCIS, A. K. (1970), 'Actions of amino acids and convulsants on bulbar reticular neurones', *Expl Brain Res.*, **11**, 111–127.

HÖSLI, L., TEBĒCIS, A. K., and FILIAS, N. (1969), 'Effects of glycine, beta-alanine and GABA, and their interaction with strychnine, on brain stem neurones', *Brain Res.*, **16**, 293–295.

HÖSLI, L., TEBĒCIS, A. K., and HAAS, H. L. (1971), 'Depressant amino acids and possible antagonists on medullary reticular neurones', *Experientia*, **27**, 732–733.

HÖSLI, L., TEBĒCIS, A. K., and SCHÖNWETTER, H. P. (1970), 'LSD and its interaction with 5-hydroxytryptamine and noradrenaline on medullary reticular neurons', *Pflügers Arch.—Eur. J. Physiol.*, **319**, 136.

HÖSLI, L., TEBĒCIS, A. K., and SCHÖNWETTER, H. P. (1971), 'A comparison of the effects of monoamines on neurones of the bulbar reticular formation', *Brain Res.*, **25**, 357–370.

HOWELLS, D. J. (1971), 'The inhibitory transmitter: GABA or its choline ester?' *J. Pharm. Pharmac.*, **23**, 794–795.

HUFFMAN, R. D., and MCFADIN, L. S. (1972), 'Suppression of presynaptic inhibition and cerebellar disfacilitation by bicuculline', *Life Sci.*, **11**, 113–121.

HUNT, C. C., and KUNO, M. (1956), 'Background discharge and evoked responses of spinal interneurones', *J. Physiol., Lond.*, **147**, 364–384.

ISRAËL, M., and WHITTAKER, V. P. (1965), 'The isolation of mossy fibre endings from the granular layer of the cerebellar cortex', *Experientia*, **21**, 325–326.

ITO, M., HIGHSTEIN, S. M., and FUKUDA, J. (1970), 'Cerebellar inhibition of the vestibulo-ocular reflex in rabbit and cat and its blockage by picrotoxin', *Brain Res.*, **17**, 524–526.

ITO, M., HIGHSTEIN, S. M., and TSUCHIYA, T. (1970), 'The postsynaptic inhibition of rabbit oculomotor neurones by secondary vestibular impulses and its blockage by picrotoxin', *Brain Res.*, **17**, 520–523.

ITO, M., HONGO, T., YOSHIDA, M., OKADA, Y., and OBATA, K. (1964), 'Antidromic and trans-synaptic activation of Deiters' neurones induced from the spinal cord', *Jap. J. Physiol.*, **14**, 638–658.

ITO, M., KAWAI, N., and UDO, M. (1968), 'The origin of cerebellar-induced inhibition of Deiters neurones. III. Localization of the inhibitory zone', *Expl Brain Res.*, **4**, 310–320.

ITO, M., OBATA, K., and OCHI, R. (1966), 'The origin of cerebellar-induced inhibition of Deiters neurones. II. Temporal correlation between the trans-synaptic activation of Purkinje cells and inhibition of Deiters neurones' *Expl Brain Res.*, **2,** 350–364.

ITO, M., UDO, M., and MANO, N. (1970), 'Long inhibitory and excitatory pathways converging onto cat reticular and Deiters' neurons and their relevance to reticulofugal axons', *J. Neurophysiol.*, **33,** 210–226.

ITO, M., UDO, M., MANO, N., and KAWAI, N. (1970), 'Synaptic action of the fastigiobulbar impulses upon neurones in the medullary reticular formation and vestibular nuclei', *Expl Brain Res.*, **11,** 29–47.

ITO, M., and YOSHIDA, M. (1966), 'The origin of cerebellar-induced inhibition of Deiters neurones. I. Monosynaptic initiation of the inhibitory postsynaptic potentials', *Expl Brain Res.*, **2,** 330–349.

ITO, M., YOSHIDA, M., and OBATA, K. (1964), 'Monosynaptic inhibition of the intracerebellar nuclei induced from the cerebellar cortex', *Experientia*, **20,** 575–576.

ITO, M., YOSHIDA, M., OBATA, K., KAWAI, N., and UDO, M. (1970), 'Inhibitory control of intracerebellar nuclei by the Purkinje cell axons', *Expl Brain Res.*, **10,** 64–80.

IVERSEN, L. L., and BLOOM, F. E. (1971), 'Autoradiographic studies of the distribution of ^{3}H-GABA in mammalian CNS', *Experientia*, **27,** 1114.

IVERSEN, L. L., and BLOOM, F. E. (1972), 'Studies of the uptake of ^{3}H-GABA and (^{3}H)glycine in slices and homogenates of rat brain and spinal cord by electron microscopic autoradiography', *Brain Res.*, **41,** 131–143.

IVERSEN, L. L., and JOHNSTON, G. A. R. (1971), 'GABA uptake in rat central nervous system: comparison of uptake in slices and homogenates and the effects of some inhibitors', *J. Neurochem.*, **18,** 1939–1950.

IVERSEN, L. L., MITCHELL, J. F., and SRINIVASAN, V. (1971), 'The release of γ-aminobutyric acid during inhibition in the cat visual cortex', *J. Physiol., Lond.*, **212,** 519–534.

IVERSEN, L. L., and NEAL, M. J. (1968), 'The uptake of (^{3}H)GABA by slices of rat cerebral cortex', *J. Neurochem.*, **15,** 1141–1149.

IWATA, N., and SAKAI, Y. (1971), 'Cholinergic mechanism in the inhibition of spinal reflexes from the superficial radial nerve of the decerebrate cat', *Experientia*, **27,** 1418–1419.

IWATA, N., SAKAI, Y., and DEGUCHI, T. (1971), 'Effects of physostigmine on the inhibition of trigeminal motoneurones by cutaneous impulses in the cat', *Expl Brain Res.*, **13,** 519–532.

JANKOWSKA, E., JUKES, M. G. M., LUND, S., and LUNDBERG, A. (1967a), 'The effect of DOPA on the spinal cord. 5. Reciprocal organization of pathways transmitting excitatory action to alpha motoneurones of flexors and extensors', *Acta physiol. scand.*, **70,** 369–388.

Jankowska, E., Jukes, M. G. M., Lund, S., and Lundberg, A. (1967b), 'The effect of DOPA on the spinal cord. 6. Half-centre organization of interneurones transmitting effects from the flexor reflex afferents', *Acta physiol. scand.*, **70,** 389–402.

Jankowska, E., and Lindström, S. (1971), 'Morphological identification of Renshaw cells', *Acta physiol. scand.*, **81,** 428–430.

Jankowska, E., Lund, S., and Lundberg, A. (1966), 'The effect of DOPA on the spinal cord. 4. Depolarization evoked in the central terminals of contralateral Ia afferent terminals by volleys in the flexor reflex afferents', *Acta physiol. scand.*, **68,** 337–341.

Jankowska, E., Lund, S., Lundberg, A., and Pompeiano, O. (1968), 'Inhibitory effects evoked through ventral reticulo-spinal pathways', *Archs. ital. Biol.*, **106,** 124–140.

Jankowska, E., and Roberts, W. J. (1972), 'Synaptic actions of single interneurones mediating reciprocal Ia inhibition of motoneurones', *J. Physiol., Lond.*, **222,** 623–642.

Jansen, J., and Brodal, A. (1954) (ed.), *Aspects of Cerebellar Anatomy.* Oslo: J. G. Tanum.

Jansen, J., and Brodal, A. (1958), 'Das Kleinhirn', in *Handbuch der mikroskopischen Anatomie des Menschen*, vol. 4/8, Nervensystem (ed. Möllendorff and Bargmann), pp. 1–323. Berlin: Springer.

Jasper, H. H., Khan, R. T., and Elliott, K. A. C. (1965), 'Amino acids released from the cerebral cortex in relation to its state of activation', *Science, N.Y.*, **147,** 1448–1449.

Jasper, H. H., and Koyama, I. (1969), 'Rate of release of amino acids from the cerebral cortex in the cat as affected by brainstem and thalamic stimulation', *Can. J. Physiol. Pharmac.*, **47,** 889–905.

Johnson, E. S., Roberts, M. H. T., Sobieszek, A., and Straughan, D. W. (1969), 'Noradrenaline sensitive cells in cat cerebral cortex', *Int. J. Neuropharmac.*, **8,** 549–566.

Johnson, E. S., Roberts, M. H. T., and Straughan, D. W. (1969), 'The responses of cortical neurones to monoamines under differing anaesthetic conditions', *J. Physiol., Lond.*, **203,** 261–280.

Johnson, E. S., Roberts, M. H. T., and Straughan, D. W. (1970), 'Amino-acid induced depression of cortical neurones', *Br. J. Pharmac.*, **38,** 659–666.

Johnson, J. L. (1972), 'Glutamic acid as a synaptic transmitter in the nervous system. A review', *Brain Res.*, **37,** 1–19.

Johnson, J. L., and Aprison, M. H. (1970), 'Distribution of glutamic acid, a transmitter candidate, and other amino acids in the dorsal sensory neuron of the cat', *Brain Res.*, **24,** 285–292.

JOHNSON, J. L., and APRISON, M. H. (1971), 'The distribution of glutamate and total free amino acids in thirteen specific regions of the cat central nervous system', *Brain Res.*, **26,** 141–148.

JOHNSTON, G. A. R. (1968), 'The intraspinal distribution of some depressant amino acids', *J. Neurochem.*, **15,** 1013–1017.

JOHNSTON, G. A. R., and CURTIS, D. R. (1972), 'γ-Aminobutyrylcholine and central inhibition', *J. Pharm. Pharmac.*, **24,** 251–252.

JOHNSTON, G. A. R., DE GROAT, W. C., and CURTIS, D. R. (1969), 'Tetanus toxin and amino acid levels in cat spinal cord', *J. Neurochem.*, **16,** 797–800.

JOHNSTON, G. A. R., and IVERSEN, L. L. (1971), 'Glycine uptake in rat central nervous system slices and homogenates: evidence for different uptake systems in spinal cord and cerebral cortex', *J. Neurochem.*, **18,** 1951–1961.

JONES, E. G., and POWELL, T. P. S. (1968), 'The ipsilateral cortical connexions of the somatic sensory areas in the cat', *Brain Res.*, **9,** 71–94.

JONES, E. G., and POWELL, T. P. S. (1969a), 'Electron microscopy of synaptic glomeruli in the thalamic relay nuclei of the cat', *Proc. R. Soc. B.*, **172,** 153–171.

JONES, E. G., and POWELL, T. P. S. (1969b), 'An electron microscopic study of the mode of termination of cortico-thalamic fibres within the sensory relay nuclei of the thalamus', *Proc. R. Soc. B.*, **172,** 173–185.

JONES, E. G., and ROCKEL, A. J. (1971), 'The synaptic organization in the medial geniculate body of afferent fibres ascending from the inferior colliculus', *Z. Zellforsch. mikrosk. Anat.*, **113,** 44–66.

JONSSON, G., FUXE, K., and HÖKFELT, T. (1972), 'On the catecholamine innervation of the hypothalamus, with special reference to the median eminence', *Brain Res.*, **40,** 271–281.

JORDAN, C. C., and WEBSTER, R. A. (1971), 'Release of acetylcholine and ^{14}C-glycine from the cat spinal cord *in vivo*', *Br. J. Pharmac.*, **43,** 441P.

JORDAN, L. M., FREDERICKSON, R. C. A., PHILLIS, J. W., and LAKE, N. (1972), 'Microelectrophoresis of 5-hydroxytryptamine: a clarification of its action on cerebral cortical neurones', *Brain Res.*, **40,** 552–558.

JORDAN, L. M., and PHILLIS, J. W. (1972), 'Acetylcholine inhibition in the intact and chronically isolated cerebral cortex', *Br. J. Pharmac.*, **45,** 584–595.

JOUVET, M. (1969), 'Biogenic amines and the states of sleep', *Science, N.Y.*, **163,** 32–41.

KANDEL, E. R. (1964), 'Electrical properties of hypothalamic neuroendocrine cells', *J. gen. Physiol.*, **47,** 691–717.

KANDEL, E. R., SPENCER, W. A., and BRINLEY, F. J., jun. (1961), 'Electrophysiology of hippocampal neurons. I. Sequential invasion and synaptic organization', *J. Neurophysiol.*, **24,** 225–242.

KANDERA, J., LEVI, G., and LAJTHA, A. (1968), 'Control of cerebral metabolite levels. II. Amino acid uptake and levels in various areas of the rat brain,' *Archs Biochem. Biophys.*, **126,** 249–260.

KÁSA, P., JOÓ, F., and CSILLIK, B. (1965), 'Histochemical localization of acetylcholinesterase in the cat cerebellar cortex', *J. Neurochem.*, **12,** 31–35.

KÁSA, P., MANN, S. P., and HEBB, C. (1970), 'Localization of choline acetyltransferase. Histochemistry at the light microscope level. Ultrastructural localization in spinal neurones', *Nature, Lond.*, **226,** 812–816.

KATAOKA, K., and DE ROBERTIS, E. (1967), 'Histamine in isolated small nerve endings and synaptic vesicles of rat brain cortex', *J. Pharmac. exp. Ther.*, **156,** 114–125.

KATZ, B. (1966), *Nerve, Muscle and Synapse*. New York: McGraw-Hill.

KAWAI, N. (1970), 'Release of 5-hydroxytryptamine from slices of superior colliculus by optic tract stimulation', *Neuropharmacology*, **9,** 395–397.

KAWAI, N., and YAMAMOTO, C. (1969), 'Effects of 5-hydroxytryptamine, LSD and related compounds on electrical activities evoked *in vitro* in thin sections from the superior colliculus', *Int. J. Neuropharmac.*, **8,** 437–449.

KAWAMURA, H., and PROVINI, L. (1970), 'Depression of cerebellar Purkinje cells by micro-iontophoretic application of GABA and related amino acids', *Brain Res.*, **24,** 293–304.

KELLERTH, J-O. (1968), 'Aspects on the relative significance of pre- and postsynaptic inhibition in the spinal cord', in *Structure and Function of Inhibitory Neuronal Mechanisms* (ed. von Euler, Skoglund and Söderberg), pp. 197–212. Oxford: Pergamon.

KELLERTH, J-O., and SZUMSKI, A. J. (1966a), 'Two types of stretch-activated post-synaptic inhibitions in spinal motoneurones as differentiated by strychnine', *Acta physiol. scand.*, **66,** 133–145.

KELLERTH, J-O., and SZUMSKI, A. J. (1966b), 'Effects of picrotoxin on stretch-activated post-synaptic inhibitions in spinal motoneurones', *Acta physiol. scand.*, **66,** 146–156.

KELLY, J. S., and DREIFUSS, J. J. (1970), 'Antidromic inhibition of identified rat supraoptic neurones', *Brain Res.*, **22,** 406–409.

KELLY, J. S., and KRNJEVIĆ, K. (1969), 'The action of glycine on cortical neurones', *Expl Brain Res.*, **9,** 155–163.

KELLY, J. S., and RENAUD, L. P. (1971), 'Post-synaptic inhibition in the cuneate blocked by GABA antagonist', *Nature, Lond.* (*New Biology*), **232,** 25–26.

KELLY, J. S., RENAUD, L., and GOTTESFELD, Z. (1971), 'The effects on inhibition of AOAA evoked increase in the cortical GABA level', *Experientia*, **27,** 1114.

KIDOKORO, Y., KUBOTA, K., SHUTO, S., and SUMINO, R. (1968), 'Reflex organization of cat masticatory muscles', *J. Neurophysiol.*, **31**, 695–708.

KIM, J. S., BAK., I. J., HASSLER, R., and OKADA, Y. (1971), 'Role of γ-aminobutyric acid (GABA) in the extrapyramidal motor system. 2. Some evidence for the existence of a type of GABA-rich strio-nigral neurons', *Expl Brain Res.*, **14**, 95–104.

KIM, J. S., HASSLER, R., KUROKAWA, M., and BAK, I. J. (1970), 'Abnormal movements and rigidity induced by harmaline in relation to striatal acetylcholine, serotonin, and dopamine', *Expl Neurol.*, **29**, 189–200.

KIRALY, J. K., and PHILLIS, J. W. (1961), 'The action of some drugs on the dorsal root potentials on the isolated toad spinal cord', *Br. J. Pharmac. chemother.*, **17**, 224–231.

KISHIDA, K., and NAKA, K. I. (1967), 'Amino acids and the spikes from the retinal ganglion cells', *Science, N.Y.*, **156**, 648–650.

KNYIHÁR, E., and CSILLIK, B. (1970), 'Localizations of inhibitors of the acetylcholine- and GABA-synthesizing systems in the rat brain. An autoradiographic assay on the distribution of ^{14}C-hemicholinium and ^{14}C-thiosemicarbazide', *Expl Brain Res.*, **11**, 1–16.

KOBAYASHI, H., and LIBET, B. (1968), 'Generation of slow postsynaptic potentials without increases in ionic conductance', *Proc. natn. Acad. Sci. U.S.A.*, **60**, 1304–1311.

KOBAYASHI, H., and LIBET, B. (1970), 'Actions of noradrenaline and acetylcholine on sympathetic ganglion cells', *J. Physiol., Lond.*, **208**, 353–372.

KOELLA, W. P. (1969), 'Serotonin and sleep', *Expl Med. Surg.*, **27**, 157–168.

KOELLE, G. B. (1954), 'The histochemical localization of cholinesterases in the nervous system of the rat', *J. comp. Neurol.*, **100**, 211–235.

KOELLE, G. B., and FRIEDENWALD, J. S. (1950), 'The histochemical localization of cholinesterase in ocular tissues', *Am. J. Ophthalmol.*, **33**, 253–256.

KOIZUMI, K., and YAMASHITA, H. (1972), 'Studies of antidromically identified neurosecretory cells of the hypothalamus by intracellular and extracellular recordings', *J. Physiol., Lond.*, **221**, 683–705.

KOJIMA, K., MIZUNO, K., and MIYAZAKI, M. (1958), 'Gamma-aminobutyric acid in ocular tissue', *Nature, Lond.*, **181**, 1200–1201.

KRAMER, S. G. (1971), 'Dopamine: a retinal neurotransmitter. I. Retinal uptake, storage, and light-stimulated release of H3-dopamine *in vivo*', *Invest. Ophthal.*, **10**, 438–452.

KRAMER, S. G., POTTS, A. M., and MANGNALL, Y. (1971), 'Dopamine: a retinal neurotransmitter. II. Autoradiographic localization of H3-dopamine in the retina', *Invest. Ophthal.*, **10**, 617–624.

KRAMER, S. Z., SHERMAN, P. A., and SEIFTER, J. (1967), 'Effects of gamma-aminobutyric acid (GABA) and sodium L-glutamate (glutamate) on the visual system and EEG of chicks', *Int. J. Neuropharmac.*, **6,** 463–472.

KRNJEVIĆ, K. (1964), 'Micro-iontophoretic studies on cortical neurons', *Int. Rev. Neurobiol.*, **7,** 41–98.

KRNJEVIĆ, K. (1967), 'Chemical transmission and cortical arousal', *Anesthesiology*, **28,** 100–105.

KRNJEVIĆ, K. (1970), 'Glutamate and γ-aminobutyric acid in brain', *Nature, Lond.*, **228,** 119–124.

KRNJEVIĆ, K. (1971), 'Microiontophoresis', in *Methods of Neurochemistry* (ed. Fried), vol. 1, ch. 4, pp. 129–172. New York: Marcel Dekker.

KRNJEVIĆ, K., and PHILLIS, J. W. (1963a), 'Iontophoretic studies of neurones in the mammalian cerebral cortex', *J. Physiol., Lond.*, **165,** 274–304.

KRNJEVIĆ, K., and PHILLIS, J. W. (1963b), 'Acetylcholine-sensitive cells in the cerebral cortex', *J. Physiol., Lond.*, **166,** 296–327.

KRNJEVIĆ, K., and PHILLIS, J. W. (1963c), 'Pharmacological properties of acetylcholine-sensitive cells in the cerebral cortex', *J. Physiol., Lond.*, **166,** 328–350.

KRNJEVIĆ, K., and PHILLIS, J. W. (1963d), 'Actions of certain amines on cerebral cortical neurones', *Br. J. Pharmac.*, **20,** 471–490.

KRNJEVIĆ, K., PUMAIN, R., and RENAUD, L. (1971a), 'Effects of Ba^{2+} and tetraethylammonium on cortical neurones', *J. Physiol., Lond.*, **215,** 223–245.

KRNJEVIĆ, K., PUMAIN, R., and RENAUD, L. (1971b), 'The mechanism of excitation by acetylcholine in the cerebral cortex', *J. Physiol., Lond.*, **215,** 247–268.

KRNJEVIĆ, K., RANDIĆ, M., and STRAUGHAN, D. W. (1965), 'Ergothioneine and central neurones', *Nature, Lond.*, **205,** 603–604.

KRNJEVIĆ, K., RANDIĆ, M., and STRAUGHAN, D. W. (1966a), 'An inhibitory process in the cerebral cortex', *J. Physiol., Lond.*, **184,** 16–48.

KRNJEVIĆ, K., RANDIĆ, M., and STRAUGHAN, D. W. (1966b), 'Nature of a cortical inhibitory process', *J. Physiol., Lond.*, **184,** 49–77.

KRNJEVIĆ, K., RANDIĆ, M., and STRAUGHAN, D. W. (1966c), 'Pharmacology of cortical inhibition', *J. Physiol., Lond.*, **184,** 78–105.

KRNJEVIĆ, K., REIFFENSTEIN, R. J., and SILVER, A. (1970), 'Chemical sensitivity of neurons in long-isolated slabs of cat cerebral cortex', *Electroenceph. clin. Neurophysiol.*, **29,** 269–282.

KRNJEVIĆ, K., and SCHWARTZ, S. (1967a), 'Some properties of unresponsive cells in the cerebral cortex', *Expl Brain Res.*, **3,** 306–319.

KRNJEVIĆ, K., and SCHWARTZ, S. (1967b), 'The action of γ-aminobutyric acid on cortical neurones', *Expl Brain Res.*, **3,** 320–336.

KRNJEVIĆ, K., and SILVER, A. (1965), 'A histochemical study of cholinergic fibres in the cerebral cortex', *J. Anat., Lond.*, **99**, 711–759.

KRNJEVIĆ, K., and SILVER, A. (1966), 'Acetylcholinesterase in the developing forebrain', *J. Anat., Lond.*, **100**, 63–89.

KRUG, M., SCHMIDT, J., and MATTHIES, H. (1970), 'Beeinflussung des Impulsmusters von spontan tätigen Neuronen der pontinen Formatio reticularis der Ratte durch Noradrenalin, Serotonin und Azetylcholin', *Acta biol. med. germ.*, **25**, 455–467.

KRŽALIĆ, Lj., MANDIĆ, V., and MIHAILOVIĆ, Lj. (1962), 'On the glutamine and γ-aminobutyric acid contents of various regions of the cat brain', *Experientia*, **18**, 368–369.

KUBÍCEK, R., and DOLÉNEK, A. (1958), 'Taurine et acides aminés dans la rétine des animaux', *J. Chromat.*, **1**, 266–268.

KUBOTA, K., KIDOKORO, Y., and SUZUKI, J. (1968), 'Postsynaptic inhibitions of trigeminal and lumbar motoneurons from the superficial radial nerve in the cat', *Jap. J. Physiol.*, **18**, 198–215.

KUHAR, M. J., GREEN, A. I., SNYDER, S. H., and GFELLER, E. (1970), 'Separation of synaptosomes storing catecholamines and gamma-aminobutyric acid in rat corpus striatum', *Brain Res.*, **21**, 405–417.

KUHAR, M. J., ROTH, R. H., and AGHAJANIAN, G. K. (1972), 'Synaptosomes from forebrains of rats with midbrain raphe lesions: selective reduction of serotonin uptake', *J. Pharmac. exp. Ther.*, **181**, 36–45.

KUHAR, M. J., SHASKAN, E. G., and SNYDER, S. H. (1971), 'The subcellular distribution of endogenous and exogenous serotonin in brain tissue: comparison of synaptosomes storing serotonin, norepinephrine, and γ-aminobutyric acid', *J. Neurochem.*, **18**, 333–343.

KUHAR, M. J., and SNYDER, S. H. (1970), 'The subcellular distribution of free H^3-glutamic acid in rat cerebral cortical slices', *J. Pharmac. exp. Ther.*, **171**, 141–152.

KUHAR, M. J., TAYLOR, K. M., and SNYDER, S. H. (1971), 'The subcellular localization of histamine and histamine methyltransferase in rat brain', *J. Neurochem.*, **18**, 1515–1527.

KUNO, M., and RUDOMIN, P. (1966), 'The release of acetylcholine from the spinal cord of the cat by antidromic stimulation of motor nerves', *J. Physiol., Lond.*, **187**, 177–193.

KUNTZMAN, R., SHORE, P. A., BOGDANSKI, D., and BRODIE, B. B. (1961), 'Microanalytical procedures for fluorometric assay of brain DOPA-5HTP decarboxylase, norepinephrine and serotonin, and a detailed mapping of decarboxylase activity in brain', *J. Neurochem.*, **6**, 226–232.

KURIYAMA, K., HABER, B., SISKEN, B., and ROBERTS, E. (1966), 'The γ-aminobutyric acid system in rabbit cerebellum', *Proc. natn. Acad. Sci. U.S.A.*, **55**, 846–852.

KURIYAMA, K., ROBERTS, E., and RUBINSTEIN, M. K. (1966), 'Elevation of γ-aminobutyric acid in brain with amino-oxyacetic acid and susceptibility to convulsive seizures in mice: a quantitative re-evaluation', *Biochem. Pharmac.*, **15**, 221–236.

KURIYAMA, K., SISKEN, B., HABER, B., and ROBERTS, E. (1968), 'The γ-aminobutyric acid system in rabbit retina', *Brain Res.*, **9**, 165–168.

LARRAMENDI, L. M. H., and VICTOR, T. (1967), 'Synapses on the Purkinje cell spines in the mouse. An electronmicroscopic study', *Brain Res.*, **5**, 15–30.

LARSON, M. D. (1969), 'An analysis of the action of strychnine on the recurrent IPSP and amino acid induced inhibitions in the cat spinal cord', *Brain Res.*, **15**, 185–200.

LATIES, A. M., and JACOBOWITZ, D. (1966), 'A comparative study of the autonomic innervation of the eye in monkey, cat, and rabbit', *Anat. Rec.*, **156**, 383–396.

LEGGE, K. F., RANDIĆ, M., and STRAUGHAN, D. W. (1966), 'The pharmacology of neurones in the pyriform cortex', *Br. J. Pharmac. Chemother.*, **26**, 87–107.

LEVIN, E., LOVELL, R. A., ELLIOTT, S. J., and ELLIOTT, K. A. C. (1961), 'The relation of gamma-aminobutyric acid to factor I in brain extracts', *J. Neurochem.*, **7**, 147–154.

LEVY, R. A., REPKIN, A. H., and ANDERSON, E. G. (1971), 'The effect of bicuculline on primary afferent terminal excitability', *Brain Res.*, **32**, 261–265.

LEWIS, P. R., and SHUTE, C. C. D. (1967), 'The cholinergic limbic system: projections to hippocampal formation, medial cortex, nuclei of the ascending cholinergic reticular system, and the subfornical organ and supra-optic crest', *Brain*, **90**, 521–540.

LEWIS, P. R., SHUTE, C. C. D., and SILVER, A. (1967), 'Confirmation from choline acetylase analyses of a massive cholinergic innervation to the rat hippocampus', *J. Physiol., Lond.*, **191**, 215–224.

LLINÁS, R. (1964), 'Mechanisms of supraspinal actions upon spinal cord activities. Pharmacological studies on reticular inhibition of alpha extensor motoneurons', *J. Neurophysiol.*, **27**, 1127–1137.

LLINÁS, R., and TERZUOLO, C. A. (1964), 'Mechanisms of supraspinal actions upon spinal cord activities. Reticular inhibitory mechanisms on alpha-extensor motoneurons', *J. Neurophysiol.*, **27**, 579–591.

LLINÁS, R., and TERZUOLO, C. A. (1965), 'Mechanisms of supraspinal actions upon spinal cord activities. Reticular inhibitory mechanisms upon flexor motoneurons', *J. Neurophysiol.*, **28**, 413–422.

LOCKE, S. (1961), 'The projection of the magnocellular medial geniculate body', *J. comp. Neurol.*, **116**, 179–193.

LOEWI, O., and HELLAUER, H. (1938), 'Über das Acetylcholin in peripheren Nerven', *Pflügers Arch.*, **240,** 769–775.

LOIZOU, L. A. (1969), 'Projections of the nucleus locus coeruleus in the albino rat', *Brain Res.*, **15,** 563–566.

LORENTE DE NÓ, R. (1947), 'Action potential of the motoneurons of the hypoglossal nucleus', *J. cell. comp. Physiol.*, **29,** 206–287.

LOVELL, R. A., ELLIOTT, S. J., and ELLIOTT, K. A. C. (1963), 'The γ-aminobutyric acid and factor I content of brain', *J. Neurochem.*, **10,** 479–488.

LOWRY, O. H., ROBERTS, N. R., and LEWIS, C. (1956), 'The quantitative histochemistry of the retina', *J. biol. Chem.*, **220,** 879–892.

LUND, S., and POMPEIANO, O. (1968), 'Monosynaptic excitation of alpha motoneurones from supraspinal structures in the cat', *Acta physiol. scand.*, **73,** 1–21.

LUNDBERG, A. (1966), 'Integration in the reflex pathway, in *Muscular Afferents and Motor Control* (ed. Granit), Nobel Symposium I, pp. 275–305. Stockholm: Almqvist and Wiksell.

MCCANCE, I. (1972), 'The role of acetylcholine in the intracerebellar nuclei of the cat', *Brain Res.*, **48,** 265–279.

MCCANCE, I., and PHILLIS, J. W. (1964), 'Discharge patterns of elements in cat cerebellar cortex, and their responses to iontophoretically applied drugs', *Nature, Lond.*, **204,** 844–846.

MCCANCE, I., and PHILLIS, J. W. (1968), 'Cholinergic mechanisms in the cerebellar cortex', *Int. J. Neuropharmac.*, **7,** 447–462.

MCCANCE, I., PHILLIS, J. W., TEBĒCIS, A. K., and WESTERMAN, R. A. (1968), 'The pharmacology of acetylcholine-excitation of thalamic neurones', *Br. J. Pharmac. Chemother.*, **32,** 652–662.

MCCANCE, I., PHILLIS, J. W., and WESTERMAN, R. A. (1968a), 'Acetylcholine-sensitivity of thalamic neurones: its relationship to synaptic transmission', *Br. J. Pharmac. Chemother.*, **32,** 635–651.

MCCANCE, I., PHILLIS, J. W., and WESTERMAN, R. A. (1968b), 'Physiological and pharmacological studies on the cerebellar projections to the feline thalamus', *Expl Neurol.*, **21,** 257–269.

MCDONALD, D. M., and RASMUSSEN, G. L. (1971), 'Ultrastructural characteristics of synaptic endings in the cochlear nucleus having acetylcholinesterase activity', *Brain Res.*, **28,** 1–18.

MCGEER, P. L., MCGEER, E. G., FIBIGER, H. C., and WICKSON, V. (1971), 'Neostriatal choline acetylase and cholinesterase following selective brain lesions', *Brain Res.*, **35,** 308–314.

MCGEER, P. L., MCGEER, E. G., and WADA, J. A. (1963), 'Central aromatic amine levels and behavior', *Archs. Neurol.*, **9,** 81–89.

MCGEER, P. L., MCGEER, E. G., WADA, J. A., and JUNG, E. (1971), 'Effects of globus pallidus lesions and Parkinson's disease on brain glutamic acid decarboxylase', *Brain Res.*, **32,** 425–431.

McGEER, E. G., WADA, J. A., TERAO, A., and JUNG, E. (1969), 'Amine synthesis in various brain regions with caudate or septal lesions', *Expl Neurol.*, **24**, 277–284.

MACINTOSH, F. C. (1941), 'The distribution of acetylcholine in the peripheral and the central nervous system', *J. Physiol., Lond.*, **99**, 436–442.

MACINTOSH, F. C. (1963), 'Synthesis and storage of acetylcholine in nervous tissue', *Can. J. Biochem. Physiol.*, **41**, 2555–2571.

MACINTOSH, F. C., and OBORIN, P. E. (1953), 'Release of acetylcholine from intact cerebral cortex', *Abstr. XIX Int. Physiol. Cong.*, pp. 580–581.

McKENZIE, G. M., and SZERB, J. C. (1968), 'The effect of dihydroxyphenylalanine, pheniprazine and dextroamphetamine on the *in vivo* release of dopamine from the caudate nucleus', *J. Pharmac. exp. Ther.*, **162**, 302–308.

McKINSTRY, D. N., and KOELLE, G. B. (1967), 'Effects of drugs on acetylcholine release from the cat superior cervical ganglion by carbachol and by preganglionic stimulation', *J. Pharmac. exp. Ther.*, **157**, 328–336.

MACLEAN, P. D. (1949), 'Psychosomatic disease and the "visceral brain". Recent developments bearing on the Papez theory of emotion', *Psychosom. Med.*, **11**, 338–353.

McLENNAN, H. (1964), 'The release of acetylcholine and of 3-hydroxytyramine from the caudate nucleus', *J. Physiol., Lond.*, **174**, 152–161.

McLENNAN, H. (1965), 'The release of dopamine from the putamen', *Experientia*, **21**, 725–726.

McLENNAN, H. (1970a), *Synaptic Transmission*, 2nd ed., pp. 1–173. Philadelphia: Saunders.

McLENNAN, H. (1970b), 'Inhibitions of long duration in the cerebral cortex. A quantitative difference between excitatory amino acids', *Expl Brain Res.*, **10**, 417–426.

McLENNAN, H. (1971), 'The pharmacology of inhibition of mitral cells in the olfactory bulb', *Brain Res.*, **29**, 177–184.

McLENNAN, H., HUFFMAN, R. D., and MARSHALL, K. C. (1968), 'Patterns of excitation of thalamic neurones by amino acids and by acetylcholine', *Nature, Lond.*, **219**, 387–388.

McLENNAN, H., MARSHALL, K. C., and HUFFMAN, R. D. (1971), 'The antagonism of glutamate action at central neurones', *Experientia*, **27**, 1116.

McLENNAN, H., and YORK, D. H. (1966), 'Cholinergic mechanisms in the caudate nucleus', *J. Physiol., Lond.*, **187**, 163–175.

McLENNAN, H., and YORK, D. H. (1967), 'The action of dopamine on neurones of the caudate nucleus', *J. Physiol., Lond.*, **189**, 393–402.

MAGHERINI, P. C., POMPEIANO, O., and THODEN, U. (1971), 'The neurochemical basis of REM sleep: a cholinergic mechanism responsible for rhythmic activation of the vestibulo-oculomotor system', *Brain Res.*, **35**, 565–569.

MAGNI, F., and WILLIS, W. D. (1963), 'Identification of reticular formation neurons by intracellular recording', *Arch. ital. Biol.*, **101**, 681–702.

MAJOROSSY, K., and RÉTHELYI, M. (1968), 'Synaptic architecture in the medical geniculate body (ventral division)', *Expl Brain Res.*, **6**, 306–323.

MALMFORS, T. (1963), 'Evidence of adrenergic neurons with synaptic terminals in the retina of rats demonstrated with fluorescence and electron microscopy', *Acta physiol. scand.*, **58**, 99–100.

MANOCHA, S. L., SHANTHA, T. R., and BOURNE, G. H. (1967), 'Histochemical mapping of the distribution of monoamine oxidase in the diencephalon and basal telencephalic centres of the brain of squirrel monkey (*Saimiri sciureus*)', *Brain Res.*, **6**, 570–586.

MARLEY, E., and VANE, J. R. (1967), 'Tryptamines and spinal cord reflexes in cats', *Br. J. Pharmac. Chemother.*, **31**, 447–465.

MARRAZZI, A. S. (1953), 'Some indications of cerebral humoral mechanisms', *Science, N.Y.*, **118**, 367–370.

MARTINI, L., and MEITES, J. (1970) (ed.), *Neurochemical Aspects of Hypothalamic Function* (Proc. Symp. 2nd Meet. Int. Soc. Neurochem., Milano, Sept., 1969). New York: Academic Press.

MASSIAN, J., and ALBE-FESSARD, D. (1963), 'Dualité des voies sensorielles afférentes contrôlant l'activité du noyau rouge', *Electroenceph. clin. Neurophysiol.*, **15**, 435–454.

MATSUSHITA, M., and IKEDA, M. (1970a), 'Olivary projections to the cerebellar nuclei in the cat', *Expl Brain Res.*, **10**, 488–500.

MATSUSHITA, M., and IKEDA, M. (1970b), 'Spinal projections to the cerebellar nuclei in the cat', *Expl Brain Res.*, **10**, 501–511.

MATUS, A. I., and DENNISON, M. E. (1971), 'Autoradiographic localisation of tritiated glycine at 'flat-vesicle' synapses in spinal cord', *Brain Res.*, **32**, 195–197.

MELLGREN, S. I., and GENESER-JENSEN, F. A. (1972), 'Distribution of monoamine oxidase in the hippocampal region of the rat', *Z. Zellforsch. mikrosk. Anat.*, **124**, 354–366.

METTLER, F. A. (1932), 'Connections of the auditory cortex of the cat', *J. comp. Neurol.*, **55**, 139–183.

MICKLE, W. A., and ADES, H. W. (1954), 'Rostral projection pathway of the vestibular system', *Am. J. Physiol.*, **176**, 243–246.

MILLER, E., HELLER, A., and MOORE, R. Y. (1969), 'Acetylcholine in rabbit visual system nuclei after enucleation and visual cortex ablation', *J. Pharmac. exp. Ther.*, **165**, 117–125.

MITCHELL, J. F. (1963), 'The spontaneous and evoked release of acetylcholine from the cerebral cortex', *J. Physiol., Lond.*, **165**, 98–116.

Mitchell, J. F., and Phillis, J. W. (1962), 'Cholinergic transmission in the frog spinal cord', *Br. J. Pharmac. Chemother.*, **19**, 534–543.

Mitchell, J. F., and Roberts, P. J. (1972), 'Evoked release of amino acids from the intact spinal cord', *Br. J. Pharmac.*, **45**, 175–176P.

Monachon, M-A., Burkard, W. P., Jalfre, M., and Haefely, W. (1972), 'Blockade of central 5-hydroxytryptamine receptors by methiothepin', *Naunyn-Schmiedeberg's Arch. Pharmac.*, **274**, 192–197.

Moore, R. Y. (1970), 'Brain lesions and amine metabolism', *Int. Rev. Neurobiol.*, **13**, 67–91.

Moore, R. Y., Bhatnagar, R. K., and Heller, A. (1971), 'Anatomical and chemical studies of a nigro-neostriatal projection in the cat', *Brain Res.*, **30**, 119–135.

Moore, R. Y., and Goldberg, J. M. (1963), 'Ascending projections of the inferior colliculus in the cat', *J. comp. Neurol.*, **121**, 109–136.

Morest, D. K. (1964), 'The neuronal architecture of the medial geniculate body of the cat', *J. Anat., Lond.*, **98**, 611–630.

Morest, D. K. (1965), 'The lateral tegmental system of the midbrain and the medial geniculate body: study with Golgi and Nauta methods in cat', *J. Anat., Lond.*, **99**, 611–634.

Morgan, R., Vrobová, G., and Wolstencroft, J. H. (1972), 'Correlation between the retinal input to lateral geniculate neurones and their relative response to glutamate and aspartate', *J. Physiol., Lond.*, **224**, 41–42P.

Morimoto, T., Takata, M., and Kawamura, Y. (1968), 'Effect of lingual nerve stimulation on hypoglossal motoneurons', *Expl Neurol.*, **22**, 174–190.

Morison, R. S., and Dempsey, E. W. (1943), 'Mechanism of thalamocortical augmentation and repetition', *Am. J. Physiol.*, **138**, 297–308.

Moss, R. L., Dyball, R. E. J., and Cross, B. A. (1971), 'Responses of antidromically identified supraoptic and paraventricular units to acetylcholine, noradrenaline and glutamate applied iontophoretically', *Brain Res.*, **35**, 573–575.

Myers, R. D. (1969), 'Temperature regulation: neurochemical systems in the hypothalamus', in *The Hypothalamus: Anatomical, Functional and Clinical Aspects*, pp. 506–523. Springfield, Ill.: Charles C. Thomas.

Myers, R. D., and Beleslin, D. B. (1970), 'The spontaneous release of 5-hydroxytryptamine and acetycholine within the diencephalon of the unanaesthetized rhesus monkey', *Expl Brain Res.*, **11**, 539–552.

Nadler, J. V., and Cooper, J. R. (1972), '*N*-Acetyl-L-aspartic acid content of human neural tumours and bovine peripheral nervous tissues', *J. Neurochem.*, **19**, 313–319.

NAVARATNAM, V., and LEWIS, P. R. (1970), 'Cholinesterase-containing neurones in the spinal cord of the rat', *Brain Res.*, **18,** 411–425.

NEAL, M. J. (1971), 'The uptake of (^{14}C)glycine by slices of mammalian spinal cord', *J. Physiol., Lond.*, **215,** 103–117.

NEAL, M. J., and IVERSEN, L. L. (1969), 'Subcellular distribution of endogenous and (^{3}H) γ-aminobutyric acid in rat cerebral cortex', *J. Neurochem.*, **16,** 1245–1252.

NEAL, M. J., and IVERSEN, L. L. (1972), 'Autoradiographic localization of ^{3}H-GABA in rat retina', *Nature, Lond.* (*New Biol.*), **235,** 217–218.

NEAL, M. J., and PICKLES, H. G. (1969), 'Uptake of ^{14}C glycine by spinal cord', *Nature, Lond.*, **222,** 679–680.

NEAL, M. J., and WHITE, R. D. (1971), 'Uptake of ^{14}C-L-glutamate by rat retina', *Br. J. Pharmac.*, **43,** 442P.

NICHOLS, C. W., JACOBOWITZ, D., and HOTTENSTEIN, M. (1967), 'The influence of light and dark on the catecholamine content of the retina and choroid', *Invest. Ophthal.*, **6,** 642–646.

NICHOLS, C. W., and KOELLE, G. B. (1968), 'Comparison of the localization of acetylcholinesterase and non-specfic cholinesterase activities in mammalian and avian retinas', *J. comp. Neurol.*, **133,** 1–16.

NICOLL, R. A. (1969), 'Inhibitory mechanisms in the rabbit olfactory bulb: dendrodendritic mechanisms', *Brain Res.*, **14,** 157–172.

NICOLL, R. A. (1971), 'Pharmacological evidence for GABA as the transmitter in granule cell inhibition in the olfactory bulb', *Brain Res.*, **35,** 137–149.

NICOLL, R. A. (1972), 'Olfactory nerves and their excitatory action in the olfactory bulb', *Expl Brain Res.*, **14,** 185–197.

NICOLL, R. A., and BARKER, J. L. (1971), 'The pharmacology of recurrent inhibition in the supraoptic neurosecretory system', *Brain Res.*, **35,** 501–511.

NISHI, S., SOEDA, H., and KOKETSU, K. (1969), 'Unusual nature of ganglionic slow EPSP studied by a voltage-clamp method', *Life Sci.* **8,** 33–42.

NOELL, W. K. (1959), 'The visual cell: electric and metabolic manifestations of its life processes', *Am. J. Ophthal.*, **48,** 347–370.

NOELL, W. K., and LASANSKY, A. (1959), 'Effects of electrophoretically applied drugs and electrical currents on the ganglion cell of the retina', *Fedn Proc. Fedn Am. Socs exp. Biol.*, **18,** 115.

NORRIS, C., GUTH, P. S., and STOCKWELL, M. S. (1972), 'Release of acetylcholine by the olivo-cochlear bundle', *Fedn Proc. Fedn Am. Socs exp. Biol.*, **31,** 580.

OBATA, K., and HIGHSTEIN, S. M. (1970), 'Blocking by picrotoxin of both vestibular inhibition and GABA action on rabbit oculomotor neurones', *Brain Res.*, **18,** 538–541.

OBATA, K., ITO, M., OCHI, R., and SATO, N. (1967), 'Pharmacological properties of the postsynaptic inhibition by Purkinje cell axons and the action of γ-aminobutyric acid on Deiters' neurones', *Expl Brain Res.*, **4**, 43–57.

OBATA, K., and TAKEDA, K. (1969), 'Release of γ-aminobutyric acid into the fourth ventricle induced by stimulation of the cat's cerebellum', *J. Neurochem.*, **16**, 1043–1047.

OBATA, K., TAKEDA, K., and SHINOZAKI, H. (1970), 'Further study on pharmacological properties of the cerebellar-induced inhibition of Deiters neurones', *Expl Brain Res.*, **11**, 327–342.

OCHI, J. (1963), 'Olfactory bulb response to antidromic olfactory tract stimulation in the rabbit', *Jap. J. Physiol.*, **13**, 113–128.

ODUTOLA, A. B. (1972), 'The organization of cholinesterase-containing systems of the monkey spinal cord', *Brain Res.*, **39**, 353–368.

OKADA, Y., NITSCH-HASSLER, C., KIM, J. S., BAK, I. J., and HASSLER, R. (1971), 'Role of γ-aminobutyric acid (GABA) in the extrapyramidal motor system. I. Regional distribution of GABA in rabbit, rat, guinea pig and baboon CNS', *Expl Brain Res.*, **13**, 514–518.

OKUMURA, N., OTSUKI, S., and KAMEYAMA, A. (1960), 'Studies on free amino acids in human brain', *J. Biochem., Tokyo*, **47**, 315–320.

O'LEARY, J. L. (1940), 'A structural analysis of the lateral geniculate nucleus of the cat', *J. comp. Neurol.*, **73**, 405–430.

OLSON, L., and FUXE, K. (1971), 'On the projections from the locus coeruleus noradrenaline neurons: the cerebellar innervation', *Brain Res.*, **28**, 165–171.

OOMURA, Y., OOYAMA, H., YAMAMOTO, T., ONO, T., and KOBAYASHI, N. (1969), 'Behavior of hypothalamic unit activity during electrophoretic application of drugs', *Ann. N.Y. Acad. Sci.*, **157**, 642–665.

OTSUKA, M., OBATA, K., MIYATA, Y., and TANAKA, Y. (1971), 'Measurement of γ-aminobutyric acid in isolated nerve cells of cat central nervous system', *J. Neurochem.*, **18**, 287–295.

PAASONEN, M. K., MACLEAN, P. D., and GIARMAN, N. J. (1957), 5-Hydroxytryptamine (serotonin, enteramine) content of structures of the limbic system', *J. Neurochem.*, **1**, 326–333.

PADJEN, A., and RANDIĆ, M. (1970), 'Some factors influencing the release of 5-hydroxyindol-3-ylacetic acid in the forebrain', *Br. J. Pharmac.*, **39**, 1–8.

PALAY, S. L. (1957), 'The fine structure of the neurohypophysis', in *Ultrastructure and Cellular Chemistry of Neural Tissue*, (ed. Waelsch), pp. 31–49. New York: Hoeber.

PARENT, A., SAINT-JACQUES, C., and POIRIER, L. J. (1969), 'Effect of interrupting the hypothalamic nervous connections on the norepinephrine and serotonin content of the hypothalamus', *Expl Neurol.*, **23**, 67–75.

PARIZEK, J., HASSLER, R., and BAK, I. J. (1971), 'Light and electron microscopic autoradiography of substantia nigra of rat after intraventricular administration of tritium labelled norepinephrine, dopamine, serotonin and the precursors', *Z. Zellforsch. mikrosk. Anat.*, **115**, 137–148.

PATON, W. D. M. (1958), 'Central and synaptic transmission in the nervous system (pharmacological aspects)', *A. Rev. Physiol.*, **20**, 431–470.

PERWEIN, J., and STRASCHILL, M. (1972), 'Die Wirkung iontophoretisch applizierten Acetylcholins auf die Aktivität retinaler Ganglienzellen der Katze. (Effect of iontophoretically acetylcholine upon the activity of retinal ganglion cells of the cat)', *Pflügers Arch.—Eur. J. Physiol.*, supp. to vol. 332, p.93.

PHILIPPU, A., HEYD, G., and BURGER, A. (1970), 'Release of noradrenaline from the hypothalamus *in vivo*', *Eur. J. Pharmac.*, **9**, 52–58.

PHILLIPS, C. G., POWELL, T. P. S., and SHEPHERD, G. M. (1963), 'Responses of mitral cells to stimulation of the lateral olfactory tract in the rabbit', *J. Physiol., Lond.*, **168**, 65–88.

PHILLIS, J. W. (1968a), 'Acetylcholinesterase in the feline cerebellum', *J. Neurochem.*, **15**, 691–698.

PHILLIS, J. W. (1968b), 'Acetylcholine release from the cerebral cortex: its role in cortical arousal', *Brain Res.*, **7**, 378–389.

PHILLIS, J. W. (1970), *The Pharmacology of Synapses*, Int. Ser. Monogr. Pure Applied Biol., Zool. Div., vol. 43. Oxford: Pergamon.

PHILLIS, J. W. (1971), 'The pharmacology of thalamic and geniculate neurons', *Int. Rev. Neurobiol.*, **14**, 1–48.

PHILLIS, J. W., and CHONG, G. C. (1965), 'Acetylcholine release from the cerebral and cerebellar cortices: its role in cortical arousal', *Nature, Lond.*, **207**, 1253–1255.

PHILLIS, J. W., and OCHS, S. (1971), 'Excitation and depression of cortical neurones during spreading depression', *Expl. Brain Res.*, **12**, 132–149.

PHILLIS, J. W., and TEBĒCIS, A. K. (1967a), 'The effects of topically applied cholinomimetic drugs on the isolated spinal cord of the toad', *Comp. Biochem. Physiol.*, **23**, 541–552.

PHILLIS, J. W., and TEBĒCIS, A. K. (1967b), 'The effects of pentobarbitone sodium on acetylcholine excitation and noradrenaline inhibition of thalamic neurones', *Life Sci.*, **6**, 1621–1625.

PHILLIS, J. W., and TEBĒCIS, A. K. (1967c), 'The responses of thalamic neurones to iontophoretically applied monoamines', *J. Physiol., Lond.*, **192**, 715–745.

PHILLIS, J. W., TEBĒCIS, A. K., and YORK, D. H. (1967a), 'The inhibitory action of monoamines on lateral geniculate neurones', *J. Physiol., Lond.*, **190**, 563–581.

Phillis, J. W., Tebēcis, A. K., and York, D. H. (1967b), 'A study of cholinoceptive cells in the lateral geniculate nucleus', *J. Physiol., Lond.*, **192,** 695–713.

Phillis, J. W., Tebēcis, A. K., and York, D. H. (1968a), 'Depression of spinal motoneurones by noradrenaline, 5-hydroxytryptamine and histamine', *Eur. J. Pharmac.*, **4,** 471–475.

Phillis, J. W., Tebēcis, A. K., and York, D. H. (1968b), 'Acetylcholine release from the feline thalamus', *J. Pharm. Pharmac.*, **20,** 476–478.

Phillis, J. W., Tebēcis, A. K., and York, D. H. (1968c), 'Histamine and some antihistamines: their actions on cerebral cortical neurones', *Br. J. Pharmac. Chemother.*, **33,** 426–440.

Phillis, J. W., and York, D. H. (1967a), 'Strychnine block of neural and drug-induced inhibition in the cerebral cortex', *Nature, Lond.*, **216,** 922–923.

Phillis, J. W., and York, D. H. (1967b), 'Cholinergic inhibition in the cerebral cortex', *Brain Res.*, **5,** 517–520.

Phillis, J. W., and York, D. H. (1968a), 'An intracortical cholinergic inhibitory synapse', *Life Sci.*, **7,** 65–69.

Phillis, J. W., and York, D. H. (1968b), 'Pharmacological studies on a cholinergic inhibition in the cerebral cortex', *Brain Res.*, **10,** 297–306.

Pin, C., Jones, B., and Jouvet, M. (1968), 'Topographie des neurones monoaminergiques du tronc cérébral du chat: étude par histofluorescence', *C.r. Séanc. Soc. Biol.*, **162,** 2136–2141.

Pohle, W., and Matthies, H. (1970), 'Die Topohistochemie von Transmittersystemen im Kortex und Hippokampus des Kaninchens', *Acta biol. med. germ.*, **25,** 447–454.

Poirier, L. J., Singh, P., Boucher, R., Bouvier, G., Olivier, A., and Larochelle, P. (1967), 'Effect of brain lesions on striatal monoamines in the cat', *Archs. Neurol.*, **17,** 601–608.

Poirier, L. J., and Sourkes, T. L. (1965), 'Influence of the substantia nigra on the catecholamine content of the striatum', *Brain*, **88,** 181–192.

Poirier, L. J., Sourkes, T. L., Bouvier, G., Boucher, R., and Carabin, S. (1966), 'Striatal amines, experimental tremor and the effect of harmaline in the monkey', *Brain*, **89,** 37–52.

Polyak, S. L. (1941), *The Retina.* Chicago: University of Chicago Press.

Pompeiano, O., and Brodal, A. (1957), 'Experimental demonstration of a somatotopical origin of rubrospinal fibres in the cat', *J. comp. Neurol.*, **108,** 225–251.

Popov, N., Pophle, W., Rösler, V., and Matthies, H. (1967), 'Regionale Verteilung von γ-Aminobuttersäure, Glutaminsäure, Asparaginsäure, Dopamin, Noradrenalin und Serotonin im Rattenhirn', *Acta biol. med. germ.*, **18,** 695–702.

PORTER, R. (1965), 'Synaptic potentials in hypoglossal motoneurones', *J. Physiol., Lond.*, **180,** 209–224.

PORTIG, P. J. and VOGT, M. (1969), 'Release into the cerebral ventricles of substances with possible transmitter function in the caudate nucleus', *J. Physiol., Lond.*, **204,** 687–715.

PRECHT, W., and YOSHIDA, M. (1971), 'Blockage of caudate-evoked inhibition of neurons in the substantia nigra by picrotoxin', *Brain Res.*, **32,** 229–233.

PURPURA, D. P., and YAHR, M. D. (1966) (ed.), *The Thalamus.* New York: Columbia University Press.

QUASTEL, D. M. J., and CURTIS, D. R. (1965), 'A central action of hemicholinium', *Nature, Lond.*, **208,** 192–194.

RALL, T. W., and GILMAN, A. G. (1970), 'The role of cyclic AMP in the nervous system', *Neurosci. Res. Prog. Bull.*, **8,** 225–323.

RALL, W., SHEPHERD, G. M., REESE, T. S., and BRIGHTMAN, M. W. (1966), 'Dendrodendritic synaptic pathway for inhibition in the olfactory bulb', *Expl Neurol.*, **14,** 44–56.

RAMWELL, P. W., SHAW, J. E., and JESSUP, R. (1966), 'Spontaneous and evoked release of prostaglandins from frog spinal cord', *Am. J. Physiol.*, **211,** 998–1004.

RANDIĆ, M., SIMINOFF, R. and STRAUGHAN, D. W. (1964), 'Acetylcholine depression of cortical neurons', *Expl Neurol.*, **9,** 236–242.

RENSHAW, B. (1941), 'Influence of discharge of motoneurons upon excitation of neighbouring motoneurons', *J. Neurophysiol.*, **4,** 167–183.

RENSHAW, B. (1946), 'Central effects of centripetal impulses in axons of spinal ventral roots', *J. Neurophysiol.*, **9,** 191–204.

RIDDELL, D., and SZERB, J. C. (1971), 'The release *in vivo* of dopamine synthesized from labelled precursors in the caudate nucleus of the cat', *J. Neurochem.*, **18,** 989–1006.

RIZZOLI, A. A. (1968), 'Distribution of glutamic acid, aspartic acid, γ-aminobutyric acid and glycine in six areas of cat spinal cord before and after transection', *Brain Res.*, **11,** 11–18.

ROBERTS, E., and EIDELBERG, E. (1960), 'Metabolic and neurophysiological roles of γ-aminobutyric acid', *Int. Rev. Neurobiol.*, **2,** 279–332.

ROBERTS, E., and FRANKEL, S. (1950), 'γ-Aminobutyric acid in brain: its formation from glutamic acid', *J. biol. Chem.*, **187,** 55–63.

ROBERTS, E., and KURIYAMA, K. (1968), 'Biochemical-physiological correlations in studies of the γ-aminobutyric acid system', *Brain Res.*, **8,** 1–35.

ROBERTS, M. H. T., and STRAUGHAN, D. W. (1967), 'Excitation and depression of cortical neurones by 5-hydroxytryptamine', *J. Physiol., Lond.*, **193,** 269–294.

ROBISON, G. A., BUTCHER, R. W., and SUTHERLAND, E. W. (1971) (ed.), *Cyclic AMP*, pp. 1–531. New York: Academic Press.

ROSE, J. E., and WOOLSEY, C. N. (1949), 'The relations of thalamic connections, cellular structure and evocable electrical activity in the auditory region of the cat', *J. comp. Neurol.*, **91,** 441–466.

ROSS, L. L., ANDREOLI, V. M., and MARCHBANKS, R. M. (1971), 'A morphological and biochemical study of subcellular fractions of the guinea pig spinal cord', *Brain Res.*, **25,** 103–119.

ROSSI, G. F., and BRODAL, A. (1957), 'Terminal distribution of spino-reticular fibres in the cat', *Archs Neurol. Psychiat., Chicago*, **78,** 439–453.

ROZSOS, I. (1958), 'The synapses of Burdach's nucleus', *Acta morph. Acad. Sci. hung.*, **8,** 105–109.

RUF, K., and STEINER, F. A. (1967), 'Steroid-sensitive single neurons in rat hypothalamus and midbrain: identification by microelectrophoresis', *Science, N.Y.*, **156,** 667–669.

RUFENER, C., and DREIFUSS, J. J. (1970), 'Selective zinc iodide-osmium tetroxide impregnation of synaptoid vesicles in the rat neurohypophysis', *Brain Res.*, **22,** 402–405.

RYALL, R. W. (1967), 'Effect of monoamines upon sympathetic preganglionic neurons', suppl. III to *Circulation Research*, vols. XX, XXI; III, 83–87.

RYALL, R. W. (1970), 'Renshaw cell mediated inhibition of Renshaw cells: patterns of excitation and inhibition from impulses in motor axon collaterals', *J. Neurophysiol.*, **33,** 257–270.

RYALL, R. W. (1971), 'Evidence for the existence of different types of receptor for glycine', *Experientia*, **27,** 1118.

RYALL, R. W. (1972), 'Neuropharmacology of the spinal cord', in *The Spinal Cord* (ed. Austin), sect. IV, pp. 113–146. Springfield, Ill.: Charles C. Thomas.

RYALL, R. W., and DE GROAT, W. C. (1972), 'The microelectrophoretic administration of noradrenaline, 5-hydroxytryptamine, acetylcholine and glycine to sacral parasympathetic preganglionic neurones', *Brain Res.*, **37,** 345–347.

RYALL, R. W., and PIERCEY, M. F. (1971), 'Excitation and inhibition of Renshaw cells by impulses in peripheral afferent nerve fibres', *J. Neurophysiol.*, **34,** 242–251.

RYALL, R. W., PIERCEY, M. F., and POLOSA, C. (1972), 'Strychnine-resistant mutual inhibition of Renshaw cells', *Brain Res.*, **41,** 119–129.

SAAVEDRA, J. M., and AXELROD, J. (1972), 'Psychotomimetic *N*-methylated tryptamines: formation in brain *in vivo* and *in vitro*', *Science, N.Y.*, **175,** 1365–1366.

SALMOIRAGHI, G. C., BLOOM, F. E., and COSTA, E. (1964), 'Adrenergic mechanisms in rabbit olfactory bulb', *Am. J. Physiol.*, **207,** 1417–1424.

SALMOIRAGHI, G. C., and STEFANIS, C. N. (1965), 'Patterns of central neurons responses to suspected transmitters', *Arch. ital. Biol.*, **103**, 705–724.

SALMOIRAGHI, G. C., and STEFANIS, C. N. (1967), 'A critique of iontophoretic studies of central nervous system neurons', *Int. Rev. Neurobiol.*, **10**, 1–30.

SALMOIRAGHI, G. C., and STEINER, F. A. (1963), 'Acetylcholine sensitivity of cat's medullary neurons', *J. Neurophysiol.*, **26**, 581–597.

SALVADOR, R. A., and ALBERS, R. W. (1959), 'The distribution of glutamic-γ-aminobutyric transaminase in the nervous system of the rhesus monkey', *J. biol. Chem.*, **234**, 922–925.

SANO, I., GAMO, T., KAKIMOTO, Y., TANIGUCHI, K., TAKESADA, M., and NISHINUMA, K. (1959), 'Distribution of catechol compounds in human brain', *Biochim. biophys. Acta*, **32**, 586–587.

SATINSKY, D. (1967), 'Pharmacological responsiveness of lateral geniculate nucleus neurons', *Int. J. Neuropharmac.*, **6**, 387–395.

SAUERLAND, E. K., and MIZUNO, N. (1969), 'Cortically induced presynaptic inhibition of trigeminal proprioceptive afferents', *Brain Res.*, **13**, 556–568.

SCHADÉ, J. P., and VAN WILGENBURG, H. (1971), 'The influence of hormones on the unit firing of neurons', in *Influence of Hormones on the Nervous System* (ed. Ford). pp. 56–62. Proc. Int. Soc. Psychoneuroendocrin. Basel: Karger.

SCHARRER, E. (1928), 'Untersuchungen über das Zwischenhirn der Fische', *Z. vergl. Physiol.*, **7**, 1.

SCHEIBEL, M. E., and SCHEIBEL, A. B. (1958), 'Structural substrates for integrative patterns in the brain stem reticular core', in *Reticular Formation of the Brain* (A Henry Ford Hospital Intern. Symposium), (ed. Jasper, Proctor, Knighton, Noshay, and Costello), pp. 31–55. Boston: Little, Brown & Co.

SCHEIBEL, M. E., and SCHEIBEL, A. B. (1966), 'Spinal motoneurons, interneurons and Renshaw cells. A Golgi study', *Arch. ital. Biol.*, **104**, 328–353.

SCHMIDT, J., KRUG., M., MAIER, R., POHLE, W., and MATTHIES, H. (1967), 'Die Wirkung von Noradrenalin, Serotonin und Azetylcholin auf die Impulsaktivität von Neuronen der pontinen Formatio reticularis der Ratte', *Acta biol. med. germ.*, **18**, 703–713.

SCHMIDT, R. F. (1963), 'Pharmacological studies on the primary afferent depolarization of the toad spinal cord', *Pflügers Arch.*, **277**, 325–346.

SCHMIDT, R. F. (1971), 'Presynaptic inhibition in the vertebrate central nervous system', *Ergebnisse der Physiologie, Biol. Chem. exp. Pharmak.*, **63**, 20–101.

SCHOLES, N. W., and ROBERTS, E. (1964), 'Pharmacological studies of the optic system of the chick; effect of γ-aminobutyric acid and pentobarbital', *Biochem. Pharmac.*, **13,** 1319–1329.

SHANK, R. P., and APRISON, M. H. (1970), 'The metabolism *in vivo* of glycine and serine in eight areas of the rat central nervous system', *J. Neurochem.*, **17,** 1461–1475.

SHAPOVALOV, A. I. (1966), 'Excitation and inhibition of spinal neurones during supraspinal stimulation', in *Muscular Afferents and Motor Control* (Proc. First Nobel Symp.) (ed. Granit), 331–348, Stockholm: Almqvist and Wiksell.

SHAW, R. K., and HEINE, J. D. (1965), 'Ninhydrin positive substances present in different areas of normal rat brain', *J. Neurochem.*, **12,** 151–155.

SHEARD, M. H., and AGHAJANIAN, G. K. (1967), 'Neural release of brain serotonin and body temperature', *Nature, Lond.*, **216,** 495–496.

SHEARD, M. H., and AGHAJANIAN, G. K. (1968), 'Stimulation of midbrain raphé neurons: behavioral effects of serotonin release', *Life Sci.*, **7,** 19–25.

SHEARD, M. H., and ZOLOVICK, A. J. (1971), 'Serotonin: release in cat brain and cerebrospinal fluid on stimulation of midbrain raphé', *Brain Res.*, **26,** 455–458.

SHENDE, M. C., and KING, R. B. (1967), 'Excitability changes of trigeminal primary afferent preterminals in brain-stem nuclear complex of squirrel monkey (*Saimiri sciureus*)', *J. Neurophysiol.*, **30,** 949–963.

SHIBUYA, T., and ANDERSON, E. G. (1968), 'The influence of chronic cord transection on the effects of 5-hydroxytryptophan, *l*-tryptophan and pargyline on spinal neuronal activity', *J. Pharmac. exp. Ther.*, **164,** 185–190.

SHIMIZU, N., MORIKAWA, N., and OKADA, M. (1959), 'Histochemical studies of monoamine oxidase of the brain of rodents', *Z. Zellforsch. mikrosk. Anat.*, **49,** 389–400.

SHUTE, C. C. D., and LEWIS, P. R. (1967), 'The ascending cholinergic reticular system: neocortical, olfactory and subcortical projections', *Brain*, **90,** 497–520.

SIGGINS, G. R., HOFFER, B. J., and BLOOM, F. E. (1971), 'Studies on norepinephrine-containing afferents to Purkinje cells of rat cerebellum. III. Evidence for mediation of norepinephrine effects by cyclic 3′, 5′- adenosine monophosphate', *Brain Res.*, **25,** 535–553.

SIGGINS, G. R., HOFFER, B. J., OLIVER, A. P., and BLOOM, F. E. (1971), 'Activation of a central noradrenergic projection to cerebellum', *Nature, Lond.*, **233,** 481–483.

SIGGINS, G. R., OLIVER, A. P., HOFFER, B. J., and BLOOM, F. E. (1971), 'Cyclic adenosine monophosphate and norepinephrine: effects on transmembrane properties of cerebellar Purkinje cells', *Science, N.Y.*, **171,** 192–194.

SILVER, A. (1967), 'Cholinesterases of the central nervous system with special reference to the cerebellum', in *Int. Rev. Neurobiol.* (ed. Pfeiffer and Smythies), vol. 10, 57–109. New York: Academic Press.

SIMS, K. L., WEITSEN, H. A., and BLOOM, F. E. (1972), 'GABA catabolism: localization of succinic semialdehyde dehydrogenase in brain motor and sensory nuclei', *Science, N.Y.*, **175,** 1479–1480.

SINGH, S. I., and MALHOTRA, C. L. (1962), 'Amino acid content of monkey brain—I. General pattern and quantitative value of glutamic acid/glutamine, γ-aminobutyric acid and aspartic acid', *J. Neurochem.*, **9,** 37–42.

SLADEK, J. R., jun. (1971), 'Differences in the distribution of catecholamine varicosities in cat and rat reticular formation', *Science, N.Y.*, **174,** 410–412.

SMITH, C. M. (1972), 'The release of acetylcholine from rabbit hippocampus', *Br. J. Pharmac.*, **45,** 172P.

SNELL, R. S. (1961), 'The histochemical localization of cholinesterase in the central nervous system', *Biblphie anat.*, **2,** 50–58.

SNYDER, S. H., KUHAR, M. J., GREEN, A. I., COYLE, J. T., and SHASKAN, E. G. (1970), 'Uptake and subcellular localization of neurotransmitters in the brain', *Int. Rev. Neurobiol.*, **13,** 127–158.

SÖDERHOLM, U. (1965), 'Histochemical localization of esterases, phosphatases and tetrazolium reductases in the motor neurones of the spinal cord of the rat and the effect of nerve division', *Acta physiol. scand.*, **65,** suppl. 256, 1–60.

SPEHLMANN, R. (1963), 'Acetylcholine and prostigmine electrophoresis at visual cortex neurons', *J. Neurophysiol.*, **26,** 127–139.

SPEHLMANN, R. (1969), 'Acetylcholine facilitation, atropine block of synaptic excitation of cortical neurons', *Science, N.Y.*, **165,** 404–405.

SPEHLMANN, R. (1971a), 'Acetylcholine and the synaptic transmission of non-specific impulses to the visual cortex', *Brain*, **94,** 139–150.

SPEHLMANN, R. (1971b), 'The effects of acetylcholine and reticular stimulation on the motor cortex', *Trans. Am. neurol. Ass.*, **96,** 309–310.

SPEHLMANN, R., DANIELS, J. C., and SMATHERS, C. C., jun. (1971), 'Acetylcholine and the synaptic transmission of specific impulses to the visual cortex', *Brain*, **94,** 125–138.

SRINIVASAN, V., NEAL, M. J., and MITCHELL, J. F. (1969), 'The effect of electrical stimulation and high potassium concentrations on the efflux of (^{3}H)γ-aminobutyric acid from brain slices', *J. Neurochem.*, **16,** 1235–1244.

STARR, M. S., and VOADEN, M. J. (1972), 'The uptake of (^{14}C) γ-aminobutyric acid by the isolated retina of the rat', *Vision Res.*, **12,** 549–557.

STEFANIS, C. (1964a), 'Hippocampal neurons: their responsiveness to microelectrophoretically administered endogenous amines', *Pharmacologist*, **6,** 171.

STEFANIS, C. (1946b), 'Electrophysiological properties of cortical motoneurons during iontophoretic application of chemical substances', *Physiologist, Lond.*, **7,** 263.

STEINER, F. A. (1968), 'Influence of microelectrophoretically applied acetylcholine on the responsiveness of hippocampal and lateral geniculate neurones', *Pflügers Arch. ges. Physiol.*, **303,** 173–180.

STEINER, F. A. (1969), 'L-Glutamic acid, gamma-aminobutyric acid and pyridoxal-5′-phosphate at the level of the single unit in rat brain', *Ann. N.Y. Acad. Sci.*, **166,** 199–209.

STEINER, F. A. (1970), 'Effects of ACTH and corticosteroids on single neurons in the hypothalamus', *Prog. Brain Res.*, **32,** 102–107.

STEINER, F. A. (1971), 'Neurotransmitter und Neuromodulatoren', *Technik und Resultate der Mikroelektrophorese im Nervensystem*, pp. 1–159. Sammlung psychiatrischer und neurologischer Einzeldarstellungen. Stuttgart: Thieme.

STEINER, F. A., and MEYER, M. (1966), 'Actions of L-glutamate, acetylcholine and dopamine on single neurones in the nuclei cuneatus and gracilis of the cat', *Experientia*, **22,** 58–59.

STEINER, F. A., PIERI, L., and KAUFMANN, L. (1968), 'Effects of dopamine and ACTH on steroid sensitive single neurones in the basal hypothalamus', *Experientia*, **24,** 1133–1134.

STEINER, F. A., and RUF, K. (1966), 'Excitatory effects of L-glutamic acid upon single unit activity in rat brain and their modification by thiosemicarbazide and pyridoxal-5′-phosphate', *Helv. physiol. Acta*, **24,** 181–192.

STEINER, F. A., RUF, K., and AKERT, K. (1969), 'Steroid-sensitive neurones in rat brain: anatomical localization and responses to neurohumours and ACTH', *Brain Res.*, **12,** 74–85.

STEINER, F. A., and WEBER, G. (1965), 'Die Beeinflussung labyrinthär erregbarer Neurone des Hirnstammes durch Acetylcholin', *Helv. physiol. Acta.*, **23,** 82–89.

STONE, J., and HOFFMANN, K.-P. (1972), 'Very slow-conducting ganglion cells in the cat's retina: a major, new functional type?', *Brain Res.*, **43,** 610–616.

STONE, T. W. (1971), 'Are noradrenaline excitations artefacts?', *Nature, Lond.*, **234,** 145–146.

STONE, T. W. (1972a), 'Cholinergic mechanisms in the rat cerebral cortex', *J. Physiol., Lond.*, **222,** 155–156P.

STONE, T. W. (1972b), 'Cholinergic mechanisms in the rat somatosensory cerebral cortex', *J. Physiol., Lond.*, **225,** 485–499.

STONE, T. W. (1972c), 'Noradrenaline effects and *p*H', *J. Pharm. Pharmac.*, **24,** 422–423.

STORM-MATHISEN, J. (1970), 'Quantitative histochemistry of acetylcholinesterase in rat hippocampal region correlated to histochemical staining', *J. Neurochem.*, **17**, 739–750.

STORM-MATHISEN, J. (1972), 'Glutamate decarboxylase in the rat hippocampal region after lesions of the afferent fibre systems. Evidence that the enzyme is localized in intrinsic neurones', *Brain Res.*, **40**, 215–235.

STORM-MATHISEN, J., and FONNUM, F. (1971), 'Quantitative histochemistry of glutamate decarboxylase in the rat hippocampal region', *J. Neurochem.*, **18**, 1105–1111.

STRASCHILL, M. (1968), 'Action of drugs on single neurons in the cat's retina', *Vision Res.*, **8**, 35–47.

STRASCHILL, M., and HOFFMANN, K-P. (1969), 'Functional aspects of localization in the cat's tectum opticum', *Brain Res.*, **13**, 274–283.

STRASCHILL, M., and PERWEIN, J. (1969), 'The inhibition of retinal ganglion cells by catecholeamines and γ-aminobutyric acid', *Pflügers Arch.—Eur. J. Physiol.*, **312**, 45–54.

STRASCHILL, M., and PERWEIN, J. (1971), 'Effect of iontophoretically applied biogenic amines and of cholinomimetic substances upon the activity of neurons in the superior colliculus and mesencephalic reticular formation of the cat', *Pflügers Arch.—Eur. J. Physiol.*, **324**, 43–55.

STRAUGHAN, D. W., and LEGGE, K. F. (1965), 'The pharmacology of amygdaloid neurones', *J. Pharm. Pharmac.*, **17**, 675–677.

STRAUGHAN, D. W., NEAL, M. J., SIMMONDS, M. A., COLLINS, G. G. S., and HILL, R. G. (1971), 'Evaluation of bicuculline as a GABA antagonist', *Nature, Lond.*, **233**, 352–354.

SUTHERLAND, E. W., ROBISON, G. A., and BUTCHER, R. W. (1968), 'Some aspects of the biological role of adenosine 3′, 5′-monophosphate (cyclic AMP)', *Circulation*, **37**, 279–306.

SUZUKI, H., and TAIRA, N. (1961), 'Effect of reticular stimulation upon synaptic transmission in cat's lateral geniculate body', *Jap. J. Physiol.*, **11**, 641–655.

SWEET, R. D., and REIS, D. J. (1971), 'Collection of (^{3}H)norepinephrine in ventriculo-cisternal perfusate during hypothalamic stimulation in cat', *Brain Res.*, **33**, 584–588.

SZENTÁGOTHAI, J., HÁMORI, J., and TÖMBÖL, T. (1966), 'Degeneration and electron microscope analysis of the synaptic glomeruli in the lateral geniculate body', *Expl Brain Res.*, **2**, 283–301.

SZERB, J. C. (1965), 'Averaged evoked potentials and cholinergic synapses in the somatosensory cortex of the cat', *Electroenceph. clin. Neurophysiol.*, **18**, 140–146.

SZERB, J. C. (1967), 'Cortical acetylcholine release and electroencephalographic arousal', *J. Physiol., Lond.*, **192**, 329–343.

TABER, E., BRODAL, A., and WALBERG, F. (1960), 'The raphe nuclei of the brain stem in the cat. I. Normal topography and cytoarchitecture and general discussion', *J. comp. Neurol.*, **114,** 161–188.

TALLAN, H. H. (1962), 'A survey of the amino acids and related compounds in nervous tissue', in *Amino Acid Pools* (ed. Holden), pp. 471–485. Amsterdam: Elsevier.

TANAKA, Y., and KATSUKI, Y. (1966), 'Pharmacological investigations of cochlear responses and of olivo-cochlear inhibition', *J. Neurophysiol.*, **29,** 94–108.

TANJI, J., and KATO, M. (1972), 'The effect of picrotoxin and strychnine upon inhibition of fusimotor neuron discharges caused by cutaneous fiber stimulation', *Experientia*, **28,** 41–42.

TAYLOR, K. M., GFELLER, E., and SNYDER, S. H. (1972), 'Regional localization of histamine and histidine in the brain of the rhesus monkey', *Brain Res.*, **41,** 171–179.

TEBĒCIS, A. K. (1967), 'Are 5-hydroxytryptamine and noradrenaline inhibitory transmitters in the medial geniculate nucleus?', *Brain Res.*, **6,** 780–782.

TEBĒCIS, A. K. (1970a), 'Properties of cholinoceptive neurones in the medial geniculate nucleus', *Br. J. Pharmac.*, **38,** 117–137.

TEBĒCIS, A. K. (1970b), 'Studies on cholinergic transmission in the medial geniculate nucleus', *Br. J. Pharmac.*, **38,** 138–147.

TEBĒCIS, A. K. (1970c), 'Effects of monoamines and amino acids on medial geniculate neurones of the cat', *Neuropharmac.*, **9,** 381–390.

TEBĒCIS, A. K. (1972a), 'Antagonism of 5-hydroxytryptamine by methiothepin shown in microelectrophoretic studies of neurones in the lateral geniculate nucleus', *Nature, Lond.* (*New Biol.*), **238,** 63–64.

TEBĒCIS, A. K. (1972b), 'Glycine as an inhibitory transmitter in the medullary reticular formation of the cat' (Proc. regional Meet. Int. Union Physiol. Sci., Sydney (Australia), 21–25 August, 1972), *Proc. Aust. Physiol. Pharmac. Soc.*, **3,** 194.

TEBĒCIS, A. K. (1972c), 'Cholinergic and non-cholinergic transmission in the medial geniculate nucleus of the cat', *J. Physiol., Lond.*, **226,** 153–172.

TEBĒCIS, A. K., and DI MARIA, A. (1972a), 'Strychnine-sensitive inhibition in the medullary reticular formation: evidence for glycine as an inhibitory transmitter', *Brain Res.*, **40,** 373–383.

TEBĒCIS, A. K., and DI MARIA, A. (1972b), 'A re-evaluation of the mode of action of 5-hydroxytryptamine on lateral geniculate neurones: comparison with catecholamines and LSD', *Expl Brain Res.*, **14,** 480–493.

TEBĒCIS, A. K., HÖSLI, L., and HAAS, H. L. (1971), 'Bicuculline and the depression of medullary reticular neurones by GABA and glycine', *Experientia*, **27,** 548.

TEBĒCIS, A. K., and PHILLIS, J. W. (1967), 'The effects of topically applied biogenic monoamines on the isolated toad spinal cord', *Comp. Biochem. Physiol.*, **23,** 553–563.

TEBĒCIS, A. K., and PHILLIS, J. W. (1969), 'The use of convulsants in studying possible functions of amino acids in the toad spinal cord', *Comp. Biochem. Physiol.*, **28,** 1303–1315.

TÖMBÖL, T. (1966), 'Short neurons and their synaptic relations in the specific thalamic nuclei', *Brain Res.*, **3,** 307–326.

TORVIK, A., and BRODAL, A. (1957), 'The origin of reticulospinal fibers in the cat. An experimental study', *Anat. Rec.*, **128,** 113–135.

TOYAMA, K., TSUKAHARA, N., and UDO, M. (1968), 'Nature of the cerebellar influences upon the red nucleus neurones', *Expl Brain Res.*, **4,** 292–309.

TSUKADA, Y., NAGATA, Y., HIRANO, S., and MATSUTANI, T. (1963), 'Active transport of amino acids into cerebral cortex slices', *J. Neurochem.*, **10,** 241–256.

UCHIZONO, K. (1965), 'Characteristics of excitatory and inhibitory synapses in the central nervous system of the cat', *Nature, Lond.*, **207,** 642–643.

UCHIZONO, K. (1966), 'Excitatory and inhibitory synapses in the cat spinal cord', *Jap. J. Physiol.*, **16,** 570–575.

UDO, M., and MANO, N. (1970), 'Discrimination of different spinal monosynaptic pathways converging onto reticular neurons', *J. Neurophysiol.*, **33,** 227–238.

UNGERSTEDT, U. (1971), 'Stereotaxic mapping of the monoamine pathways in the rat brain', *Acta physiol. scand.*, suppl. 347, 1–48.

UNO, M., YOSHIDA, M., and HIROTA, I. (1970), 'The mode of cerebello-thalamic relay transmission investigated with intracellular recording from cells of the ventrolateral nucleus of cat's thalamus', *Expl Brain Res.*, **10.** 121–139.

VAN GELDER, N. M. (1965a), 'The histochemical demonstration of γ-aminobutyric acid metabolism by reduction of a tetrazolium salt', *J. Neurochem.*, **12,** 231–237.

VAN GELDER, N. M. (1965b), 'A comparison of γ-aminobutyric acid metabolism in rabbit and mouse nervous tissue', *J. Neurochem.*, **12,** 239–244.

VAN HARREVELD, A. (1959), 'Compounds in brain extracts causing spreading depression of cerebral cortical activity and contraction of crustacean muscle', *J. Neurochem.*, **3,** 300–315.

VAN HARREVELD, A. (1966), *Brain Tissue Electrolytes*. Washington: Butterworths.

VAN HARREVELD, A., and FIFKOVÁ, E. (1970), 'Glutamate release from the retina during spreading depression', *J. Neurobiol.*, **2,** 13–29.

VAN HARREVELD, A., and FIFKOVÁ, E. (1971), 'Effects of glutamate and other amino acids on the retina', *J. Neurochem.*, **18,** 2145–2154.

VASTOLA, E. F. (1959), 'After-positivity in lateral geniculate body', *J. Neurophysiol.*, **22**, 258–272.

VOADEN, M. J., and STARR, M. S. (1972), 'The efflux of radioactive GABA from rat retina *in vitro*', *Vision Res.*, **12**, 559–566.

VOGT, M. (1954), 'The concentration of sympathin in different parts of the central nervous system under normal conditions and after the administration of drugs', *J. Physiol., Lond.*, **123**, 451–481.

VOGT, M. (1969), 'Release from brain tissue of compounds with possible transmitter function: interaction of drugs with these substances', *Br. J. Pharmac.*, **37**, 325–337.

VON BAUMGARTEN, R., BLOOM, F. E., OLIVER, A. P., and SALMOIRAGHI, G. C. (1963), 'Response of individual olfactory nerve cells to microelectrophoretically administered chemical substances', *Pflügers Arch. ges. Physiol.*, **277**, 125–140.

VON VOIGTLANDER, P. F., and MOORE, K. E. (1971a), 'The release of H^3-dopamine from cat brain following electrical stimulation of the substantia nigra and caudate nucleus', *Neuropharmacology*, **10**, 733–741.

VON VOIGTLANDER, P. F., and MOORE, K. E. (1971b), 'Nigro-striatal pathway: stimulation-evoked release of (^3H) dopamine from caudate nucleus', *Brain Res.*, **35**, 580–583.

VON VOIGTLANDER, P. F., and MOORE, K. E. (1971c), '*In vivo* electrically evoked release of (^3H) noradrenaline from cat brain', *J. Pharm. Pharmac.*, **23**, 381–382.

WAGNER, A., and KLEE, M. R. (1968), 'Pharmacological investigation of polysynaptic IPSPs of cat's flexor motoneurons', *Proc. int. Un. Physiol. Sci.*, **7**, 457.

WALBERG, E. (1965), 'An electron microscopic study of terminal degeneration in the inferior olive of the cat', *J. comp. Neurol.*, **125**, 205–222.

WALLACH, D. P. (1961), 'Studies on the GABA pathway. I. The inhibition of γ-aminobutyric acid-α-ketoglutaric acid transaminase *in vitro* and *in vivo* by U-7524 (amino-oxyacetic acid)', *Biochem. Pharmac.*, **5**, 323–331.

WALTHER, J. B., and RASMUSSEN, G. L. (1960), 'Descending connections of auditory cortex and thalamus of the cat', *Fedn Proc. Fedn Am. Socs exp. Biol.*, **19**, 291.

WATANABE, T. (1971), 'Effect of picrotoxin on two-tone inhibition of auditory neurons in the cochlear nucleus', *Brain Res.*, **28**, 586–590.

WATANABE, T., and SIMADA, Z. (1970), 'Collicular auditory interneuron and its blocking by dihydro-β-erythroidine hydrobromide', *Proc. Japan Acad.*, **46**, 983–988.

WATANABE, T., and SIMADA, Z. (1971), 'Picrotoxin: effect on collicular auditory neurons', *Brain Res.*, **28**, 582–585.

WEDNER, H. J., HOFFER, B. J., STEINER, A. L., PARKER, C. W., and BLOOM, F. E. (1972), 'Localizing cyclic AMP intracellularly by immunofluorescence', *Fedn Proc. Fedn Am. Socs exp. Biol.*, **31,** 513.

WEIGHT, F. F. (1968), 'Cholinergic mechanisms in recurrent inhibition of motoneurons', in *Psychopharmacology*, A Review of Progress 1957–1967 (ed. Efron), pp. 69–75. Washington: U.S. Public Health Service.

WEIGHT, F. F. (1971), 'Mechanisms of synaptic transmission', *Neurosci. Res.*, **4,** 1–27.

WEIGHT, F. F., and SALMOIRAGHI, G. C. (1966a), 'Responses of spinal cord interneurons to acetylcholine, norepinephrine and serotonin administered by microelectrophoresis', *J. Pharmac. exp. Ther.*, **153,** 420–427.

WEIGHT, F. F., and SALMOIRAGHI, G. C. (1966b), 'Adrenergic responses of Renshaw cells', *J. Pharmac. exp. Ther.*, **154,** 391–397.

WEIGHT, F. F., and SALMOIRAGHI, G. C. (1967), 'Motoneurone depression by norepinephrine', *Nature, Lond.*, **213,** 1229–1230.

WEIGHT, F. F., and VOTAVA, J. (1970), 'Slow synaptic excitation in sympathetic ganglion cells: evidence for synaptic inactivation of potassium conductance', *Science, N.Y.*, **170,** 755–758.

WERMAN, R. (1966), 'A review—criteria for identification of a central nervous system transmitter', *Comp. Biochem. Physiol.*, **18,** 745–766.

WERMAN, R. (1972a), 'CNS cellular level: membranes', *A. Rev. Physiol.*, **34,** 337–374.

WERMAN, R. (1972b), 'Amino acids as central neurotransmitters', in 'Neurotransmitters', *Res. Publ. Assoc. Res. nerv. ment. Dis.*, **50,** 147–180.

WERMAN, R., and APRISON, M. H. (1970), 'The current status of the glycine story', in *Proc. Fourth Int. Cong. Pharmac.* (ed. Eigenmann), vol. 5, pp. 52–56. Basel: Schwabe.

WERMAN, R., DAVIDOFF, R. A., and APRISON, M. H. (1968), 'Inhibitory action of glycine on spinal neurons in the cat', *J. Neurophysiol.*, **31,** 81–95.

WESTECKER, M. E. (1969), 'The time course of facilitation and inhibition in the olfactory bulb, investigated with double pulse stimulation of the lateral olfactory tract', *Brain Res.*, **16,** 527–529.

WESTECKER, M. E. (1970), 'Excitatory and inhibitory interactions in the olfactory bulb involving dendrodendritic synapses between mitral cells and granular cells', *Pflügers Arch.—Eur. J. Physiol.*, **317,** 173–186.

WHITFIELD, I. C., and ALLANSON, J. T. (1958), 'A study of the effect of some neurally active drugs on inhibition in the auditory pathway', *Arch. ital. Biol.*, **96,** 29–37.

WHITTAKER, V. P., and SHERIDAN, M. N. (1965), 'The morphology and acetylcholine content of isolated cerebral cortical synaptic vesicles', *J. Neurochem.*, **12,** 363–372.

WIDÉN, L., and AJMONE MARSAN, C. (1960), 'Effects of corticipetal and corticifugal impulses upon single elements of the dorsolateral geniculate nucleus', *Expl Neurol.*, **2,** 468–502.

WILLIS, W. D., and WILLIS, J. C. (1966), 'Properties of interneurons in the ventral spinal cord', *Arch. ital. Biol.*, **104,** 354–386.

WILSON, V. J., and YOSHIDA, M. (1969a), 'Comparison of effects of stimulation of Deiters' nucleus and medial longitudinal fasciculus on neck, forelimb and hindlimb motoneurons', *J. Neurophysiol.*, **32,** 743–758.

WILSON, V. J., and YOSHIDA, M. (1969b), 'Monosynaptic inhibition of neck motoneurons by the medial vestibular nucleus', *Expl Brain Res.*, **9,** 365–380.

WILSON, V. J., YOSHIDA, M., and SCHOR, R. H. (1970), 'Supraspinal monosynaptic excitation and inhibition of thoracic back motoneurons', *Expl Brain Res.*, **11,** 282–295.

WOFSEY, A. R., KUHAR, M. J., and SNYDER, S. H. (1971), 'A unique synaptosomal fraction, which accumulates glutamic and aspartic acids, in brain tissue'. *Proc. natn. Acad. Sci. U.S.A.*, **68,** 1102–1106.

WOLFGRAM, F. J. (1954), 'Relative amounts of choline acetylase and cholinesterases in dorsal and ventral roots of cattle', *Am. J. Physiol.*, **176,** 505–507.

WOLSTENCROFT, J. H. (1964), 'Reticulospinal neurones', *J. Physiol., Lond.*, **174,** 91–108.

WOODWARD, D. J., HOFFER, B. J., SIGGINS, G. R., and OLIVER, A. P. (1971), 'Inhibition of Purkinje cells in the frog cerebellum. II. Evidence for GABA as the inhibitory transmitter', *Brain Res.*, **33,** 91–100.

WOOLLARD, H. H., and HARPMAN, J. A. (1940), 'The connexions of the inferior colliculus and of the dorsal nucleus of the lateral lemniscus', *J. Anat., Lond.*, **74,** 441–458.

YAGI, K., AZUMA, T., and MATSUDA, K. (1966), 'Neurosecretory cell: capable of conducting impulse in rats', *Science, N.Y.*, **154,** 778–779.

YAMAMOTO, C. (1967), 'Pharmacological studies of norepinephrine, acetylcholine and related compounds on neurons in Deiters' nucleus and the cerebellum', *J. Pharmac. exp. Ther.*, **156,** 39–47.

YAMAMOTO, C., YAMAMOTO, T., and IWAMA, K. (1963), 'The inhibitory systems in the olfactory bulb studied by intracellular recording', *J. Neurophysiol.*, **26,** 403–415.

YAMASHITA, H., KOIZUMI, K., and BROOKS, C. MCC. (1970), 'Electrophysiological studies of neurosecretory cells in the cat hypothalamus', *Brain Res.*, **20,** 462–466.

YORK, D. H. (1967), 'The inhibitory action of dopamine on neurones of the caudate nucleus', *Brain Res.*, **5,** 263–266.

YORK, D. H. (1970), 'Possible dopaminergic pathway from substantia nigra to putamen', *Brain Res.*, **20,** 233–249.

YORK, D. H. (1972), 'Dopamine receptor blockade—a central action of chlorpromazine on striatal neurones', *Brain Res.*, **37,** 91–99.

YOSHIDA, M., and PRECHT, W. (1971), 'Monosynaptic inhibition of neurons of the substantia nigra by caudato-nigral fibers', *Brain Res.*, **32,** 225–228.

YOSHIDA, M., RABIN, A., and ANDERSON, M. (1971), 'Two types of monosynaptic inhibition of pallidal neurons produced by stimulation of the diencephalon and substantia nigra', *Brain Res.*, **30,** 235–239.

ZETLER, G., and SCHLOSSER, L. (1955), 'Über die Verteilung von Substanz P und Cholinacetylase im Gehirn', *Arch. exp.. Path. Pharmak.*, **224,** 159–175.

ZIEGLGÄNSBERGER, W., and HERZ, A. (1971), 'Changes of cutaneous receptive fields of spino-cervical-tract neurones and other dorsal horn neurones by microelectrophoretically administered amino acids', *Expl Brain. Res.*, **13,** 111–126.

SUBJECT INDEX